Miller's Guide to Framing & Roofing

Home Reference Series

Miller's Guide to Framing & Roofing

MARK R. MILLER
Professor
The University of Texas at Tyler
Tyler, Texas

REX MILLER
Professor Emeritus
State University College at Buffalo
Buffalo, New York

McGraw-Hill

New York   Chicago   San Francisco   Lisbon   London
Madrid   Mexico City   Milan   New Delhi   San Juan
Seoul   Singapore   Sydney   Toronto

Library of Congress Cataloging-in-Publication Data

Miller, Mark R.
    Miller's guide to framing and roofing / Mark R. Miller, Rex Miller.
        p.   cm.
    Includes index.
    ISBN 0-07-145144-7
    1. Framing (Building) 2. Roofs. 3. Roofing. I. Title: Guide to framing and roofing. II.
Miller, Rex. III. Title.

    TH2301.M53   2005
    694'.2—dc22

                                                                2005047871

1 2 3 4 5 6 7 8 9 0   QPD/QPD   0 1 0 9 8 7 6 5

ISBN 0-07-145144-7

*The sponsoring editor for this book was Larry Hager, the editing supervisor was
Caroline Levine, and the production supervisor was Sherri Souffrance. The art director for
the cover was Handel Low. It was set in ITC Century Light by Kim J. Sheran of McGraw-
Hill Professional's Hightstown, N.J., composition unit.*

McGraw-Hill books are available at special quantity discounts to use as premiums and sales
promotions, or for use in corporate training programs. For more information, please write to
the Director of Special Sales, McGraw-Hill Professional, Two Penn Plaza, New York, NY
10121. Or contact your local bookstore.

Contents

6 Installing Windows and Doors

Preface

The purpose of this book is to aid those who are interested in the construction of a home, for either themselves or others.

The book has something for the apprentice, the seasoned carpenter, and the do-it-yourselfer. Although it contains some theory of the principles of construction and various methods for accomplishing a given task, the main emphasis of the text is on the practical, everyday application of good building practices.

Many illustrations are included to show the variety of parts and techniques used presently as well as parts found in older homes that may need replacement or repair. Obviously, not all related problems will be presented here because there is a great deal of ingenuity required of the worker on the job. For standard procedure, however, the International Council of Codes (ICC) does give guidance to those who are interested in meeting Code requirements.

Keep in mind that the Code, as it is commonly called, changes from time to time. In fact, every three years or so a revised Code is published. The Code is in a constant state of flux and is changed as conditions warrant. Therefore, it is possible that information related to various specifications will change. New materials and techniques are developed constantly. It is important that anyone planning to build a home anywhere be aware of the local zoning and building requirements.

MARK R. MILLER
REX MILLER

Acknowledgments

Every book is the product of many people, especially in the field of carpentry. All involved make it a "good" one worth the time it takes to read. This book is no exception, for there were many people who contributed to its creation. There were many manufacturers, builders, construction persons, distributors, and editors and printers involved. The following are some of those who contributed their time, expertise, and materials to make this a better representative of the construction field. To them we owe much. Thanks, for your aid and opinions.

AFM
American Plywood Association
American Polysteel Forms, Inc.
American Wood-Preservers Association
Black and Decker
David White Instruments, Inc.
Duofast, Inc.

Forest Products Laboratory
Fox & Jacobs, Inc.
Georgia-Pacific
IRL Daffin
Jet, Inc.
Kurmas Wood Foundations
Lennox Furnace Company
Millers Falls—a Division of Ingersoll-Rand Co.
Portland Cement Corporation
Proctor Products, Inc.
Richmond Screw Anchor, Inc.
Rockwell International, Inc.
Rub-R-Wall, Inc.
Skrobarczk Properties
Southern Pine Council—Richard Wallace
southernpinecouncil.com
Stablia™
Stanley Tools, Inc.
Universal Form Clamp, Inc.

Miller's Guide to Framing & Roofing

1

CHAPTER

Safety Measures and Carpentry Tools

BECAUSE CARPENTRY INVOLVES ALL KINDS of challenging jobs, it is an exciting industry. You will have to work with hand tools, power tools, and all types of building materials. You can become very skilled at your job. You get a chance to be proud of what you do. You can stand back and look at the building you just helped erect and feel great about a job well done.

One of the exciting things about being a carpenter is watching a building come up. You actually see it grow from the ground up. Many people work with you to make it possible to complete the structure. Being part of a team can be rewarding, too.

This book will help you do a good job in carpentry, whether you are remodeling an existing building or starting from the ground up. Because it covers all the basic construction techniques, it will aid you in making the right decisions.

You have to do something over and over again to gain skill. When reading this book, you might not always get the idea the first time. Go over it again until you understand. Then go out and practice what you just read. This way you can see for yourself how the instructions actually work. Of course, no one can learn carpentry by merely reading a book. You have to read, reread, and then do. This "do" part is the most important. You have to take the hammer or saw in hand and actually do the work. There is nothing like good, honest sweat from a hard day's work. At the end of the day you can say "I did that" and be proud that you did.

This chapter should help you build these skills:

• Select personal protective gear
• Work safely as a carpenter
• Measure building materials
• Lay out building parts
• Cut building materials
• Fasten materials
• Shape and smooth materials
• Identify basic hand tools
• Recognize common power tools

SAFETY

Figure 1-1 shows a carpenter using one of the latest means of driving nails: the compressed-air-driven nail driver, which drives nails into the wood with a single stroke. The black cartridge that appears to run up near the carpenter's leg is a part of the nailer. It holds the nails and feeds them as needed.

Fig. 1-1 *This carpenter is using an air-driven nail driver to nail these framing members.* (Duo-Fast)

As for safety, notice the carpenter's shoes. They have rubber soles for gripping the wood. This will prevent a slip through the joists and a serious fall. The steel toes in the shoes prevent damage to a foot from falling materials. The soles of the shoes are very thick to prevent nails from going through. The hard hat protects the carpenter's head from falling lumber, shingles, or other building materials. The carpenter's safety glasses cannot be seen in Fig. 1-1, but they are required equipment for the safe worker.

Other Safety Measures

To protect the eyes, it is best to wear safety glasses. Make sure your safety glasses are of tempered glass. They will not shatter and cause eye damage. In some instances you should wear goggles. This prevents splinters and other flying objects from entering the eye from under or around the safety glasses. Ordinary glasses aren't always the best, even if they are tempered glass. Just become aware of the possibilities of eye damage whenever you start a new job or procedure. See Fig. 1-2 for a couple of types of safety glasses.

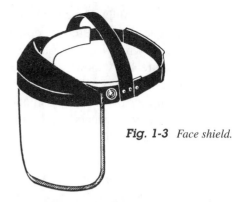

Fig. 1-3 Face shield.

Fig. 1-2 Safety glasses.

Sneakers are used only by roofers. Sneakers, sandals, and dress shoes do not provide enough protection for the carpenter on the job. Only safety shoes should be worn on the job.

Gloves Some types of carpentry work require the sensitivity of the bare fingers. Other types do not require the hands or fingers to be exposed. In cold or even cool weather, gloves may be in order. Gloves are often needed to protect your hands from splinters and rough materials. It's only common sense to use gloves when handling rough materials.

Probably the best gloves for carpenter work are a lightweight type. A suede finish to the leather improves the gripping ability of the gloves. Cloth gloves tend to catch on rough building materials. They may be preferred, however, if you work with short nails or other small objects.

Body protection Before you go to work on any job, make sure your entire body is properly protected. The hard hat comes in a couple of styles. Under some conditions the face shield is better protection. See Fig. 1-3.

Is your body covered with heavy work clothing? This is the first question to ask before going onto the job site. Has as much of your body as practical been covered with clothing? Has your head been properly protected? Are your eyes covered with approved safety glasses or face shield? Are your shoes sturdy, with safety toes and steel soles to protect against nails? Are gloves available when you need them?

General Safety Rules

Some safety procedures should be followed at all times. This applies to carpentry work especially:

- Pay close attention to what is being done.
- Move carefully when walking or climbing.
- (Take a look at Fig. 1-4. This type of made-on-the-job ladder can cause trouble.) Use the leg muscles when lifting.

Fig. 1-4 A made-on-the-job ladder.

- Move long objects carefully. The end of a carelessly handled 2 × 4 can damage hundreds of dollars worth of glass doors and windows. Keep the workplace neat and tidy. Figure 1-5A shows a cluttered working area. It would be hard to walk along here without tripping. If a dumpster is used for trash and debris, as in Fig. 1-5B, many accidents can be prevented. Sharpen or replace dull tools.

(A)

(B)

Fig. 1-5 *(A) Cluttered work site; (B) A work area can be kept clean if a large dumpster is kept nearby for trash and debris.*

- Disconnect power tools before adjusting them.
- Keep power tool guards in place.
- Avoid interrupting another person who is using a power tool.
- Remove hazards as soon as they are noticed.

Safety on the Job

A safe working site makes it easier to get the job done. Lost time due to accidents puts a building schedule behind. This can cost many thousands of dollars and lead to late delivery of the building. If the job is properly organized and safety is taken into consideration, the smooth flow of work is quickly noticed. No one wants to get hurt. Pain is no fun. Safety is just common sense. If you know how to do something safely, it will not take any longer than if you did it in an unsafe manner. Besides, why would you deliberately do something that is dangerous? All safety requires is a few precautions on the job. Safety becomes a habit once you get the proper attitude established in your thinking. Some of these important habits to acquire are:

- Know exactly what is to be done before you start a job.
- Use a tool only when it can be used safely. Wear all safety clothing recommended for the job. Provide a safe place to stand to do the work. Set ladders securely. Provide strong scaffolding.
- Avoid wet, slippery areas.
- Keep the working area as neat as practical.
- Remove or correct safety hazards as soon as they are noticed. Bend protruding nails over. Remove loose boards.
- Remember where other workers are and what they are doing.
- Keep fingers and hands away from cutting edges at all times.
- Stay alert!

Safety Hazards

Carpenters work in unfinished surroundings. While a house is being built, there are many unsafe places around the building site. You have to stand on or climb ladders, which can be unsafe. You may not have a good footing while standing on a ladder. You may not be climbing a ladder in the proper way. Holding onto the rungs of the ladder is very unsafe. You should always hold onto the outside rails of the ladder when climbing.

There are holes that can cause you to trip. They may be located in the front yard where the water or sewage lines come into the building. There may be holes for any number of reasons. These holes can cause you all kinds of problems, especially if you fall into them or turn your ankle.

The house in Fig. 1-6 is almost completed. However, if you look closely you can see that some wood has been left on the garage roof. This wood can slide down and hit a person working below. The front porch has not been poured. This means stepping out of the front door can be a rather long step. Other debris around the yard can be a source of trouble. Long sliv-

Fig. 1-6 *Even when a house is almost finished, there can still be hazards. Wood left on a roof could slide off and hurt someone, and without the front porch, it is a long step down.*

ers of flashing can cause trouble if you step on them and they rake your leg. You have to watch your every step around a construction site.

Outdoor work Much of the time carpentry is performed outdoors. This means you will be exposed to the weather, so dress accordingly. Wet weather increases the accident rate. Mud can make a secure place to stand hard to find. Mud can also cause you to slip if you don't clean it off your shoes. Be very careful when it is muddy and you are climbing on a roof or a ladder.

Tools Any tool that can cut wood can cut flesh. You have to keep in mind that although tools are an aid to the carpenter, they can also be a source of injury. A chisel can cut your hand as easily as the wood. In fact, it can do a quicker job on your hand than on the wood it was intended for. Saws can cut wood and bones. Be careful with all types of saws, both hand and electric. Hammers can do a beautiful job on your fingers if you miss the nailhead. The pain involved is intensified in cold weather. Broken bones can be easily avoided if you keep your eye on the nail while you're hammering. Besides that, you will get the job done more quickly. And, after all, that's why you are there—to get the job done and do it right the first time. Tools can help you do the job right. They can also cause you injury. The choice is up to you.

In order to work safely with tools you should know what they can do and how they do it. The next few pages are designed to help you use tools properly.

USING CARPENTER TOOLS

A carpenter is lost without tools. This means you have to have some way of containing them. A toolbox is

very important. If you have a place to put everything, then you can find the right tool when it is needed. A toolbox should have all the tools mentioned here. In fact, you will probably add more as you become more experienced. Tools have been designed for every task. All it takes is a few minutes with a hardware manufacturer's catalog to find just about everything you'll ever need. If you can't find what you need, the manufacturers are interested in making it.

Measuring Tools

Folding rule When using the folding rule, place it flat on the work. The 0 end of the rule should be exactly even with the end of the space or board to be measured. The correct distance is indicated by the reading on the rule.

A very accurate reading may be obtained by turning the edge of the rule toward the work. In this position, the marked graduations of the face of the rule touch the surface of the board. With a sharp pencil, mark the exact distance desired. Start the mark with the point of the pencil in contact with the mark on the rule. Move the pencil directly away from the rule while making the mark.

One problem with the folding rule is that it breaks easily if it is twisted. This happens most commonly when it is being folded or unfolded. The user may not be aware of the twisting action at the time. You should keep the joints oiled lightly. This makes the rule operate more easily.

Pocket tape Beginners may find the pocket tape (Fig. 1-7) the most useful measuring tool for all types of work. It extends smoothly to full length. It returns quickly to its compact case when the return button is pressed. Steel tapes are available in a variety of lengths. For most carpentry a rule 6, 8, 10, or 12 feet long is used.

Fig. 1-7 *Tape measure.* (Stanley Tools)

Longer tapes are available. They come in 20-, 50-, and 100-foot lengths. See Fig. 1-8. This tape can be extended to 50 feet to measure lot size and the location of a house on a lot. It has many uses around a building site. A crank handle can be used to wind it up once you are finished with it. The hook on the end of the tape makes it easy for one person to use it. Just hook the tape over the end of a board or nail and extend it to your desired length.

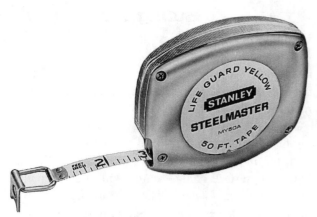

Fig. 1-8 *A longer tape measure.*

Saws

Carpenters use a number of different saws. These saws are designed for specific types of work. Many are misused. They will still do the job, but they would do a better job if used properly. Handsaws take quite a bit of abuse on a construction site. It is best to buy a good-quality saw and keep it lightly oiled.

Standard skew-handsaw This saw has a wooden handle. It has a 22-inch length. A 10-point saw (with 10 teeth per inch) is suggested for crosscutting. Crosscutting means cutting wood *across* the grain. The 26-inch-length, 5½-point saw is suggested for ripping, or cutting *with* the wood grain.

Figure 1-9 shows a carpenter using a handsaw. This saw is used in places where the electric saw cannot be used. Keeping it sharp makes a difference in the quality of the cut and the ease with which it can be used.

Backsaw The backsaw gets its name from the piece of heavy metal that makes up the top edge of the cutting part of the saw. See Fig. 1-10. It has a fine tooth configuration. This means it can be used to cut crossgrain and leave a smoother finished piece of work. This type of saw is used by finish carpenters who want to cut trim or molding.

Miter box As you can see from Fig. 1-11A, the miter box has a backsaw mounted in it. This box can be adjusted using the lever under the saw handle (see arrow). You can adjust it for the cut you wish. It can cut from 90° to 45°. It is used for finish cuts on moldings and trim materials. The angle of the cut is determined by the location of the saw in reference to the bed of the box. Release the clamp on the bottom of the saw support to adjust the saw to any degree desired. The wood is held with one hand against the fence of the box and the bed. Then the saw is used by the other hand. As you can see from the setup, the cutting should take place when the saw is pushed forward. The backward movement of the saw should be made with the pressure on the saw released slightly. If you try to cut on the backward movement, you will just pull the wood away from the fence and damage the quality of the cut.

Coping saw Another type of saw the carpenter can make use of is the coping saw (Fig. 1-12). This one can cut small thicknesses of wood at any curve or angle desired. It can be used to make sure a piece of paneling fits properly or a piece of molding fits another piece in the corner. The blade is placed in the frame with the teeth pointing toward the handle. This means it cuts only on the downward stroke. Make sure you properly support the piece of wood being cut. A number of blades can be obtained for this type of saw. The number of teeth in the blade determines the smoothness of the cut.

Hammers and Other Small Tools

There are a number of different types of hammers. The one the carpenter uses is the *claw* hammer. It has claws that can extract nails from wood if they have been put in the wrong place or have bent while being driven. Hammers can be bought in 20-ounce, 24-ounce, 28-ounce, and 32-ounce weights for carpentry work. Most carpenters prefer a 20-ounce. You have to work with a number of different weights to find out which will work best for you. Keep in mind that the hammer should be of tempered steel. If the end of the hammer has a tendency to splinter or chip off when it hits a nail, the pieces can hit you in the eye or elsewhere, causing serious damage. It is best to wear safety glasses whenever you use a hammer.

Nails are driven by hammers. Figure 1-13 shows the gage, inch, and penny relationships for the common box nail. The *d* after the number means *penny*. This is a measuring unit inherited from the English in the colonial days. There is little or no relationship between

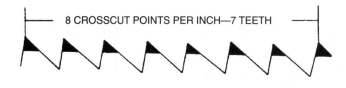

8 CROSSCUT POINTS PER INCH—7 TEETH

6 RIP POINTS PER INCH—5 TEETH

THE NUMBER OF POINTS PER INCH ON A HANDSAW
DETERMINES THE FINENESS OR COARSENESS
OF CUT. MORE POINTS PRODUCE A FINER CUT.

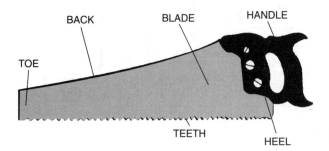

BACK   BLADE   HANDLE

TOE

TEETH   HEEL

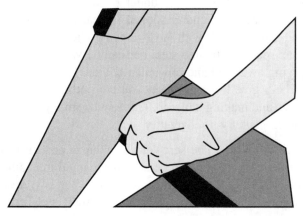

Fig. 1-9 *Using a handsaw.*

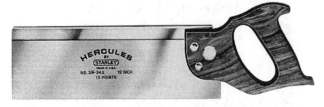

Fig. 1-10 *Backsaw.* (Stanley Tools)

Fig. 1-11A *Miter box.* (Stanley Tools)

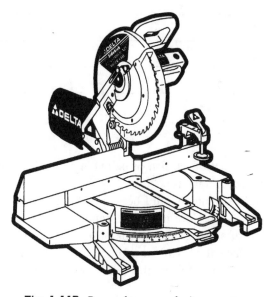

Fig. 1-11B *Powered compound miter saw.* (Delta)

Fig. 1-12 *Coping saw.* (Stanley Tools)

penny and inches. If you want to be able to talk about it intelligently, you'll have to learn both inches and penny. The gage is nothing more than the American Wire Gage number for the wire that the nails were made from originally. Finish nails have the same measuring unit (penny) but do not have the large, flat heads.

Nail set Finish nails are driven below the surface of the wood by a nail set. The nail set is placed on the head of the nail. The large end of the nail set is struck by the hammer. This causes the nail to go below the surface of the wood. Then the hole left by the countersunk nail is filled with wood filler and finished off with

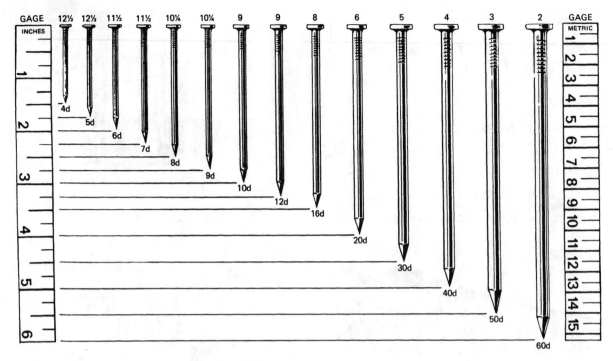

Fig. 1-13 *Nails. (Forest Products Laboratory)*

a smooth coat of varnish or paint. Figure 1-14 shows the nail set and its use.

The carpenter would be lost without a hammer. See Fig. 1-15. Here the carpenter is placing sheathing on rafters to form a roof base. The hammer is used to drive the boards into place, since they have to overlap slightly. Then the nails are driven by the hammer also.

In some cases a hammer will not do the job. The job may require a hatchet. See Fig. 1-16. This device can be used to pry and to drive. It can pry boards loose when they are improperly installed. It can sharpen posts to be driven at the site. The hatchet can sharpen the ends of stakes for staking out the site. It can also withdraw nails. This type of tool can also be used to drive stubborn sections of a wall into place when they are erected for the first time. The tool has many uses.

Scratch awl An awl is a handy tool for a carpenter. It can be used to mark wood with a scratch mark and to produce pilot holes for screws. Once it is in your tool box, you can think of a hundred uses for it. Since it does have a very sharp point, it is best to treat it with respect. See Fig. 1-17.

Wrecking bar This device (Fig. 1-18) has a couple of names, depending on which part of the country you are in at the time. It is called a wrecking bar in some parts and a crowbar in others. One end has a chisel-

sharp flat surface to get under boards and pry them loose. The other end is hooked so that the slot in the end can pull nails with the leverage of the long handle. This specially treated steel bar can be very helpful in prying away old and unwanted boards. It can be used to help give leverage when you are putting a wall in place and making it plumb. This tool has many uses for the carpenter with ingenuity.

Screwdrivers The screwdriver is an important tool for the carpenter. It can be used for many things other than turning screws. There are two types of screwdrivers. The standard type has a straight slot-fitting blade at its end. This type is the most common of screwdrivers. The Phillips-head screwdriver has a cross or X on the end to fit a screw head of the same design. Figure 1-19 shows the two types of screwdrivers.

Squares

In order to make corners meet and standard sizes of materials fit properly, you must have things square. That calls for a number of squares to check that the two walls or two pieces come together at a perpendicular.

Try square The *try square* can be used to mark small pieces for cutting. If one edge is straight and the handle part of the square (Fig. 1-20) is placed against this straight edge, then the blade can be used to mark the

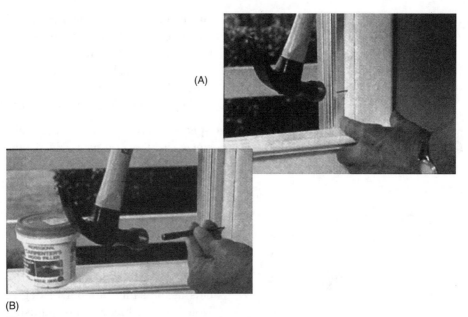

(A)

(B)

Fig. 1-14 *(A) Driving a nail with a hammer. (B) Finishing the job with a nail set to make sure the hammer doesn't leave an impression in the soft wood of the window frame.*

Fig. 1-15 *Putting on roof sheathing. The carpenter is using a hammer to drive the board into place.*

wood perpendicular to the edge. This comes in handy when you are cutting 2 × 4s and want them to be square.

Framing square   The framing square is a very important tool for the carpenter. It allows you to make square

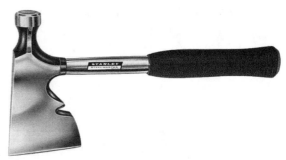

Fig. 1-16 *Hatchet.* (Stanley Tools)

Fig. 1-17 *Scratch awl.* (Stanley Tools)

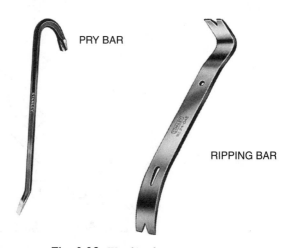

PRY BAR

RIPPING BAR

Fig. 1-18 *Wrecking bars.* (Stanley Tools)

Fig. 1-19 *Two types of screwdrivers.*

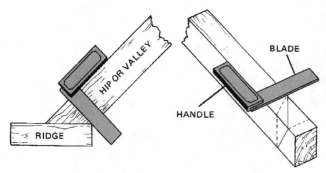

Fig. 1-20 *Use of a try square.* (Stanley Tools)

cuts in dimensional lumber. This tool can be used to lay out rafters and roof framing. See Fig. 1-21. It is also used to lay out stair steps.

Later in this book you will see a step-by-step procedure for using the framing square. The tools are described as they are called for in actual use.

Speed square Another alternative to the framing square is the compact speed square, as shown in Fig. 1-22. A speed square measures only 7 inches in length from either side of the 90° angle it forms. It is used for quickly laying out rafters and other roof framing members.

Step-by-step instructions on how to use the speed square will be discussed in Chap. 4.

Bevel A bevel can be adjusted to any angle to make cuts at the same number of degrees. See Fig. 1-23. Note how the blade can be adjusted. Now take a look at Fig. 1-24. Here you can see the overhang of rafters. If you want the ends to be parallel with the side of the house, you can use the bevel to mark them before they are cut off. Simply adjust the bevel so the han-

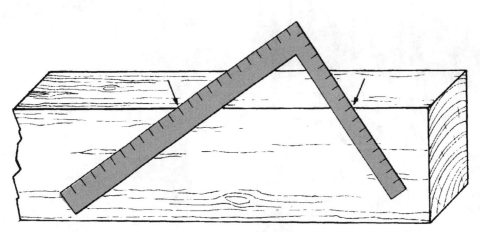

Fig. 1-21 *Framing square.* (Stanley Tools)

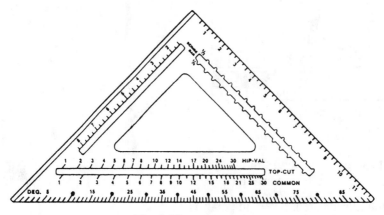

Fig. 1-22 *Speed square.*

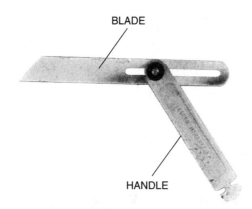

Fig. 1-23 Bevel. (Stanley Tools)

Fig. 1-24 Rafter overhang cut to a given angle.

dle is on top of the rafter and the blade fits against the soleplate below. Tighten the screw and move the bevel down the rafter to where you want the cut. Mark the angle along the blade of the bevel. Cut along the mark, and you have what you see in Fig. 1-24. It is a good device for transferring angles from one place to another.

Chisel Occasionally you may need a wood chisel. It is sharpened on one end. When the other end is struck with a hammer, the cutting end will do its job. That is, of course, if you have kept it sharpened. See Fig. 1-25.

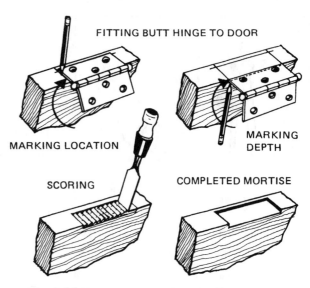

FITTING BUTT HINGE TO DOOR

MARKING LOCATION

MARKING DEPTH

SCORING

COMPLETED MORTISE

Fig. 1-25 Using a wood chisel to complete a mortise.

The chisel is commonly used in fitting or hanging doors. It is used to remove the area where the door hinge fits. Note how it is used to score the area (Fig. 1-25); it is then used at an angle to remove the ridges. A great deal of the work with the chisel is done by using the palm of the hand as the force behind the cutting edge. A hammer can be used. In fact, chisels have a metal tip on the handle so the force of the hammer blows will not chip the handle. Other applications are up to you, the carpenter. You'll find many uses for the chisel in making things fit.

Plane Planes (Fig. 1-26) are designed to remove small shavings of wood along a surface. One hand holds the knob in front and the other the handle in back. The blade is adjusted so that only a small sliver of wood is removed each time the plane is passed over the wood. It can be used to make sure that doors and windows fit properly. It can be used for any number of wood smoothing operations.

Dividers and compass Occasionally a carpenter must draw a circle. This is done with a compass. The compass shown in Fig. 1-27 can be converted to a divider

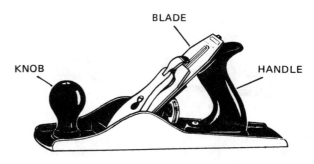

BLADE

KNOB

HANDLE

Fig. 1-26 Smooth plane. (Stanley Tools)

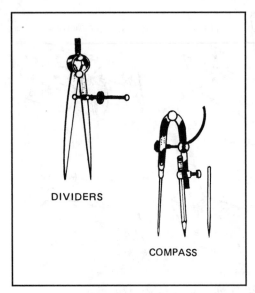

Fig. 1-27 *Dividers and compass*

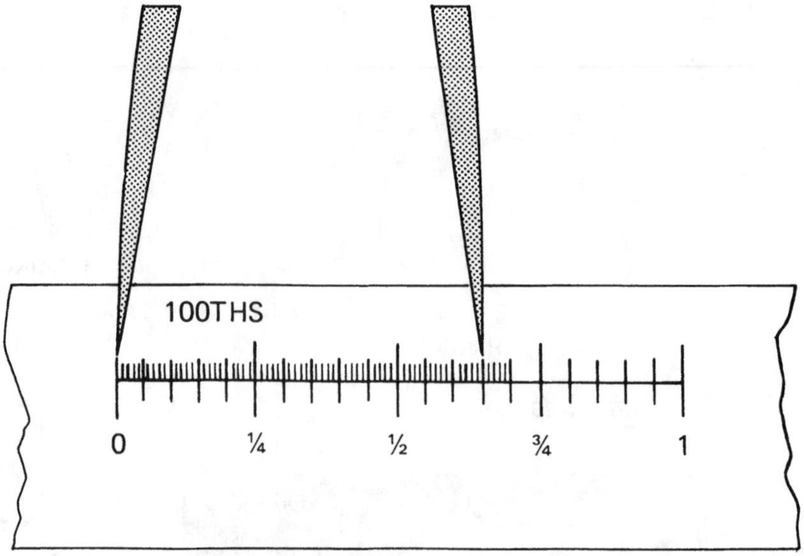

Fig. 1-28 *Dividers being used to transfer hundredths of an inch.*

by removing the pencil and inserting a straight steel pin. The compass has a sharp point that fits into the wood surface. The pencil part is used to mark the circle circumference. It is adjustable to various radii.

The dividers in Fig. 1-27 have two points made of hardened metal. They are adjustable. It is possible to use them to transfer a given measurement from the framing square or measuring device to another location. See Fig. 1-28.

Level In order to have things look as they should, a level is necessary. There are a number of sizes and shapes available. This one shown in Figure 1-29 is the most common type used by carpenters. The bubbles in the glass tubes tell you if the level is obtained. In Fig. 1-30 the carpenter is using the level to make sure the window is in properly before nailing it into place permanently.

If the vertical and horizontal bubbles are lined up between the lines, then the window is plumb, or vertical. A plumb bob is a small, pointed weight. It is attached to a string and dropped from a height. If the bob is just above the ground, it will indicate the vertical direction by its string. Keeping windows, doors,

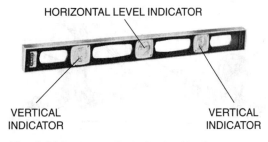

Fig. 1-29 *A commonly used type of level.* (Stanley Tools)

Fig. 1-30 *Using a level to make sure a window is placed properly before nailing.* (Andersen)

and frames square and level makes a difference in fitting. It is much easier to fit prehung doors into a frame that is square. When it comes to placing panels of 4-×-8-foot plywood sheathing on a roof or on walls, squareness can make a difference as to fit. Besides, a square fit and a plumb door and window look better than those that are a little off. Figure 1-31 shows three plumb bobs.

Files A carpenter finds use for a number of types of files. The files have different surfaces for doing different jobs. Tapping out a hole to get something to fit may be just the

Fig. 1-31 *Plumb bobs.* (Stanley Tools)

job for a file. Some files are used for sharpening saws and touching up tool cutting edges. Figure 1-32 shows different types of files. Other files may also be useful. You can acquire them later as you develop a need for them.

Clamps C clamps are used for many holding jobs. They come in handy when placing kitchen cabinets by holding them in place until screws can be inserted and properly seated. This type of clamp can be used for an extra hand every now and then when two hands aren't enough to hold a combination of pieces till you can nail them. See Fig. 1-33.

Cold chisel It is always good to have a cold chisel around. It is very much needed when you can't remove a nail. Its head may have broken off and the nail must be removed. The chisel can cut the nail and permit the separation of the wood pieces. See Fig. 1-34.

If a chisel of this type starts to "mushroom" at the head, you should remove the splintered ends with a grinder. Hammering on the end can produce a mushrooming effect. These pieces should be taken off since they can easily fly off when hit with a hammer. That is another reason for using eye protection when using tools.

Caulking gun In times of energy crisis, the caulking gun gets plenty of use. It is used to fill in around windows and doors and everywhere there may be an air leak. There are many types of caulk being made today. Another chapter will cover the details of the caulking compounds and their uses.

This gun is easily operated. Insert the cartridge and cut its tip to the shape you want. Puncture the thin plastic film inside. A bit of pressure will cause the caulk to come out the end. The long rod protruding from the end of the gun is turned over. This is so the serrated edge will engage the hand trigger. Remove the pressure from the cartridge when you are finished. Do this by rotating the rod so that the serrations are not engaged by the trigger of the gun.

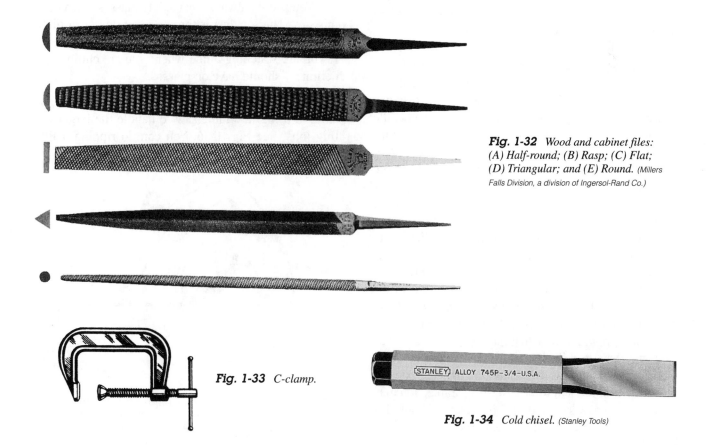

Fig. 1-32 *Wood and cabinet files:*
(A) Half-round; (B) Rasp; (C) Flat;
(D) Triangular; and (E) Round. (Millers
Falls Division, a division of Ingersol-Rand Co.)

Fig. 1-33 *C-clamp.*

Fig. 1-34 *Cold chisel.* (Stanley Tools)

Table 1-1 *Size of Extension Cords for Portable Tools*

Cord Length, Feet	Full-Load Rating of the Tool in Amperes at 115 Volts					
	0 to 2.0	2.10 to 3.4	3.5 to 5.0	5.1 to 7.0	7.1 to 12.0	12.1 to 16.0
	Wire Size (AWG)					
25	18	18	18	16	14	14
50	18	18	18	16	14	12
75	18	18	16	14	12	10
100	18	16	14	12	10	8
200	16	14	12	10	8	6
300	14	12	10	8	6	4
400	12	10	8	6	4	4
500	12	10	8	6	4	2
600	10	8	6	4	2	2
800	10	8	6	4	2	1
1000	8	6	4	2	1	0

If the voltage is lower than 115 volts at the outlet, have the voltage increased or use a much larger cable than listed.

Power Tools

The carpenter uses many power tools to aid in getting the job done. The quicker the job is done, the more valuable the work of the carpenter becomes. This is called productivity. The more you are able to produce, the more valuable you are. This means the contractor can make money on the job. This means you can have a job the next time there is a need for a good carpenter. Power tools make your work go faster. They also help you to do a job without getting fatigued. Many tools have been designed with you in mind. They are portable and operate from an extension cord.

The extension cord should be the proper size to take the current needed for the tool being used. See Table 1-1. Note how the distance between the outlet and the tool using the power is critical. If the distance is great, then the wire must be larger in size to handle the current without too much loss. The higher the number of the wire, the smaller the diameter of the wire. The larger the size of the wire (diameter), the more current it can handle without dropping the voltage.

Some carpenters run an extension cord from the house next door for power before the building site is furnished power. If the cord is too long or has the wrong size wire, it drops the voltage below 115. This means the saws or other tools using electricity will draw more current and therefore drop the voltage more. Every time the voltage is dropped, the device tries to obtain more current. This becomes a self-defeating phenomenon. You wind up with a saw that has little cutting power. You may have a drill that won't drill into a piece of wood without stalling. Of course the damage done to the electric motor is in some cases irreparable. You may have to buy a new saw or drill. Double-check Table 1-1 for the proper wire size in your extension cord.

Portable saw This is the most often used and abused of carpenter's equipment. The electric portable saw, such as the one shown in Fig. 1-35, is used to cut all 2 × 4s and other dimensional lumber. It is used to cut off rafters. This saw is used to cut sheathing for roofs. It is used for almost every sawing job required in carpentry.

This saw has a guard over the blade. The guard should always be left intact. Do not remove the saw guard. If not held properly against the wood being cut, the saw can kick back and into your leg.

You should always wear safety glasses when using this saw. The sawdust is thrown in a number of directions, and one of these is straight up toward your eyes. If you are watching a line where you are cutting, you definitely should have on glasses.

Table saw If the house has been enclosed, it is possible to bring in a table saw to handle the larger cutting jobs. See Fig. 1-36. You can do ripping a little

Fig. 1-35 *Portable power saw. The favorite power tool of every carpenter. Note the blade should not extend more than ⅛" below the wood being cut. Also note the direction of the blade rotation.*

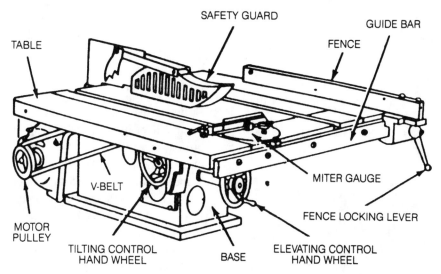

Fig. 1-36 *Table saw.* *(Power Tool Division, Rockwell International)*

more safely with this type of saw because it has a rip fence. If a push stick is used to push the wood through and past the blade, it is safe to operate. Do not remove the safety guard. This saw can be used for both crosscut and rip. The blade is lowered or raised to the thickness of the wood. It should protrude about ¼ to ½ inch above the wood being cut. This saw usually requires a 1-horsepower motor. This means it will draw about 6.5 amperes to run and over 35 to start. It is best not to run the saw on an extension cord. It should be wired directly to the power source with circuit breakers installed in the line.

Radial arm saw This type of saw is brought in only if the house can be locked up at night. The saw is expensive and too heavy to be moved every day. It should have its own circuit. The saw will draw a lot of current when it hits a knot while cutting wood. See Fig. 1-37.

Fig. 1-37 *Radial arm saw.* *(DeWalt)*

In this model the moving saw blade is pulled toward the operator. In the process of being pulled toward you, the blade rotates so that it forces the wood being cut against the bench stop. Just make sure your left hand is in the proper place when you pull the blade back with your right hand. It takes a lot of care to operate a saw of this type. The saw works well for cutting large-dimensional lumber. It will crosscut or rip. This saw will also do miter cuts at almost any angle. Once you become familiar with it, the saw can be used to bevel crosscut, bevel miter, bevel rip, and even cut circles. However, it does take practice to develop some degree of skill with this saw.

Router The router has a high-speed type of motor. It will slow down when overloaded. It takes the beginner some time to adjust to *feeding* the router properly. If you feed it too fast, it will stall or burn the edge you're routing. If you feed it too slowly, it may not cut the way you wish. You will have to practice with this tool for some time before you're ready to use it to make furniture. It can be used for routing holes where needed. It can be used to take the edges off laminated plastic on countertops. Use the correct bit, though. This type of tool can be used to the extent of the carpenter's imagination. See Fig. 1-38.

Saw blades There are a number of saw blades available for the portable, table, or radial saw. They may be standard steel types or they may be carbide tipped. Carbide-tipped blades tend to last longer. See Fig. 1-39.

Combination blades (those that can be used for both crosscut and rip) with a carbide tip give a smooth finish. They come in 7–7¼-inch diameter with 24 teeth. The arbor hole for mounting the blade on the saw is ¾ to ⅝ inch. A safety combination blade is also made in

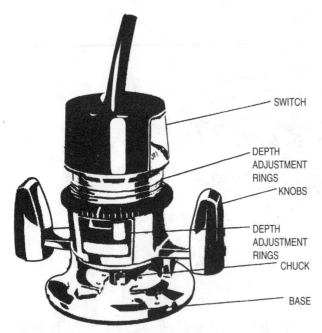

Fig. 1-38 The hand-held router has many uses in carpentry.

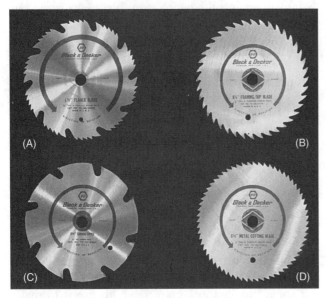

Fig. 1-39 Saw blades. (A) Planer blade; (B) Framing rip blade; (C) Carbide tipped; (D) Metal cutting blade. (Black & Decker)

10-inch-diameter size with 10 teeth and the same arbor hole sizes as the combination carbide-tipped blade.

The planer blade is used to crosscut, rip, or miter hard or soft woods. It is 6½ or 10 inches in diameter with 50 teeth. It too can fit anything from ¾- to ⅝-inch arbors.

If you want a smooth cut on plywood without the splinters that plywood can generate, you had better use a carbide-tipped plywood blade. It is equipped with 60 teeth and can be used to cut plywood, Formica, or laminated countertop plastic. It can also be used for straight cutoff work in hard or soft woods. Note the shape of the

saw teeth to get some idea as to how each is designed for a specific job. You can identify these after using them for some time. Until you can, mark them with a grease pencil or marking pen when you take them off. A teflon coated blade works better when cutting treated lumber.

Saber saw   The saber saw has a blade that can be used to cut circles in wood. See Fig. 1-40. It can be used to cut around any circle or curve. If you are making an inside cut, it is best to drill a starter hole first. Then, insert the blade into the hole and follow your mark. The saber saw is especially useful in cutting out holes for heat ducts in flooring. Another use for this type of saw is cutting holes in roof sheathing for pipes and other protrusions. The saw blade is mounted so that it cuts on the upward stroke. With a fence attached, the saw can also do ripping.

Drill   The portable power drill is used by carpenters for many tasks. Holes must be drilled in soleplates for anchor bolts. Using an electric power drill (Fig. 1-41) is faster and easier than drilling by hand. This drill is capable of drilling almost any size hole through dimensional lumber. A drill bit with a carbide tip enables the carpenter to drill in concrete as well as bricks. Carpenters use this type of masonry hole to insert anchor bolts in concrete that has already hardened. Electrical boxes have to be mounted in drilled holes in the brick and concrete. The job can be made easier and can be more efficiently accomplished with the portable power hand drill.

The drill has a tough, durable plastic case. Plastic cases are safer when used where there is electrical work in progress.

Fig. 1-40 Saber saw.

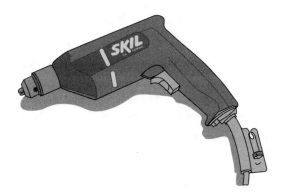

Fig. 1-41 *Hand-held portable drill.*

Fig. 1-43 *A cordless drill and a cordless saw using matching batteries.*

Carpenters are now using cordless electric drills (Fig. 1-42). Cordless drills can be moved about the job without the need for extension cords. Improved battery technology has made the cordless drills almost as powerful as regular electric drills. The cordless drill has numbers on the chuck to show the power applied to the shaft. Keep in mind that the higher the number, the greater the torque. At low power settings, the chuck will slip when the set level of power is reached. This allows the user to set the drill to drive screws.

Figure 1-43 shows a cordless drill and a cordless saw. This cordless technology is now used by carpenters and do-it-yourselfers. Cordless tools can be obtained in sets that use the same charger system (Fig. 1-44). An extra set of batteries should be kept charging at all times and then "swapped out" for the discharged ones. This way no time is lost waiting for the battery to reach full charge. Batteries for cordless tools are rated by battery voltage. High voltage gives more power than low voltage.

As a rule, battery-powered tools do not give the full power of regular tools. However, most jobs don't require full power. Uses for electric drills are limited

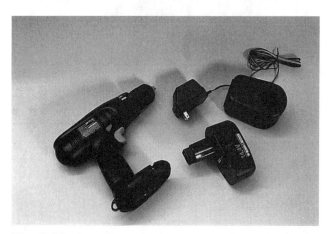

Fig. 1-44 *One charger can be used to charge saw and drill batteries of same voltage.*

only by the imagination of the user. The cordless feature is very handy when mounting countertops on cabinets. Sanding discs can be placed in the tool and used for finishing wood. Wall and roof parts are often screwed in place rather than nailed. Using the drill with special screwdriver bits can make the job faster than nailing.

Sanders The belt sander shown in Fig. 1-45 and the orbital sanders shown in Figs. 1-46A and B can do almost any required sanding job. The carpenter needs the sander occasionally. It helps align parts properly, especially those that don't fit by just a small amount. The sander can be used to finish off windows, doors, counters, cabinets, and floors. A larger model of the belt sander is used to sand floors before they are sealed and varnished. The orbital or vibrating sanders are used primarily to put a very fine finish on a piece of wood. Sandpaper is attached to the bottom of the sander. The sander is held by hand over the area to be sanded. The operator has to

Fig. 1-42 *A cordless hand drill with variable torque*

Fig. 1-45 *Belt sander.* (Black & Decker)

(A)

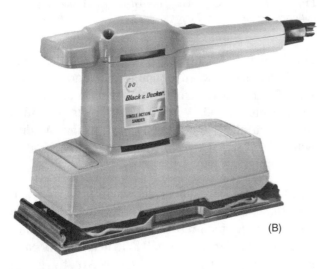

(B)

Fig. 1-46 *Orbital sanders: (A) dual action and (B) single action.*
(Black & Decker)

remove the sanding dust occasionally to see how well the job is progressing.

Nailers One of the greatest tools the carpenter has acquired recently is the nailer. See Fig. 1-47. It can drive nails or staples into wood better than a hammer. The nailer is operated by compressed air. The staples and nails are especially designed to be driven by the machine. See Tables 1-2 and 1-3 for the variety of fasteners used with this type of machine. The stapler or nailer can also be used to install siding or trim around a window.

The tool's low air pressure requirements (60 to 90 pounds per square inch) allow it to be moved from place to place. Nails for this machine (Fig. 1-47) are from 6d to 16d. It is magazine-fed for rapid use. Just pull the trigger.

Fig. 1-47 *Air-powered nailer.* (Duo-Fast)

2
CHAPTER

Building Floor Frames

IN THIS UNIT YOU WILL LEARN HOW TO BUILD frame floors. You will learn how to make floors over basements and crawl spaces. You will also learn how to make openings for stairs and other things. Overall, you will learn how to:

- Connect the floor to the foundation
- Place needed girders and supports
- Lay out the joist spacings
- Measure and cut the parts
- Put the floor frame together
- Lay the subflooring
- Build special framing
- Alter a standard floor frame to save energy

INTRODUCTION

Floors form the base for the rest of the building. Floor frames are built over basements and crawl spaces. Houses built on concrete slabs do not have floor frames. However, multilevel buildings may have both slabs and floor frames.

First the foundation is laid. Then the floor frame is made of posts, beams, sill plates, joists, and a subfloor. When these are put together they form a level platform. The rest of the building is held up by this platform. The first wooden parts are called the *sill plates*. The sill plates are laid on the edges of the foundations. Often, additional supports are needed in the middle of the foundation area. See Fig. 2-1. These are called *midfloor supports* and may take several forms. These supports may be made of concrete or masonry. Wooden posts and metal columns are also used. Wooden timbers called *girders* are laid across the central supports. Floor joists then reach (span) from the sill on the foundation to the central girder. The floor

joists support the floor surface. The joists are supported by the sill and girder. These in turn rest on the foundation.

Two types of floor framing are used on multistory buildings. The most common is the platform type. In platform construction each floor is built separately. The other type is called the balloon frame. In balloon frames the wall studs reach from the sill to the top of the second floor. Floor frames are attached to the long wall studs. The two differ on how the wall and floor frames are connected. These will be covered in detail later.

SEQUENCE

The carpenter should build the floor frame in this sequence:

1. Check the level of foundation and supports.
2. Lay sill seals, termite shields, etc.
3. Lay the sill.
4. Lay girders.
5. Select joist style and spacing.
6. Lay out joists for openings and partitions.
7. Cut joists to length and shape.
8. Set joists.
   a. Lay in place.
   b. Nail opening frame.
   c. Nail regular joists.
9. Cut scabs, trim joist edges.
10. Nail bridging at tops.
11. Lay subfloor.
12. Nail bridging at bottom.
13. Trim floor at ends and edges.
14. Cut special openings in floor.

SILL PLACEMENT

The sill is the first wooden part attached to the foundation. However, other things must be done before the sill is laid. When the anchors and foundation surface are adequate, a seal must be placed on the foundation. The seal may be a roll of insulation material or caulking. If a metal termite shield is used, it is placed over the seal. Next, the sill is prepared and fitted over the anchor bolts onto the foundation. See Fig. 2-2.

The seal forms a barrier to moisture and insects. Roll-insulation-type material, as in Fig. 2-3, may be used. The roll should be laid in one continuous strip with no joints. At corners, the rolls should overlap about 2 inches.

Fig. 2-1 *The piers and foundation walls will help support the floor frame.*

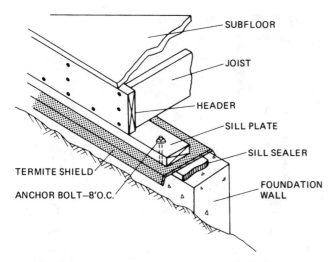

Fig. 2-2 *Section showing floor, joists, and sill placement.*

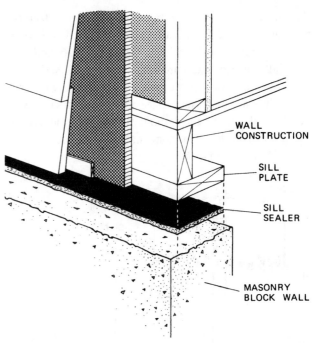

Fig. 2-3 *A seal fills in between the top of the foundation wall and the sill. It helps conserve energy by making the sill more weathertight. (Conwed)*

To protect against termites, two things are often done. A solid masonry top is used. Metal shields are also used. Some foundation walls are built of brick or concrete block that have hollow spaces. They are sealed with mortar or concrete on the top. A solid concrete foundation provides the best protection from termite penetration.

However, termites can penetrate cracks in masonry. Termites can enter a crack as small as 1/64 inch in width. Metal termite shields are used in many parts of the country. Figure 2-4 shows a termite shield installed.

Anchor the Sill

The sill must be anchored to the foundation. The anchors keep the frame from sliding from the foundation. They also keep the building from lifting in high winds. Three methods are used to anchor the sill to the foun-

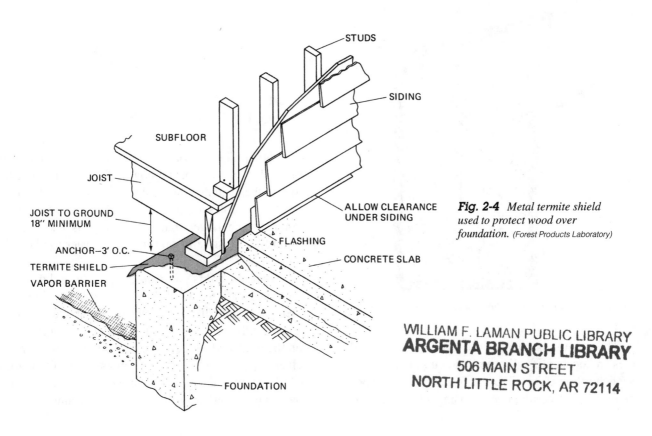

Fig. 2-4 *Metal termite shield used to protect wood over foundation. (Forest Products Laboratory)*

dation. The first uses bolts embedded in the foundation, as in Fig. 2-5. Sill straps are also used and so are special drilled bolt anchors. See Figs. 2-6 and 2-7. Special masonry nails are also sometimes used; however, they are not recommended for anchoring exterior walls. It is not necessary to use many anchors per wall. Anchors should be used about every 4 feet, depending upon local codes. Anchors may not be required on walls shorter than 4 feet.

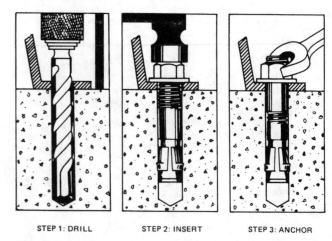

STEP 1: DRILL   STEP 2: INSERT   STEP 3: ANCHOR

Fig. 2-7 Anchor holes may be drilled after the concrete has set.
(Hilti-Fastening Systems)

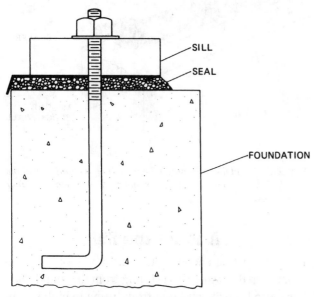

Fig. 2-5 Anchor bolt in foundation.

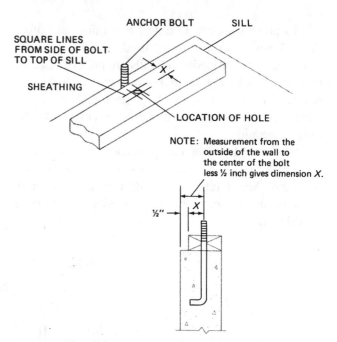

NOTE: Measurement from the outside of the wall to the center of the bolt less ½ inch gives dimension X.

Fig. 2-8 Locate the holes for anchor bolts.

Fig. 2-6 Anchor straps or clips can be used to anchor the sill.

The anchor bolts must fit through the sill. The holes are located first. Washers and nuts are taken from the anchor bolts. The sill board is laid next to the bolts. See Fig. 2-8. Lines are marked using a framing square as a guide. The sheathing thickness is subtracted from one-half the width of the board. This distance is used to find the center of the hole for each anchor. The centers for the holes are then marked. As a rule the hole is bored ¼ inch larger than the bolt. This leaves some room for adjustments and makes it easier to place the sill.

Next, the sill is put over the anchors and the spacing and locations are checked. All sills are fitted and then removed. Sill sealer and termite shields are laid, and the sills are replaced. The washers and nuts are put on the bolts and tightened. The sill is checked for levelness and straightness. Low spots in the foundation can be shimmed with wooden wedges. However, it is best to use grout or mortar to level the foundation.

Special masonry nails may be used to anchor interior walls on slabs. These are driven by sledge hammers or by nail guns. The nail mainly prevents side slippage of the wall. Figure 2-9 shows a nail gun application.

Fig. 2-9 *A nail gun can be used to drive nails in the slab and for toenailing.* (Duo-Fast)

Setting Girders

Girders support the joists on one end. Usually the girder is placed halfway between the outside walls. The distance between the supports is called the *span*. The span on most houses is too great for joists to reach from wall to wall. Central support is given by girders.

Determine girder location Plans give the general spacing for supports and girders. Spans up to 14 feet are common for 2-×-10-inch or 2-×-12-inch lumber. The girder is laid across the leveled girder supports. A chalk line may be used to check the level. The support may be shimmed with mortar, grout, or wooden wedges. The supports are placed to equalize the span. They also help lower expense. The piers shown in Fig. 2-1 must be leveled for the floor frame.

The girder is often built by nailing boards together. Figure 2-10 shows a built-up girder. Girders are often made of either 2-×-10-inch or 2-×-12-inch lumber. Joints in the girder are staggered. The size of the girder and joists is also given on the plans.

There are several advantages to using built-up girders. First, thin boards are less expensive than thick ones. The lumber is more stable because it is drier. There is less shrinkage and movement of this type of girder. Wooden girders are also more fire-resistant than steel girders. Solid or laminated wooden girders take a long time to burn through. They do not sag or break until they have burned nearly through. Steel, on the other hand, will sag when it gets hot. It only takes a few minutes for steel to get hot enough to sag.

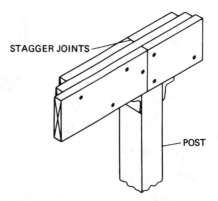

Fig. 2-10 *Built-up wood girder.* (Forest Products Laboratory)

The ends of girders can be supported in several ways. Figure 2-11A and B shows two methods. The ends of girders set in walls should be cut at an angle. See Fig. 2-12. In a fire, the beam may fall free. If the ends are cut at an angle, they will not break the wall.

Metal girders should have a wooden sill plate on top. This board forms a nail base for the joists. Basement girders are often supported by post jacks. See Fig. 2-13. Post jacks are used until the basement floor is finished. A support post, called a *lolly column*, may be built beneath the girder. It is usually made from 2 × 4 lumber. Walls may be built beneath the girder. In many areas this is done so that the basement may later be finished out as rooms.

JOISTS

Joists are the supports under the floor. They span from the sill to the girder. The subfloor is laid on the joists.

Lay Out the Joists

Joists are built in two basic ways. The first is the *platform method*. The platform method is the more common method today. The other method is called *balloon framing*. It is used for two-story buildings in some areas. However, the platform method is more common for multistory buildings.

Joist spacing The most common spacing is 16 inches. This makes a strong floor support. It also allows the carpenter to use standard sizes. However, 12-inch and 24-inch spacing are also used. The spacing depends upon the weight the floor must carry. Weight comes from people, furniture, and snow, wind, and rain. Local building codes will often tell what joist spacings should be used.

Joist spacings are given by the distance from the center of one board to the center of the next. This is called the distance *on centers*. For a 16-inch spacing, it would be written, *16 inches O. C.*

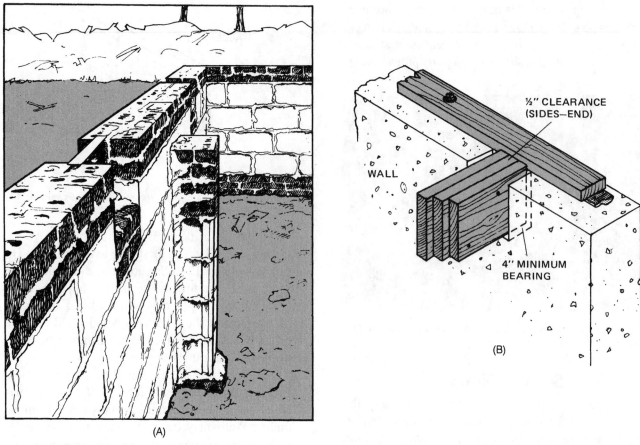

Fig. 2-11 *Two methods of supporting girder ends: (A) Projecting post; (B) Recessed pocket.* (Forest Products Laboratory)

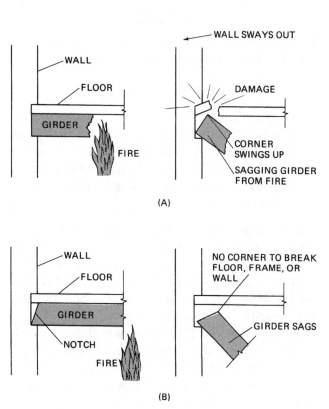

Fig. 2-12 *For solid walls, girder ends must be cut at an angle.*

Fig. 2-13 *A post jack supports the girder until a column or a wall is built.*

Modular spacings are 12, 16, or 24 inches O.C. These modules allow the carpenter to use standard-size sheet materials easily. The standard-size sheet is 48 × 96 inches (4 × 8 feet). Any of the modular sizes divides evenly into the standard sheet size. By using modules, the amount of cutting and fitting is greatly reduced. This is important since sheet materials are used on subfloors, floors, outside walls, inside walls, roof decks, and ceilings.

Joist layout for platform frames The position of the floor joists may be marked on a board called a *header*. The header fits across the end of the joists. See Fig. 2-14.

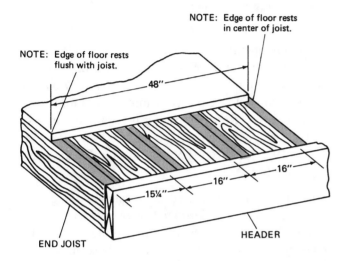

Fig. 2-14 *The positions of the floor joists may be marked on the header.*

Joist spacing is given by the distance between centers. However, the center of the board is a hard mark to use. It is much easier to mark the edge of a board. After all, if the centers are spaced right, the edges will be too!

The header is laid flat on the foundation. The end of the header is even with the end joist. The distance from the end of the header to the edge of the first joist is marked. However, this distance is not the same as the O.C. spacing. See Fig. 2-15. The first distance is always ¾ inch less than the spacing. This lets the edge of the flooring rest flush, or even, with the outside edge of the joist on the outside wall. This will make laying the flooring quicker and easier.

The rest of the marks are made at the regular O.C. spacing. See Fig. 2-15. As shown, an X indicates on which side of the line to put the joist.

Mark a pole It is faster to transfer marks than to measure each one. The spacing can be laid out on a board first. The board can then be used to transfer spacings. This board is called a *pole*. A pole saves time because measurements are done only once. To transfer the marks, the pole is laid next to a header. Use a square to

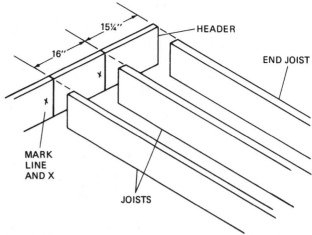

Fig. 2-15 *The first joist must be spaced ¾ inch less than the O.C. spacing used. Mark from end of header.*

project the spacing from the pole to the header. A square may be used to check the "square" of the line and mark.

Joists under walls Joists under walls are doubled. There are two ways of building a double joist. When the joist supports a wall, the two joists are nailed together. See Fig. 2-16. Pipes or vents sometimes go through the floors and walls. Then, a different method is used. See Fig. 2-17. The joists are spaced approximately 4 inches

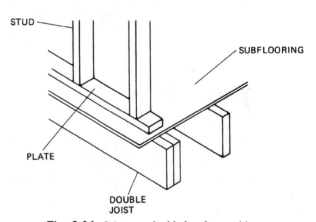

Fig. 2-16 *Joists are doubled under partitions.*

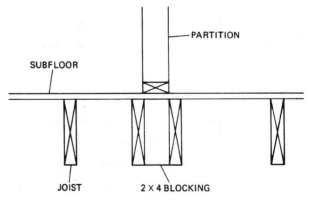

Fig. 2-17 *Double joists under a partition are spaced apart when pipes must go between them.*

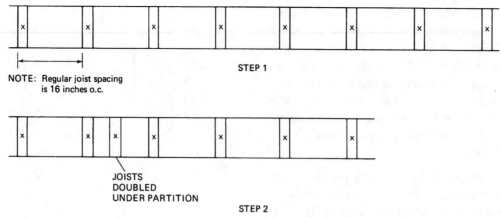

STEP 1

NOTE: Regular joist spacing is 16 inches o.c.

JOISTS DOUBLED UNDER PARTITION

STEP 2

Fig. 2-18 *Header joist layout. Next, add the joists for partitions.*

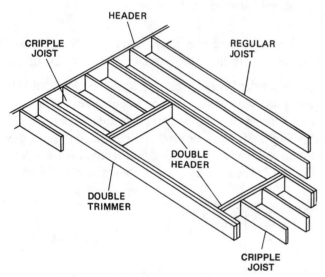

HEADER

CRIPPLE JOIST

REGULAR JOIST

DOUBLE HEADER

DOUBLE TRIMMER

CRIPPLE JOIST

Fig. 2-19 *Frame parts for a floor opening.*

apart. This space allows the passage of pipes or vents. Figure 2-18 shows the header layout pole with a partition added. Special blocking should be used in the double joist. Two or three blocks are used. The blocking serves as a fire stop and as bracing.

Joists for openings Openings are made in floors for stairs and chimneys. Double joists are used on the sides of openings. They are called *double trimmers.* Double trimmers are placed without regard for regular joist spacings. Regular spacing is continued on each end of the opening. Short "cripple" joists are used. See Fig. 2-19. A pole can show the spacing for the openings. See Fig. 2-20.

Girder spacing The joists are located on the girders also. Remember, marks do not show centerlines of the joists. Centerlines are hard to use, so marks show the edge of a board. These marks are easily seen.

Balloon layout Balloon framing is different. See Fig. 2-21. The wall studs rest on the sill. The joists and the studs are nailed together as shown. However, the end joists are nailed to the end wall studs.

The first joist is located back from the edge. The distance is the same as the wall thickness. The second joist is located by the first wall stud.

The wall stud is located first. The first edge of the stud is ¾ inch less than the O.C. spacing. For 16 inches O.C., the stud is 15¼ inches from the end. A 2-inch

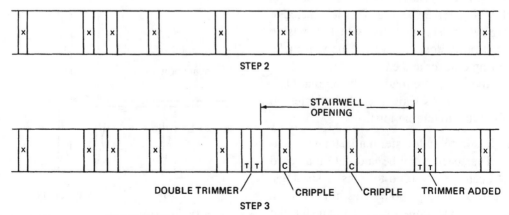

STEP 2

STAIRWELL OPENING

DOUBLE TRIMMER

CRIPPLE

CRIPPLE

TRIMMER ADDED

STEP 3

Fig. 2-20 *Add the trimmers for the opening to the layout pole.*

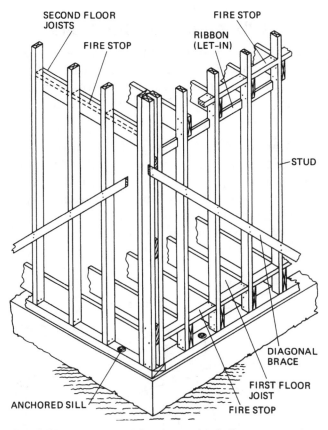

Fig. 2-21 *Joist and stud framing used in balloon construction.*
(Forest Products Laboratory)

stud will be 1½ inches thick. Thus, the edge of the first joist will be 16¾ inches from the edge.

Engineered Wood Joists

An alternative to solid dimensional lumber, which can warp as the humidity in your house changes throughout the year, are engineered wood joists. Various types and sizes are illustrated in Fig. 2-22. Engineered wood products can be used for floor joists and rafters. They are made in the shape of an I-beam. The top and bot-

tom sections are made of laminated or solid lumber that is grooved to allow for a thinner middle section. This middle section is typically composed of 3/8-inch-thick *oriented strand board* (OSB). The OSB is glued and pressed into the top and bottom grooved sections.

The rim joists used for this type of construction resemble solid dimensional lumber, however, they are manufactured from laminated veneer lumber for extra strength and increased dimensional stability. The main advantages of engineered wood products over standard solid dimensional lumber are as follows:

1. They are lighter in weight.
2. They have better strength to weight ratios.
3. Knockouts can be removed for cross ventilation and wiring;
4. They are more stable with no shrinkage-related call backs.
5. They can be manufactured in a continuous 30-foot length.
6. They are less expensive when reduced labor and material costs are considered.

The engineered wood joists are made in several sizes depending upon load and span. Sizes range from 9½ inches tall with 1½-inch-wide top and bottom flanges to 16 inches tall with 3-inch-wide top and bottom flanges (refer to Fig. 2-23). Installation of floor sheathing is done in the same manner as if you were using traditional solid lumber floor joists (Fig 2-24).

Because of the irregular "I" shape of the joists, they typically are not used around the perimeter as rim joists. The solid shape laminated veneer rim joists create a smooth side in which sheathing or siding can be attached easily. In addition, metal joist hangers also can be easily attached to the solid laminated veneer board. Because the size and types of joists vary among manufacturers, charts, tables, and building techniques brochures are provided wherever you purchase these products. Building techniques for joists are quite similar

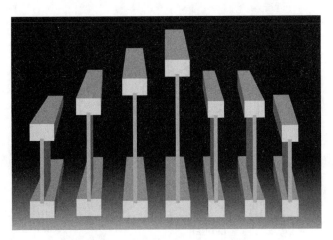

Fig. 2-22 *Types of engineered wood joists.*

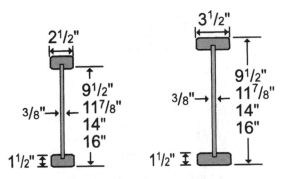

Fig. 2-23 *Sizes of engineered wood joists*

Fig. 2-24 *Installing engineered wood joists.*

Other Ways to Cut Joists

Ends of joists may be cut in other ways. The ends may be aligned and joined. Ends are cut square for some systems. For others, the ends are notched. Metal girders are sometimes used. Then, joists are cut to rest on metal girders.

End-joined joists The ends of the joists are cut square to fit together. The ends are then butted together as in Fig. 2-26. A gusset is nailed (10d) on each side to hold the joists together. Gussets may be made of either plywood or metal. This method saves lumber. Builders use it when they make several houses at one time.

to those used when framing with traditional solid dimensional lumber, which will be discussed in depth later in this chapter.

Cut Joists

The joists span, or reach, from the sill to the girder. Note that joists do not cover the full width of the sill. Space is left on the sill for the joist header. See Fig. 2-4. For lumber 2 inches thick, the spacing would be 1½ inches. Joists are cut so that they rest on the girder. Four inches of the joist should rest on the girder.

The quickest way to cut joists is to cut each end square. Figure 2-25 shows square-cut ends. The joists overlap across the girder. The ends rest on the sill with room for the joist header. This way the header fits even with the edge.

It is sometimes easier to put the joist header on after the joists are toenailed and spaced. Or, the joist header may be put down first. The joists are then just butted next to the header. But it is very important to carefully check the spacing of the joists before they are nailed to either the sill or the header.

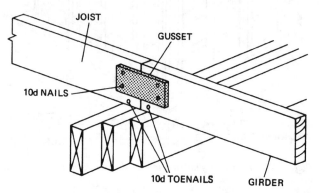

Fig. 2-26 *Joists may be butt-joined on the girder. Gussets may be used to hold them. Plywood subfloor may also hold the joists.*

Notched and lapped joists Girders may be notched and lapped. See Fig. 2-27. This connection has more interlocking but takes longer and costs more. First, the notch is cut on the end of the joists. Next a 2-×-4-inch joist support is nailed (16d) on the girder. Nails should be staggered 6 or 9 inches apart. The joists are then laid in place. The ends overlap across the girder.

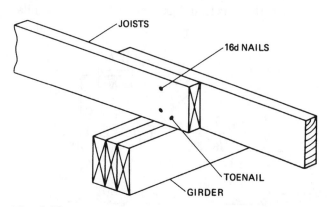

Fig. 2-25 *Joists may overlap on the girder. The overlap may be long or short.*

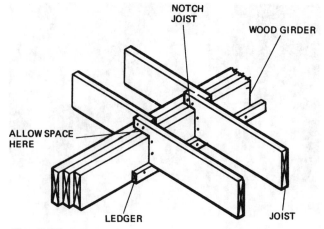

Fig. 2-27 *Joists can be notched and lapped on the girder.* (Forest Products Laboratory)

Joist-girder butts This is a quick method. With it the top of the joist can be even with the top of the girder.

A 2-×-4-inch ledger is nailed to the girder with 16d common nails. See Fig. 2-28. The joist rests on the ledger and not on the girder. This method is not as strong.

Also, a board is used to join the girder ends. The board is called a *scab*. The scab also makes a surface for the subfloor. It is a 2-×-4-inch board. It is nailed with three 16d nails on each end.

Joist hangers Joist hangers are metal brackets. See Fig. 2-29. The brackets hold up the joist. They are nailed (10d) to the girder. The joist ends are cut square. Then, the joist is placed into the hanger. It is also nailed with 10d nails as in Fig. 2-29. Using joist hangers saves time. The carpenter need not cut notches or nail up ledgers.

Joists for metal girders Joists must be cut to fit into metal girders. See Fig. 2-30. A 2-×-4-inch board is first bolted to the metal girder. The ends of the joist are then beveled. This lets the joist fit into the metal girder. The joist rests on the board. The board also is a nail base for the joist. The tops of the joists must be scabbed. The scabs are made of 2-×-4-inch boards. Three 16d nails are driven into each end.

Setting the Joists

Two jobs are involved in setting joists. The first is laying the joists in place. The second is nailing the joists. The carpenter should follow a given sequence.

Lay the joists in place First, the header is toenailed in place. Then the full-length joists are cut. Then they are laid by the marks on the sill or header. Each side of the joist is then toenailed (10d) to the sill. See Fig. 2-31. Joists next to openings are not nailed. Next, the ends of the joists are toenailed (10d) on the girder. Then the overlapped ends of the joists are nailed (16d) together.

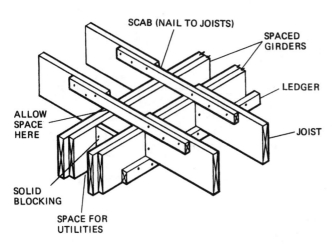

Fig. 2-28 *Joists may be butted against the girder. Note that the girder may be spaced for pipes, etc.* (Forest Products Laboratory)

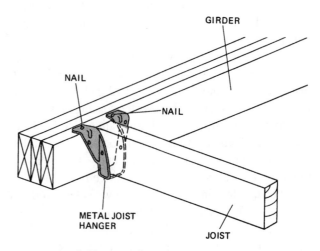

Fig. 2-29 *Using metal joist hangers saves time.*

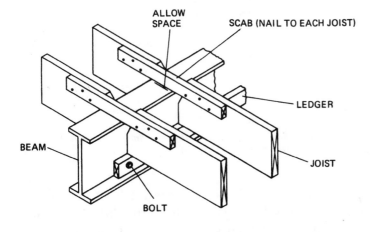

(A)

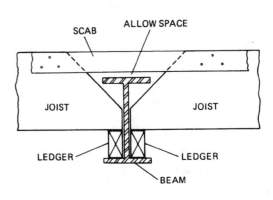

(B)

Fig. 2-30 *Systems for joining joists to metal girders.* (Forest Products Laboratory)

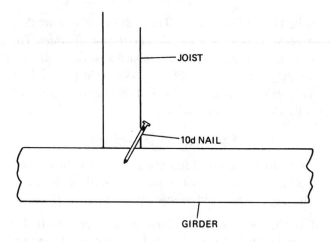

Fig. 2-31 *The joists are toenailed to the girders.*

See Fig. 2-25. These nails are driven at an angle, as in Fig. 2-32.

Nail opening frame A special sequence must be used around the openings. The regular joists next to the opening should not be nailed down. The opening joists are nailed (16d) in place first. These are called *trimmers*.

Then, the first *headers* for openings are nailed (16d) in place. Note that two headers are used. For 2-×-10-inch joists, three nails are used. For 2-×-12 inch lumber, four nails are used. Figure 2-33 shows the spacing of the nails.

Next, short cripple or tail joists are nailed in place. They span from the first header to the joist header. Three 16d nails are driven at each end. Then, the second header is nailed (16d) in place. See Fig. 2-34.

The double trimmer joist is now nailed (16d) in place. These pieces are nailed next to the opening. The nails are alternated top and bottom. See Fig. 2-35. This finishes the opening. The regular joists next to the opening are nailed in place. Finally, the header is nailed to the joist ends. Three 16d nails are driven into each joist.

Fire Stops

Fire stops are short pieces nailed between joists and studs. See Fig. 2-36. They are made of the same boards

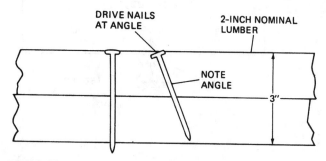

Fig. 2-32 *When nailing joists together, drive nails at an angle. This holds better and the ends do not stick through.*

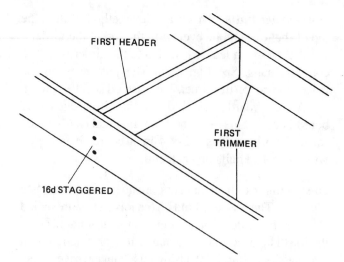

Fig. 2-33 *Nailing the first parts of an opening.*

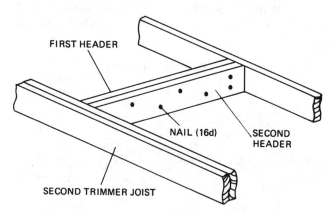

Fig. 2-34 *Add the second header and trimmer joists.*

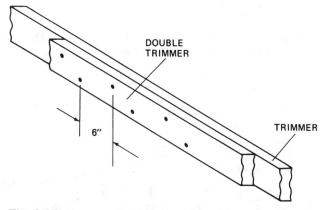

Fig. 2-35 *Stagger nails on double trimmer. Alternate nails on top and bottom.*

as the joists. Fire stops keep fire from spreading between walls and floors. They also help keep joists from twisting and spreading. Fire stops are usually put at or near the girder. Two 16d nails are driven at each end of the stop. Stagger the boards slightly as shown. This makes it easy to nail them in place.

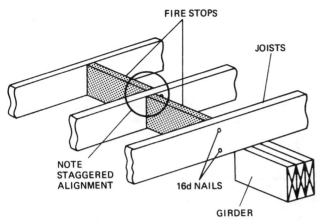

Fig. 2-36 *Fire stops are nailed in. They keep fire from spreading between walls and floors.*

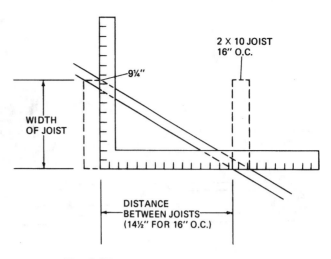

Fig. 2-37 *Use a square to lay out bridging.*

Bridging

Bridging is used to keep joists from twisting or bending. Bridging is centered between the girder and the header. For most spans, center bridging is adequate. For joist spans longer than 16 feet, more bridging is used. Bridging should be put in every 8 feet. This must be done to comply with most building codes.

Most bridging is cut from boards. It may be cut from either 1-inch or 2-inch lumber. Use the framing square to mark the angles as in Fig. 2-37. With this method, the angle may be found.

A radial arm saw may be used to cut multiple pieces. See Fig. 2-38. Also, a jig can be built to use a portable power saw. See Fig. 2-39.

Special steel bridging is also used. Figure 2-40 shows an example. Often, only one nail is needed in each end. Steel bridging meets most codes and standards.

All the bridging pieces should be cut first. Nails are driven into the bridging before it is put up. Two 8d or 10d nails are driven into each end. Next, a chalk line is strung across the tops of the joists. This gives a line for the bridging.

The bridging is nailed at the top first. This lets the carpenter space the joists for the flooring when it is laid. The bottoms of the bridging are nailed after the flooring is laid.

The bridging is staggered on either side of the chalk line. This prevents two pieces of bridging from being nailed at the same spot on a joist. To nail them both at the same place would cause the joist to split. See Fig. 2-41.

SUBFLOORS

The last step in making a floor frame is laying the subflooring. Subflooring is also called *underlayment*. The subfloor is the platform that supports the rest of the

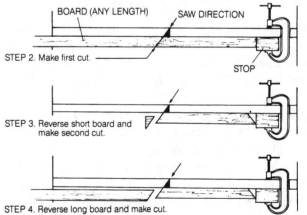

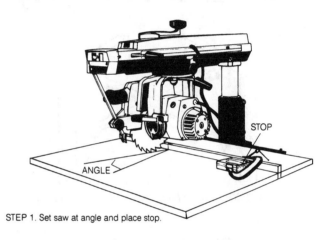

STEP 1. Set saw at angle and place stop.

STEP 2. Make first cut.

STEP 3. Reverse short board and make second cut.

STEP 4. Reverse long board and make cut.

Fig. 2-38 *Radial arm saw set up for bridging.*

structure. It is covered with a finish floor material in the living spaces. This may be of wood, carpeting, tile, or stone. However, the finish floor is added much later.

Several materials are commonly used for subflooring. The most common material is plywood. Plywood should be C—D grade with waterproof or exterior glues. Other materials used are chipboard, fiberboard, and boards.

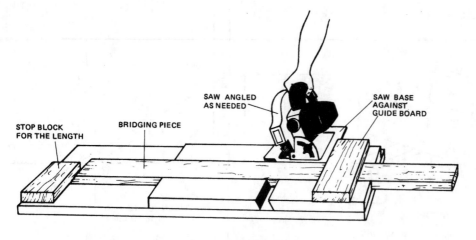

Fig. 2-39 A jig may be built to cut bridging.

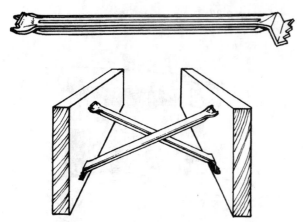

Fig. 2-40 Most building codes allow steel bridging.

Plywood Subfloor

Plywood is an ideal subflooring material. It is quickly laid and takes little cutting and trimming. It may be either nailed or glued to the joists. Plywood is very flat and smooth. This makes the finished floor smooth and easy to lay. Builders use thicknesses from ½- to ¾-inch plywood. The most common thicknesses are ½ and ⅝ inch. The FHA minimum is ½-inch-thick plywood.

Plywood as subflooring has fewer squeaks than boards. This is because fewer nails are required. The squeak in floors is caused when nails work loose. Table 2-1 shows minimum standards for plywood use.

Chipboard and Fiberboard

As a rule, plywood is stronger than other types of underlayment. However, both fiberboard and chipboard are also used. Chipboard underlayment is used more often. The minimum thickness for chipboard or fiberboard is ⅝ inch. This thickness must also be laid over 16-inch joist spacing. Both chipboard and fiberboard are laid in the same manner as plywood. In any case, the ends of the large sheets are staggered. See Fig. 2-42.

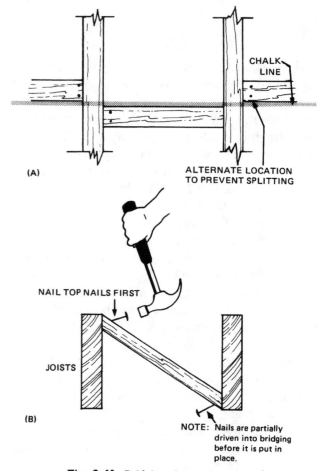

Fig. 2-41 Bridging pieces are staggered.

Laying Sheets

The same methods are used for any sheet subflooring. Nails are used most often, but glue is also used. An outside corner is used as the starting point. The long grain or sheet length is laid across the joists. See Fig. 2-43. The ends of the different courses are staggered. This prevents the ends from all lining up on one joist. If they did, it could weaken the floor. By staggering the end joints, each layer adds strength to the total floor.

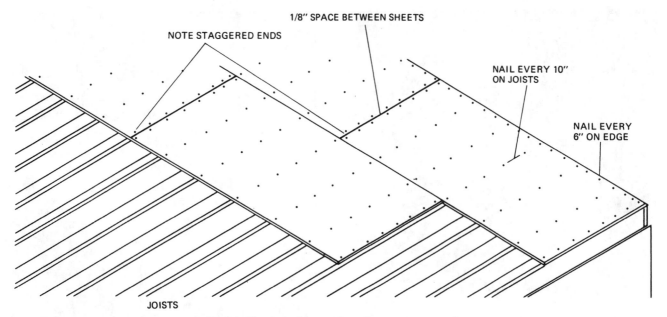

Fig. 2-42 *The ends of subfloor sheets are staggered.*

Table 2-1 *Minimum Flooring Standards*

Single-Layer (Resilient) Floor			
Joists, Inches O.C.	Minimum Thickness, Inches	Common Thickness, Inches	Minimum Index
12	19/32	5/8	24/12
16	5/8	5/8 or 3/4	32/16
24	3/4	3/4	48/24
Subflooring with Finish Floor Layer Applied			
Joists, Inches O.C.	Minimum Thickness, Inches	Common Thickness, Inches	Minimum Index
12	1/2	1/2 or 5/8	32/16
16	1/2	5/8	32/16
24	3/4	3/4	48/24

NOTES: 1. C-C grade underlayment plywood.
2. Each piece must be continuous over two spans.
3. Sizes can vary with span and depth of joists in some locations.

The carpenter must allow for expansion and contraction. To do this, the sheets are spaced slightly apart. A paper match cover may be used for spacing. Its thickness is about the correct space.

Nailing The outside edges are nailed first with 8d nails. Special "sinker" nails may be used. The outside nails should be driven about 6 inches apart. Nails are driven into the inner joists about 10 inches apart. See Fig. 2-44. Power nailers can be used to save time, cost, and effort. See Fig. 2-45.

Gluing Gluing is now widely used for subflooring. Modern glues are strong and durable. Glues, also called adhesives, are quickly applied. Glue will not squeak as will nails. Figure 2-46 shows glue being applied to floor joists. Floors are also laid with tongue-and-groove joints (Fig. 2-47). Buffer boards are used to protect the edges of the boards as the panels are put in place. (See Fig. 2-48.)

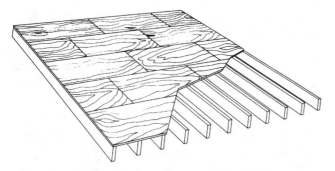

Fig. 2-43 *The long grain runs across the joists.*

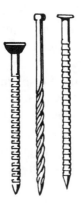

Fig. 2-44 *Flooring nails. Note the "sinker" head on the first nail.*

Fig. 2-45 *Using power nailers saves time and effort.* (Duo-Fast)

Fig. 2-46 *Subflooring is often glued to the joists. This makes the floor free of squeaks.* (American Plywood Association)

Fig. 2-47 *Plywood subflooring may also have tongue-and-groove joints. This is stronger.* (American Plywood Assocation)

Fig. 2-48 *A buffer board is used to protect the edges of tongue-and-groove panels.* (American Plywood Association)

Board Subflooring

Boards are also used for subflooring. There are two ways of using boards. The older method lays the boards diagonally across the joists. Figure 2-49 shows this. This way takes more time and trimming. It takes a longer time to lay the floor, and more material is wasted by trimming. However, diagonal flooring is still used. It is preferred where wood board finish flooring will be used. This way the finish flooring may be laid at right angles to the joists. Having two layers that run in different directions gives greater strength.

Today, board subflooring is often laid at right angles to the joists. This is appropriate when the finish floor will be sheets of material.

Either way, two kinds of boards are used. Plain boards are laid with a small space between the boards. It allows for expansion. End joints must be made over a joist for support. See Fig. 2-50. Grooved boards are also used. See Fig. 2-51. End joints may be made at any point with grooved boards.

Nailing Boards are laid from an outside edge toward the center. The first course is laid and nailed with 8d nails. Two nails are used for boards 6 inches wide or less. Three nails are used for boards wider than 6 inches.

The boards are nailed down untrimmed. The ends stick out over the edge of the floor. This is done for both grooved and plain boards. After the floor is done,

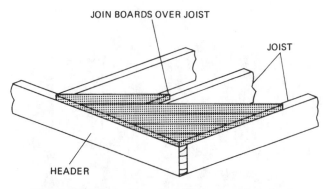

Fig. 2-49 *Diagonal board subfloors are still used today.*

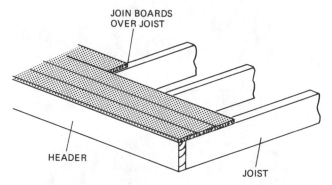

Fig. 2-50 *Plain board subfloor may be laid across joists. Joints must be made over a joist.*

Fig. 2-51 *Grooved flooring is laid across joists. Joints can be made anywhere.*

the ends are sawed off. They are sawed off even with the floor edges.

SPECIAL JOISTS

The carpenter should know how to make special joists. Several types of joists are used in some buildings. Special joists are used for overhangs and sunken floors. A sunken floor is any floor lower than the rest. Sunken floors are used for special flooring such as stone. Floors may also be lowered for appearance. Special joists are also used to recess floors into foundations. This is done to make a building look lower. This is called the *low-profile* building.

Overhangs

Overhangs are called *cantilevers*. They are used for special effects. Porches, decks, balconies, and projecting windows are all examples. Figure 2-52A shows an example of projecting windows. Figure 2-52B is a different type of bay. However, both rest on overhanging floor joist systems. Overhangs are also used for "garrison" style houses. When a second floor extends over the wall of the first, it is called a garrison style. See Fig. 2-53A and B.

The longest projection without special anchors is 24 inches. Windows and overhangs seldom extend 24 inches. A balcony, however, would extend more than 24 inches. Thus, a balcony would need special anchors.

Overhangs with joist direction Some overhangs project in the same direction as the floor joists. Little extra framing is needed for this. This is the easiest way to build overhangs. In this method, the joists are simply made longer. Blocking is nailed over the sill with 16d nails. Figure 2-54 shows blocking and headers for this type of overhang. Here the joists rest on the sill. Some overhangs extend over a wall instead of a foundation. Then, the double top plate of the lower wall supports the joists.

Overhangs at angles to joist direction Special construction is needed to frame this type of overhang. It is similar to framing openings in the floor frame. Stringer joists form the base for the subflooring. Stringer joists must be nailed to the main floor joists. See Fig. 2-55. They must be inset twice the distance of the overhang. Two methods of attaching the stringers are used. The first method is to use a wooden ledger. However, this ledger is placed on the top. See Fig. 2-56. The other method uses a metal joist hanger. Special anchors are needed for large overhangs such as rooms or decks.

Sunken Floors

Subfloors are lowered for two main reasons. A finish floor may be made lower than an adjoining finish floor for appearance. Or the subfloor may be made lower to accommodate a finish floor of a different material. The different flooring could be stone, tile, brick, or concrete. These materials are used for appearance or to drain

(A)

(B)

Fig. 2-52 *(A) A bay window rests on overhanging floor joists.*
(American Plywood Association); *(B) A different type of bay. Both types rest on overhanging floor joist systems.*

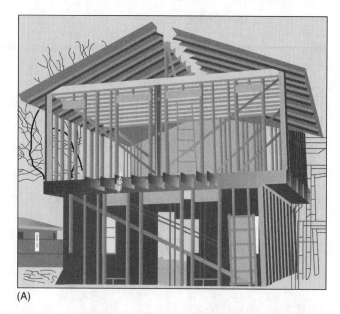

(A)

(B)

Fig. 2-53 *(A) Joist protections from the overhang for a garrison type second story; (B) The finished house.*

water. However, they are thicker than most finish floors. To make the floor level, special framing is done to lower the subfloor.

The sunken portion is framed like a special opening. First, header joists are nailed (16d) in place. See Fig. 2-57. The headers are not as deep, or wide, as the main joists. This lowers the floor level. To carry the load

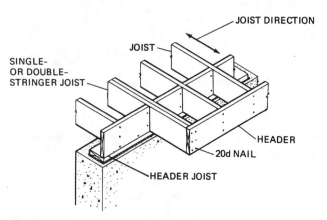

Fig. 2-54 *Some overhangs simply extend the regular joist. (Forest Products Laboratory)*

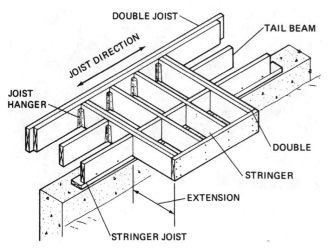

Fig. 2-55 *Frame for an overhang at an angle to the joists.* (Forest Products Laboratory)

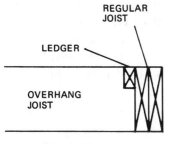

Fig. 2-56 *A top ledger "let in" is a good anchor.*

with thinner boards, more headers are used. The headers are added by spacing them closer together. Double joists are nailed (16d) after the headers.

Low Profiles

The lower profile home has a regular size frame. However, the subfloor and walls are joined differently. Figure 2-58 shows the arrangement. The sill is below the

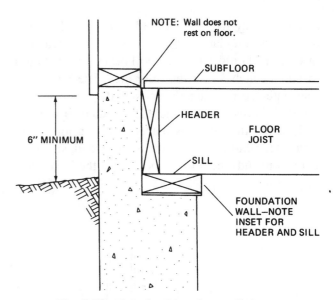

Fig. 2-58 *Floor detail for a low-profile house.*

top of the foundation. The bottom plate for the wall is attached to the foundation. The wall is not nailed to the subfloor. This makes the joists below the common foundation level. The building will appear to be lower than normal.

ENERGY FACTORS

Most energy is not lost through the floor. The most heat is lost through the ceiling. This is because heat rises. However, energy can be saved by insulating the floor. In the past, most floors were not insulated. Floors over basements need not be insulated. Floors over enclosed basements are the best energy savers.

Floors over crawl spaces should be insulated. The crawl space should also be totally enclosed. The foundation should have ventilation ports. But, they should be closed in winter. The most energy is saved by insulating certain areas. Floors under over-

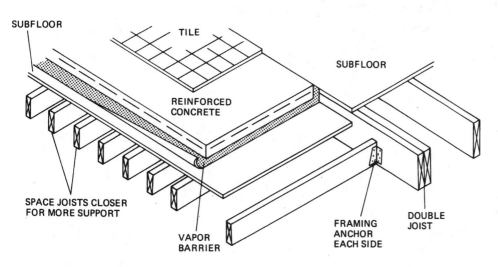

Fig. 2-57 *Details of frame for a sunken floor.*

hangs and bay windows should be insulated. Floors next to the foundation should also be insulated. The insulation should start at the sill or header. It should extend 12 inches into the floor area. See Fig. 2-59. The outer corners are the most critical areas. But, for the best results, the whole floor can be insulated. Roll or bat insulation is placed between joists and supported. Supports are made of wood strips or wire. Nail (6d) them to the bottom of the joists. See Fig. 2-60.

Fig. 2-60 *Insulation between floor joists should be supported.*

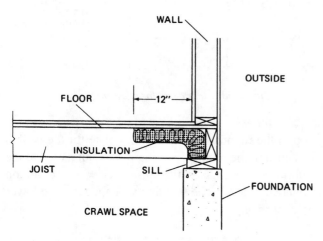

Fig. 2-59 *Insulate the outside floor edges.*

Moisture Barriers

Basements and slabs must have moisture barriers beneath them. Moisture barriers are not needed under a floor over a basement. However, floors over crawl spaces should have moisture barriers. The moisture barrier is laid over the subfloor. See Fig. 2-61. A moisture barrier may be added to older floors below the joists. This may be held in place by either wooden strips or wires. Six-mil plastic or builder's felt is used. See Fig. 2-62.

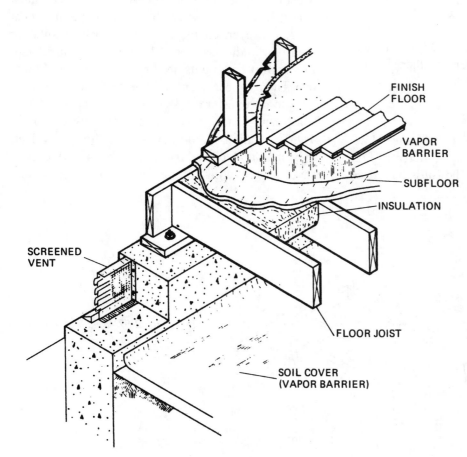

Fig. 2-61 *A moisture barrier is laid over the subfloor above a crawl space.* (Forest Products Laboratory)

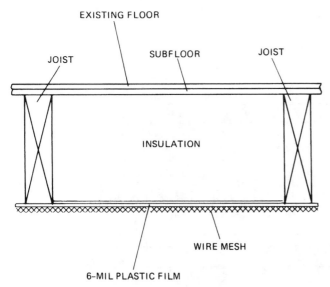

Fig. 2-62 *A moisture barrier may also be added beneath floors.*

Energy Plenums

A *plenum* is a space for controlled air. The air is pressurized a little more than normal. Plenum systems over crawl spaces allow air to circulate beneath floors. This maximizes the heating and cooling effects. Figure 2-63 shows how the air is circulated. Doing this keeps the temperature more even. Even temperatures are more efficient and comfortable.

The plenum must be carefully built. Insulation is used in special areas. See Fig. 2-64. A hatch is needed for plenum floors. The hatch gives access to the plenum area. Access is needed for inspection and servicing. There are no outside doors or vents to the plenum.

The plenum arrangement offers an advantage to the builder: A plenum house can be built more cheaply. There are several reasons for this. The circulation system is simpler. No ducts are built beneath floors or in attics. Common vents are cut in the floors of all the rooms. The system forces air into the sealed plenum

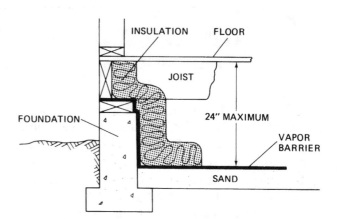

Fig. 2-64 *Section view of the energy plenum.*

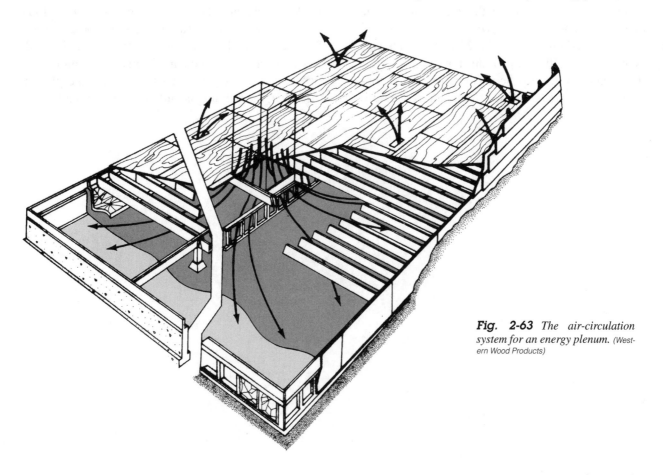

Fig. 2-63 *The air-circulation system for an energy plenum.* (Western Wood Products)

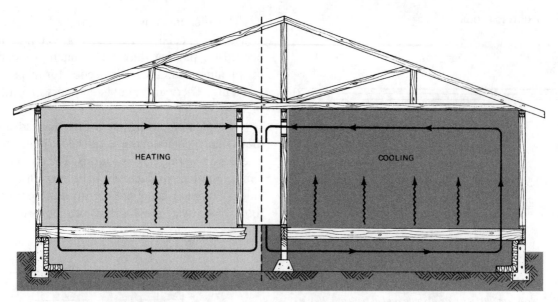

Fig. 2-65 *Conditioned air is circulated through the plenum.* (Western Wood Products)

(see Fig. 2-65). The air does not lose energy in the insulated plenum. The forced air then enters the various rooms from the plenum. The blower unit is in a central portion of the house. The blower can send air evenly from a central area. The enclosed louvered space lets the air return freely to the blower.

Rough plumbing is brought into the crawl space first. Then the foundation is laid. Fuel lines and cleanouts are located outside the crawl space. The minimum clearance in the crawl space should be 18 inches. The maximum should be 24 inches. This size gives the greatest efficiency for air movement.

Foundation walls may be masonry or poured concrete. Special treated plywood foundations are also used. Proper drainage is essential. After the foundation is built, the sill is anchored. Standard sills, seals, and termite shields may be used. The plenum area must be covered with sand. Next, a vapor barrier is laid over the ground and extends up over the sill. See Fig. 2-64. This completely seals the plenum area (Fig. 2-65). Then insulation is laid. Either rigid or batt insulation may be used. It should extend from the sill to about 24 inches inside the plenum. The most energy loss occurs at foundation corners. The insulation covers these corners. Then, the floor joists are nailed to the sill. The joist header is nailed (16d) on and insulated. Subflooring is then nailed to the joists. This completes the plenum. After this, the building is built as a normal platform.

3
CHAPTER

Framing
Walls

HOW TO BUILD A WALL FRAME IS THE TOPIC of this unit. How to cut the parts for the wall is covered, then how to connect the wall to the rest of the building. Why the parts are made as they are is also explained. You will learn how to:

• Lay out wall sections
• Measure and cut the parts
• Assemble and erect wall sections
• Join wall sections together
• Change a standard wall frame to save energy and materials

INTRODUCTION

There are two ways of framing a building. The most common is called the *western platform method*. The other is the *balloon method*. In most buildings today, the western platform method is used.

In the western platform method, walls are put up after the subfloor has been laid. Walls are started by making a frame. The frame is made by nailing boards to the tops and bottoms of other boards. The top and bottom boards are called *plates*. The vertical boards are called *studs*. The frame must be made very strong because it holds up the roof. After the frame is put together, it is raised and nailed in place. The wall frames are put up one at a time. A roof can be built next. Then the walls are covered and windows and doors are installed. Figure 3-1 shows a wall frame in place. Note that the roof is not on the building yet.

Fig. 3-1 *Wall frames are put up after the floor is built.*

Sometimes the first covering for the wall is added before the wall is raised. This first covering is called *sheathing*. It is very easy to nail the sheathing on the frame while the frame is flat on the floor. Doing this makes the job quicker and reduces problems with

keeping the frame square. Another advantage is that no scaffolds will be needed to reach all the areas. This reduces the possibility of accidents and saves time in moving scaffolds. However, most builders still nail sheathing on after the wall is raised.

The wall is attached to the floor in most cases. When the floor is built over joists, the wall is nailed to the floor. See Fig. 3-2. When the wall is built on a slab, it is anchored to the slab. See Fig. 3-3. As a rule, walls for both types of floors are made the same way.

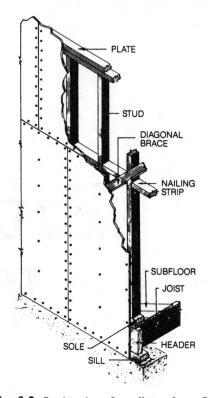

Fig. 3-2 *Section view of a wall on a frame floor.*

Walls that help hold up the roof, or the next floor, are made first. As a rule, all outside, or exterior, walls do this. These walls are also referred to as *load-bearing* walls.

Inside walls are called *partitions*. They can be load-bearing walls, too. However, not all partitions carry loads. Interior walls that do not carry loads may be built after the roof is up. Interior walls that do not carry loads are also called *curtain walls*.

After the walls are put up, the roof is built. The walls are then covered. The first cover is the sheathing. Putting on the sheathing after the roof lets a builder get the building waterproofed or weatherproofed a little sooner. The siding is put on much later.

Wall sections are made one at a time. The longest outside walls are made first. The end walls are made next. However, the sequence can be changed to fit the

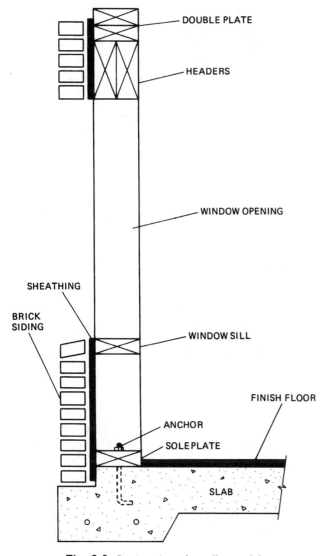

Fig. 3-3 *Section view of a wall on a slab.*

job. There are many ways to make walls. This chapter will show the most common method.

SEQUENCE

The general sequence for making wall frames is:

1. Lay out the longest outside wall section.
2. Cut the parts.
3. Nail the parts together.
4. Raise the wall.
5. Brace it in place.
6. Lay out the next wall.
7. Repeat the process.
8. Join the walls.
9. Do all outside walls.
10. Do all inside walls.

11. Build the roof.
12. Sheath the outside walls.
13. Install outside doors and windows.
14. Cut soleplates from inside door openings.

WALL LAYOUT

All the parts of the wall must be planned. The carpenter must know beforehand what parts to cut and where to nail. The carpenter plans the construction by making a *wall layout* on boards. One layout is done on the top and bottom boards (plates). Another layout is done for the wall studs.

Plate Layout

Soleplates   First, select pieces of 2-inch lumber for the bottom of the wall. The bottom part of the wall is called a *soleplate*. The soleplate is laid along the edges of the floor where the walls will be. No soleplate is put across large door openings. Sometimes, the soleplate can be made across small doors. After the wall is put up, the soleplate is sawed out.

Top plate   After a soleplate has been laid, another piece is laid beside it. It will be the top part of the wall. This piece is called a *top plate*. The soleplate and the top plate are laid next to each other with a flat side up. Figure 3-4 shows how the soleplate and the top plate are laid so that you can measure and mark both the top and bottom plates at the same time. Because it keeps both top and bottom locations aligned, this method ensures accuracy.

Sole- and top plates are often spliced. The splice must occur over the center of a full stud. See Fig. 3-5. Otherwise, the wall section will be weakened.

Stud Layout

Several types of studs are used in walls. The studs that run from the soleplate to the top plate are called *full studs*. Studs that run from the soleplate to the top of a rough opening are called *trimmer studs*. Short studs that run from either plate to a header or a sill are called *cripple studs*. Figure 3-6 shows a part of a wall section.

Spacing full studs   Most full studs are spaced a standard distance apart. This standard distance is an even part of the sizes of plywood, sheathing, and other building materials. The studs act both as roof support and as a nailing base for the sheathing. The most common standard distance is 16 inches. Studs spaced 16 inches apart are said to be 16 inches on *centers* (O.C.). Another common spacing is 24 inches O.C. However,

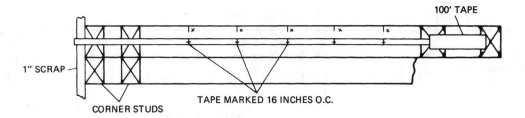

1" SCRAP

100' TAPE

CORNER STUDS

TAPE MARKED 16 INCHES O.C.

NAIL A PIECE OF 1 INCH SCRAP AT THE END. HOOK
THE TAPE OVER THE END. MARK THE INTERVAL.
THEN MAKE A SMALL X TO THE RIGHT OF THE MARK.

Fig. 3-4 Lay sole- and top
plates for marking.

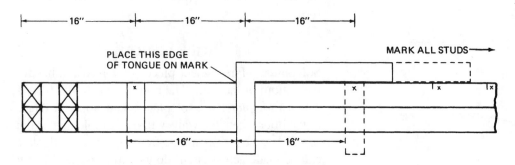

16" 16" 16"

PLACE THIS EDGE
OF TONGUE ON MARK

MARK ALL STUDS ➞

16" 16"

USE A 2 INCH SCRAP TO MARK WIDTHS. THEN
MARK STUD LOCATIONS.

Fig. 3-5 Top plates are spliced over a stud.

lines on centers are not used to show where boards are placed. An easier way is shown in Fig. 3-4. The stud is located by using the left end of the wall as a starting point. Nail a 1-inch scrap piece there. Then the distance between centers is measured from the block on the outside corner of the wall. A mark is made, and the X is made to the right of the mark.

After all of the marks are made, a square is used to line in the locations. This method is easier because the measurement is taken from the side of the stud. The distance is the same whether it is center-to-center or side-to-side.

Space the rough openings The next step in spacing is to find where the windows and doors are to be made. The openings for windows and doors are framed no matter what the stud spacing is. The openings are called rough openings (R.O.). The size of rough openings may be shown in different ways. The actual size may be shown in the plan. However, many plans show *window schedules*, which often list the window as a number, like "2442." This means that the window sash opening is 2 feet 4 inches wide and 4 feet 2 inches high. The rough openings are larger. Wooden windows require larger R.O.s than metal ones. It is best to refer to specifications. However, when they are not available, carpenters can use a rule of thumb: Wooden windows are usually written, for example, 24 × 42. The R.O. should be 3 inches wider (2'7") and 4 inches higher (4'6"). Metal frames are usually written without the X. The R.O. is 2 inches wider and 3 inches higher. For a 2442 metal window, the R.O. would be 2 feet 6 inches wide and 4 feet 5 inches high.

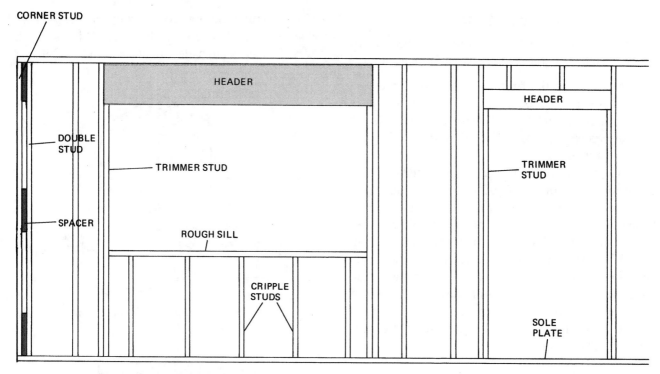

Fig. 3-6 *A wall section and parts. Note the large header over the opening. It is sometimes larger than needed, to eliminate the need for top cripples. This saves labor costs because it takes longer to cut and nail cripples.*

To find the locations of rough openings, measure the distance from the corner or end of the wall to the center of the rough opening. Make a mark called a *centerline* on the soleplate. From the centerline measure half of the rough opening width on each side. Make another mark at each side of the centerline for the rough opening. Mark a line, as in Fig. 3-7, to show the thickness of a stud. This thickness goes on the outside of the opening. Note that the distance between the lines must be the width of the opening. Mark a T in the regular stud spaces. This tells you that a *trimmer* stud is placed there. On the outside of the space for the trimmer stud, lay out another thickness. Mark an X from one corner of the opening to the other as shown in the figure. The X is used to show where a full stud is placed. The T is used to show where a trimmer stud is placed. The trimmer stud does not extend from the sole- to the top plate. Thus, it is shown only on the soleplate.

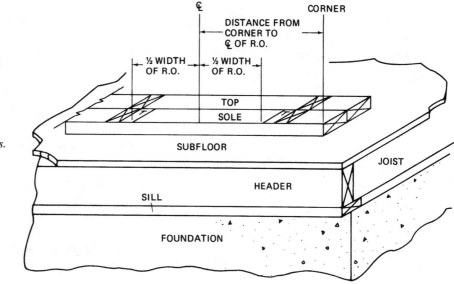

Fig. 3-7 *Locate rough openings.*

Corner Studs

A strong way of nailing the walls together is needed. To accomplish this, a double stud is used for one corner.

Corner studs on the first wall The spacing of the studs starts at one corner. Another stud space is marked at the second corner. The regular spacing is not used at corners. A stud is laid out at each corner, or end, of the wall.

To make the corner stronger, another stud is placed in the corner section. See Fig. 3-8A. The second corner stud is spaced from the first with spacer blocks. This is called a *built-up corner*. It is done to give the corner greater strength and to make a nail base on the inside. A nail base is needed on the inside to nail the inside wall covering in place. After the end stud is marked on the corner, mark the thickness of one more stud on the plate. Mark it with an S for *spacer*. Next, lay out the stud as shown in Fig. 3-8B. Mark it with an X to indicate a full stud.

Corner studs for the second wall The second wall does not need double studs. It is laid out in the regular way. The walls are nailed together as in Fig. 3-9.

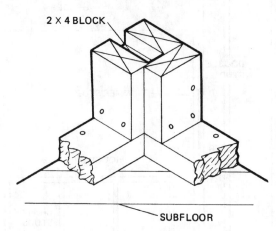

Fig. 3-9 *The second wall is nailed to the double corner.* (Forest Products Laboratory)

Partition Studs

As mentioned earlier, inside walls are called partitions. The partitions must be solidly nailed to the outside walls. To make a solid nail base, special studs are built into the exterior walls. See Fig. 3-10. The most common method is to place two studs as shown in Fig. 3-10. The studs are placed 1½ inches apart. The space is made just like corners on the first wall. The corner of the partition can now rest on the two studs. The two special studs act as a nail base for the wall. Also, ¾ inch of each stud is exposed. This makes a nail base for the interior wall covering. Other methods of joining partitions to the outside walls are shown in Figs. 3-11 and 3-12.

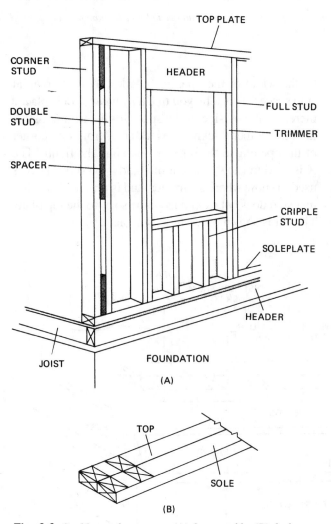

Fig. 3-8 *Stud layout for corners: (A) the assembly; (B) the layout.*

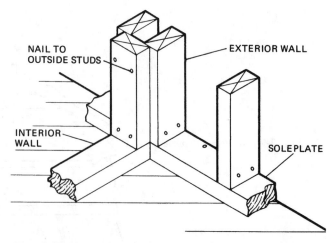

Fig. 3-10 *Partition walls are nailed to special studs in outside walls.* (Forest Products Laboratory)

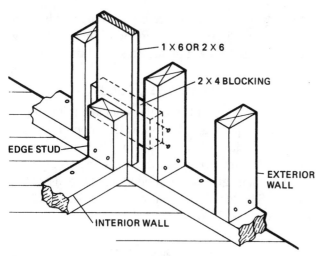

Fig. 3-11 *Another way to join partitions to the outside walls.* (Forest Products Laboratory)

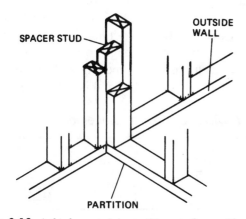

Fig. 3-12 *A third way to join partitions to the outside walls.*

Find Stud Length

Before cutting any studs, the carpenter must find the proper lengths. There are two ways of doing this. One way lets the carpenter measure the length. In the other way, the carpenter must compute it.

Make a master stud pattern A carpenter can make a master stud pattern. This is the way that stud lengths are measured. A 2-inch board just like the studs is used. A side view of the wall section is drawn. The side view is full-size and will show the stud lengths.

The carpenter starts by laying out the distance between the floor and the ceiling height. The distances in between are then shown. Rough openings are added. This then shows the lengths of trimmers and cripples. The pattern also lets the carpenter check the measurements before cutting the pieces. The stud pattern is also called a story pole or rod. As a rule, it is done for only one floor of the house. Figure 3-13 shows a master stud pattern.

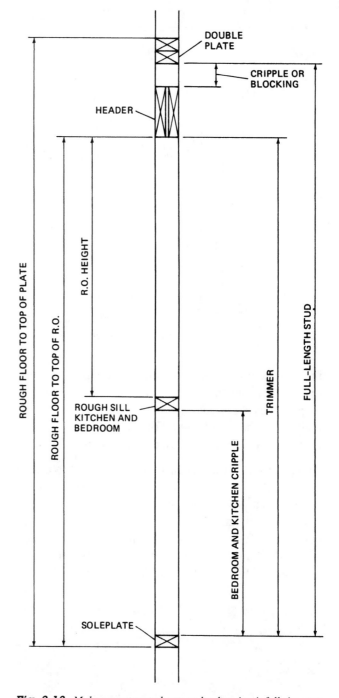

Fig. 3-13 *Make a master stud pattern by drawing it full size on a board.*

Compute stud length Another way to find stud lengths is to compute them. The carpenter must do some arithmetic and check it very carefully. This method is shown in Fig. 3-14.

The usual finished floor-to-ceiling distance is 8 feet ½ inch. The thicknesses of the finish floor and ceiling material are added to this dimension. The floor and ceiling thicknesses are commonly ½ inch each. The distance is written all in inches and adds up to 97½

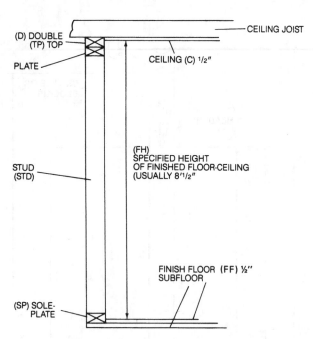

Fig. 3-14 *Computing a stud length.*

inches. The thickness of the sole- and top plates is subtracted. For one soleplate and a double top plate, this thickness is 4½ inches (3 times 1½ inches). The remainder is the stud length. For this example, the stud length is 93 inches. This is a commonly used length.

Using precut studs Sometimes the studs are cut to length at the mill and delivered to the site. When this is done, the carpenter does not make any measurements or cuts for full studs. The standard length for such precut studs is 93 inches. The carpenter who orders such materials should be very careful to specify "precision end trim" (P.E.T.) for lumber. See Fig. 3-15.

Fig. 3-15 *Precut studs ready at a site.*

Frame Rough Openings

The locations of full and trimmer studs for the rough opening are shown on the soleplates. The lengths of these can be found from the size of the rough opening.

The story pole is used for reference. The width of the R.O. sets the distance between the trimmers. See Fig. 3-6. A full stud is used on the outside of the trimmer.

Trimmer Studs

Trimmer studs extend from the soleplate to the top of the rough opening. They provide support for the header. The header must support the wall over the opening. It is important that the header be solidly held. The trimmers give solid support to the ends of the headers. The length of trimmers is the distance from the soleplate to the header.

Header Size

The size of the header is determined by the width of the rough opening. Table 3-1 shows the size for a typical opening width. As in Fig. 3-3, two header pieces are used over the rough opening. In some cases, the headers may be large enough to completely fill the space between the rough opening and the top plate. See Fig. 3-16. However, doing this will make the wall

Table 3-1 *Size of Lumber for Headers*

Width of Rough Opening	Minimum Size Lumber	
	One Story	Two or More
3'0"	2 × 4	2 × 4
3'6"	2 × 4	2 × 6
4'0"	2 × 6	2 × 6
4'6"	2 × 6	2 × 6
5'0"	2 × 6	2 × 6
6'0"	2 × 6	2 × 8
8'0"	2 × 8	2 × 10
10'0"	2 × 10	2 × 12
12'0"	2 × 12	2 × 12

Fig. 3-16 *Solid headers may be used instead of cripples.*

section around the header shrink and expand at a different rate than the rest of the wall. A solid header section is also harder to insulate.

The sill The bottom of the rough opening is framed in by a sill. The sill does not carry a load. Thus, it can simply be nailed between the trimmer studs at the bottom of the rough opening. The sill does not require solid support beneath the ends; however, it is common to use another short trimmer for this. See Fig. 3-17.

Fig. 3-17 A trimmer may be used under both the header and the sill.

Cripples Cripple studs are short studs. They join the header to the top plate. They also join the sill to the soleplate. They continue the regular stud spacing. This makes a nail base for sheathing and wallboard.

CUTTING STUDS TO LENGTH

After the wall parts are laid out, carpenters must find how many studs are needed. They must do this for each length. A great deal of time can be saved if all studs are cut at one time. Saws are set and full studs are cut first. Then the settings are changed and all trimmer studs are cut. Next, cripple studs are cut. Headers and sills should be cut last.

Cutting Tips

Most cuts to length are done with power saws. Two types are used. Radial arm saws can be moved to a location. See Fig. 3-18. Portable circular saws can be used almost anyplace. See Fig. 3-19. With either

Fig. 3-18 A radial arm saw can be used on the site.

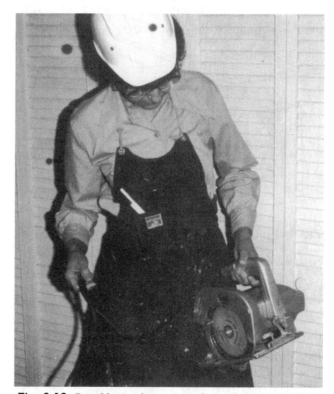

Fig. 3-19 Portable circular saws can be used almost anywhere.

type of saw, special setups can be used. Pieces can be cut to the right length without measuring each one.

Portable circular saw Sawing several pieces to the same length is done with a special jig. See Fig. 3-20. Two pieces of stud lumber are nailed to a base. Enough space is left between them for another stud. A stop block is nailed at one end. A guide board is then nailed across the two outside pieces. Care should be taken because the guide is for the saw frame. The blade cuts a few inches away.

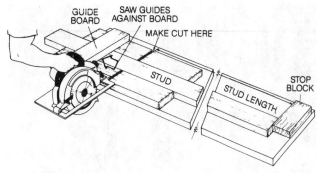

Fig. 3-20 *A jig for cutting studs with a portable circular saw.*

Radial arm saw A different method is used on a radial arm saw. No marking is done. Figure 3-21 shows how to set the stop block. This is quicker and easier. The piece to be cut is simply moved to touch the stop block. This sets the length. The piece is held against the back guide of the saw. The saw is then pulled through the work.

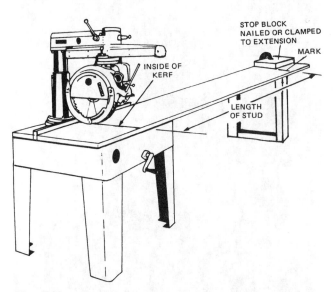

Fig. 3-21 *A stop block is used to set the length with a radial arm saw. This way all the studs can be cut without measuring. This saves much time and money.*

WALL ASSEMBLY

Once all pieces are cut, the wall may be assembled. Headers should be assembled before starting. See Figs. 3-22 and 3-23. The soleplate is moved about 4 inches away from the edge of the floor. It is laid flat with the stud markings on top. Then it is turned on edge. The markings should face toward the middle of the house. Then the top plate is moved away from the soleplate. The distance should be more than the length of a full stud. The soleplate and top plate markings should be aligned. They must point toward each other.

Fig. 3-22 *Use ½-inch plywood for spaces between headers.*

Fig. 3-23 *Nailers can be used to assemble headers.*

The straightest studs are selected for the corners. They are put at the corners just as they will be nailed. Next, a full stud is laid at every X location between the sole- and top plates. Figure 3-24 shows carpenters doing this.

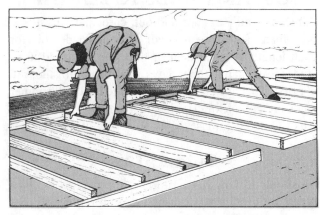

Fig. 3-24 *Laying the studs in place on the subfloor.*

Nailing Studs to Plates

All studs are laid in position. The soleplate and top plate are tapped into position. Make sure that each stud is within the marks.

Corner studs Before nailing, corner spacers are cut. Three spacers are used at each corner. Each spacer should be about 16 inches long. Spacers are put between the two studs at the corners. One spacer should be at the top, one should be at the bottom, and one should be in the center. Two 16d nails should be used on each side. See Fig. 3-25. After the corner studs are nailed together, they are nailed to the plates. Two 16d nails are used for each end of each stud.

Full studs All the full studs are nailed in place at the X marks. Two 16d nails are used at each end. Figures 3-26 and 3-27 show how.

Trimmer studs Next, the trimmer studs are laid against the full studs. The spacing of the rough openings is checked. The trimmers are then nailed to the studs from the trimmer side. See Fig. 3-28. Use 10d nails. The nails should be staggered and 16 inches apart. Staggered means that one nail is near the top edge and the next nail is near the bottom edge. See Fig. 3-29.

Headers and sills The headers are nailed in place next. The headers are placed flush with the edges of the studs. The headers are nailed in place with 16d nails. The nails are driven through the studs into the ends of the headers.

The sills are nailed in place after the headers. Locate the position of the sill. Toenail the ends of the sill to the trimmers. Also, another trimmer may be used. See Fig. 3-30.

Cripple studs Next, the cripples are laid out. They are nailed in place from the soleplate with two 16d nails. Then, two 16d nails are used to nail the sill to the ends of the cripples. Finally, the top cripples are nailed into place. See Fig. 3-31.

CORNER BRACES

Corner braces should be put on before the wall is raised. The bracing prevents the wall frame from swaying sideways. Two methods are commonly used.

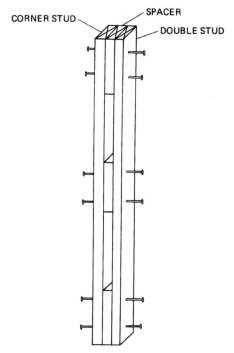

Fig. 3-25 *Nailing spacers into corner studs. Use two 16d nails on each side of spacer.*

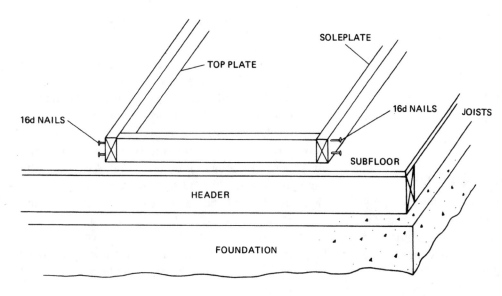

Fig. 3-26 *Nailing full studs to plates.*

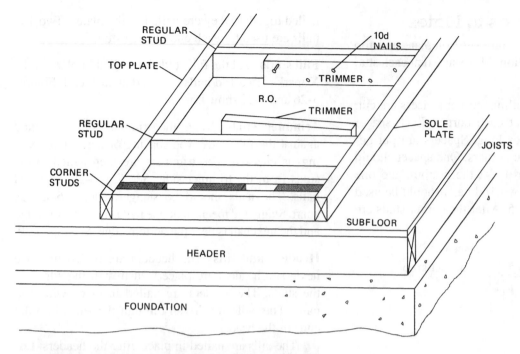

Fig. 3-27 Trimmers are nailed into the rough opening.

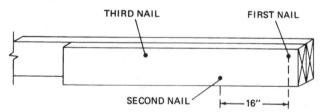

Fig. 3-28 Nailing studs for wall frames.

Fig. 3-29 In stagger nailing, one nail is near the top and the other is near the bottom.

Plywood Corner Braces

The first method uses plywood sheets as shown in Fig. 3-32. It is best to nail the plywood on before the wall is erected. Plywood bracing costs more but takes little time and effort. If used, it should be the same thickness as the wall sheathing. Plywood should not be used where "energy" sheathing is used. This will be discussed later.

Diagonal Corner Braces

The second method is to make a board brace. The brace is made from 1-×-4-inch lumber. It is "let in" or recessed into the studs. The angle may be any angle, but 45° is common. Braces are "let in" to the studs so that the outside surface of the wall will stay flat. See Fig. 3-33.

To make a diagonal brace, select the piece of wood to be used and nail it temporarily in place across the outside. Mark the layout on each stud. Also mark the angle on the ends of the brace. Remove the brace and gage the depth of the cut across each stud. Make the cuts with a saw. Knock out the wood between the cuts with a chisel. Then, trim off the ends of the brace. Place the brace in the "let in" slots. Check for proper fit. If the fit is good, nail the brace in place. Use two 8d nails at each stud.

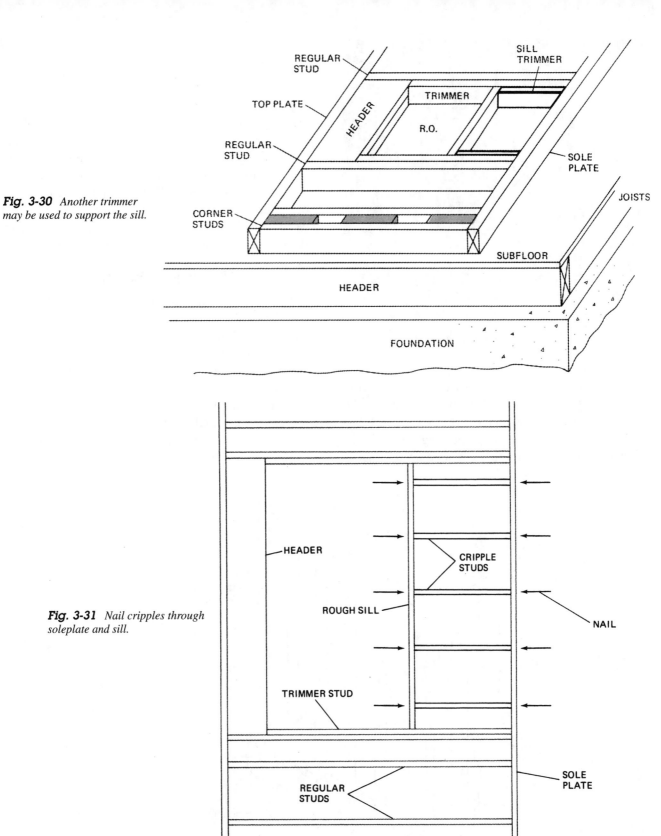

Fig. 3-30 *Another trimmer may be used to support the sill.*

Fig. 3-31 *Nail cripples through soleplate and sill.*

ERECT THE WALLS

Once the wall section is assembled, it is raised upright. The wall may be raised by hand. See Fig. 3-34. Special wall jacks may also be used, as in Fig. 3-35. Figure 3- 36 shows another method, raising the wall with a fork-lift unit. This is common when a large wall must be placed over anchor bolts on a slab. For slabs, the anchor holes in the soleplate should be drilled before the

Fig. 3-32 Plywood is used for corner bracing.

wall is raised (Fig. 3-37). See the section on anchoring sills in Chapter 2. Walls that have been raised must be braced. The brace is temporary. The brace is used to hold the wall upright at the proper angle. Wind will easily blow walls down if they are not braced. Bracing

is usually kept until the roof is on. See Figs. 3-38 and 3-39.

Wall Sheathing

Wall sheathing may be put on a wall before it is raised. The advantage is that this reduces the amount of lifting and holding. However, it slows down getting the roof up. It is also harder to make extra openings for vents and other small objects. Wall sheathing may be nailed or stapled. Manufacturer's recommendations should be followed. Sheathing is made from wood, plywood, fiber, fiberboard, plastic foam, or gypsum board.

To Raise the Wall

Before the wall is raised, a line should be marked. The line should show the inside edge of the wall. It is made with the chalk line on the subfloor. This shows the position of the wall.

Before the wall is raised, all needed equipment should be ready. Enough people or equipment to raise the wall should be ready. Also the tools to plumb and brace the wall should be ready.

When all are ready, the wall may be raised. The top edge of the wall is picked up first. Lower parts are

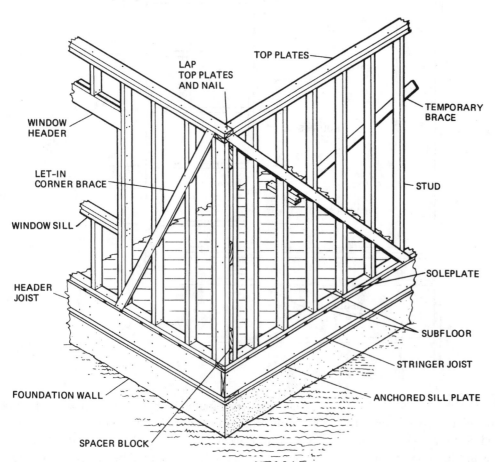

Fig. 3-33 Boards are used for corner bracing. (Forest Products Laboratory)

Fig. 3-34 *Raising a wall section by hand.* (American Plywood Association)

Fig. 3-35 *Raising a wall with wall jacks.* (Proctor Products)

Fig. 3-36 *Raising a wall with a fork-lift unit.*

Fig. 3-37 *Wall sections are nailed to slabs after being erected.* (Fox and Jacobs)

Fig. 3-38 *Once raised, walls are held in place with temporary braces.*

grasped to raise it into a vertical position. See Figs. 3-34 through 3-36. The wall is held firmly upright. As it is held, the wall is pushed even with the chalk line on the floor.

Some walls are built erect, not raised into position. As a rule, these walls are built on slabs. The layout and cutting are the same.

However, the soleplate is bolted to the slab first. Next, studs are toenailed to the soleplate. The top plate

Fig. 3-39 *Nailing a temporary brace.* (Fox and Jacobs)

Fig. 3-40 *Once walls have been pulled into place, they are nailed to the subfloor.* (Proctor Products)

can be put on next. Then the wall is braced. Openings and partition bases may then be built.

Put Up a Temporary Brace

After the wall is positioned on the floor, it is nailed in place. See Fig. 3-40. Then one end is plumbed for vertical alignment. A special jig can be used for the level. See Fig. 3-41. After the wall is plumb, a brace is nailed. See Fig. 3-39.

Each wall is treated in the same manner. After the first wall, each added wall will form a corner. The corners are joined by nailing as in Fig. 3-42. It is important that the corners be plumb.

After several walls have been erected, the double plate is added at the top. Figure 3-42 shows a double plate in place. Note that the double plate overlaps on corners to add extra strength.

INTERIOR WALLS

Interior walls, or partitions, are made in much the same manner as outside walls. However, the carpenter must remember that studs for inside walls may be longer. This is because inside walls are often curtain walls. A curtain wall does not help support the load of the roof. Because of this a double plate might not be used. But, the top of the wall must be just as high.

Most builders wish to make the building weathertight quickly. Therefore the first partitions that are made are load-bearing partitions.

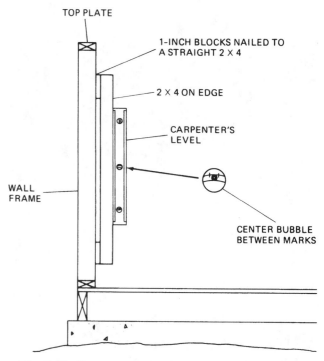

Fig. 3-41 *Use a spanner jig to plumb corners with the level.*

When roof trusses are used, load-bearing partitions may not be necessary. Trusses distribute loads so that inside support is not needed. However, many carpenters make all walls alike. This lets them cut all studs the same. It also lets them use any wall for support.

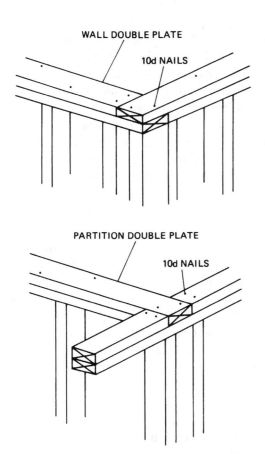

Fig. 3-42 *Double plates are added after several walls have been erected.*

Locate Soleplates for Partitions

To locate the partitions, the centerlines are determined from the plans. The centerline is then marked on the floor with the chalk line. Plates are then laid out.

Studs

Studs and headers are cut. The partition walls are done in the same way as outside walls. As a rule, the longer partitions are done first. Then the short partition walls are done. Last are the shortest walls for the closets.

Corners

Corners and wall intersections are made just as for outside walls. The size and amount of blocking can be reduced. The purpose of blocking is to provide nail surfaces. These are needed at inside and outside corners. They are a base for nailing wall covering.

Headers and Trimmers

Headers are not required for rough openings in curtain walls. Often, openings are framed with single boards. See Fig. 3-43. The header for inside walls is much like a sill. Trimmers are optional. They provide more sup-

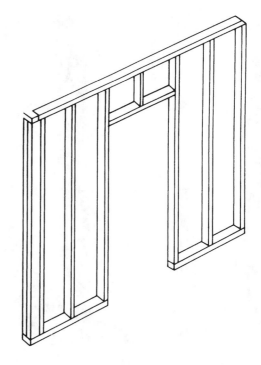

Fig. 3-43 *Single boards are also used as headers on inside partitions.*

port. They are recommended when single-board headers are used.

Soleplate

The soleplate is not cut out for door openings. It is made as one piece. It is cut away after the wall is raised. See Fig. 3-44.

Special Walls

Several conditions may call for special wall framing. Walls may need to be thicker to enclose plumbing. Drain pipes may be wider than the standard 3½-inch-thick wall. Thickness may be added in two ways: Wider studs may be used, or extra strips may be nailed to the edges of studs. Other special needs are soundproofing and small openings.

Soundproofing

Most household noise is transmitted by sound waves vibrating through the air. Blocking the air path with a standard interior wall or ceiling reduces the sound somewhat, but not completely.

The reason? Vibrations still travel through solids, particularly when the materials provide a continuous path. Standard interior walls and metal air ducts allow sound waves to continue their transmission.

So an effective sound control system must not only block the sound path, it needs to break the vibration

Fig. 3-44 *At door openings, the partition soleplates are cut out after the wall is up.*

a structure's ability to resist sound transmission. The higher the STC rating, the greater the structure's ability to limit the transmission of sound.

By incorporating combinations of all three elements for proper sound control, it is possible to raise the STC rating of a standard interior wall from 4 to 12 decibels, depending on the wall's construction.

Special insulation is available for sound control. When installed inside interior walls, the acoustic batts provide extra material to help block the sound path. The fiberglass insulation has tiny air pockets that absorb sound energy.

To break the vibration path, the insulation manufacturer recommends two options: One approach is to install staggered wall studs. The sole- and top plates are made wider. Regular studs are used. The studs are staggered from one side to the other. See Fig. 3-45. Insulation can now be woven between the studs. Sound is transmitted better through solid objects. But, with staggered studs, there is no solid part from wall to wall. In this way, there is no solid bridge for easy sound passage.

Another alternative is to install resilient channels between drywall and studs. For floors and ceilings, resilient channels should be installed in addition to the insulation batts. Figure 3-46A illustrates this using wood studs, 16 inches on center with ½-inch drywall on each side and one thickness of fiberglass acoustic batt, which is 3.5 inches thick. Figure 3-46B shows a better option. Single wood studs are used 16 inches on

path. Further noise reduction is accomplished with a system that absorbs the sound.

The Sound Transmission Class (STC) rating is a measurement—expressed in decibels—that describes

WALL DETAIL	DESCRIPTION	STC RATING
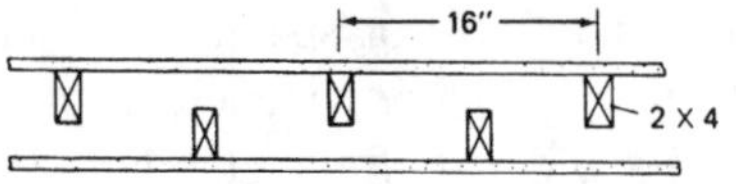 ⊢— 16″ —⊣ 2 X 4	1/2-INCH GYPSUM WALLBOARD	45
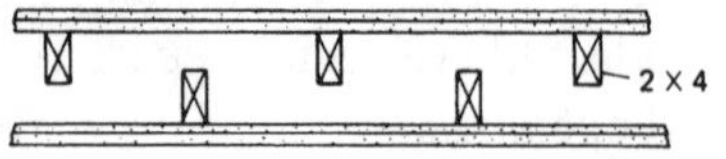 2 X 4	5/8-INCH GYPSUM WALLBOARD (DOUBLE LAYER EACH SIDE)	53
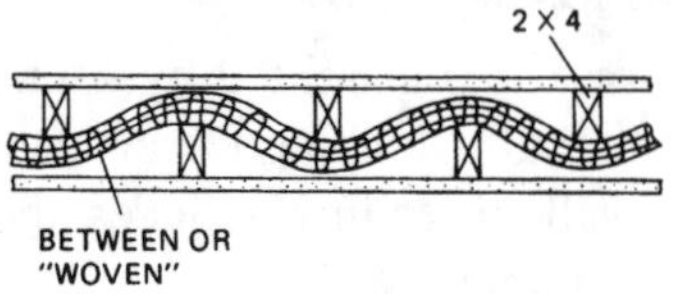 2 X 4 BETWEEN OR "WOVEN"	1/2-INCH GYPSUM WALLBOARD 1 1/2-INCH FIBROUS INSULATION	60
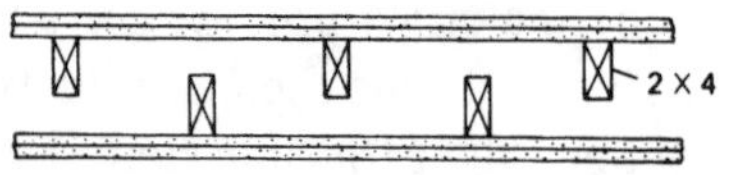 2 X 4	1/2-INCH SOUND DEADENING BOARD (NAILED) 1/2-INCH GYPSUM WALLBOARD (LAMINATED)	51

Fig. 3-45 *Sound insulation of double walls.* (Western Wood Products)

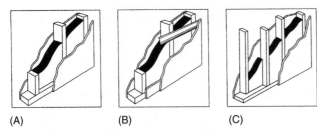

(A)   (B)   (C)

Fig. 3-46 *(A) Single wood studs, 16 inches on center. (Owens-Corning); (B) Single wood studs, 16 inches on center with resilient channel. (Owens-Corning); (C) Staggered wood studs, 16 inches on center. (Owens-Corning)*

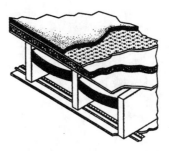

Fig. 3-48 *Standard carpet and pad with ⅜-inch particle board surface, ⅜-inch plywood subfloor, and a single sheet of ½-inch gypsum board mounted on resilient channels spaced 24 inches. (Owens-Corning)*

center with a resilient channel and ½-inch drywall on each side. A 3.5-inch-thick insulation batt is used. Resilient channels help reduce noise by dissipating sound energy and decreasing sound transmission through the framing. For the wall, install resilient channels spaced 24 inches horizontally over 16 inches on-center framing. Another, better option is shown in Fig. 3-46C. Here, staggered wood studs placed 16 inches on-center with ½-inch gypsum board on each side with one thickness of fiberglass acoustic batt, 3.5 inches thick. Keep in mind that the insulation batts inserted between the walls are not intended as a thermal barrier, but for sound control.

There are other sources of sound or noise transmission that also must be addressed if the problem is to be minimized. Figure 3-47 shows the electrical wiring that can cause sound transmission if holes for the wire are not caulked. Do not place light switches and outlets back to back. Place wall fixtures a minimum of 24 inches apart. Light switches should be spaced at least 36 inches apart. Note also how elastic caulk is used to fill in around the electrical receptacle box.

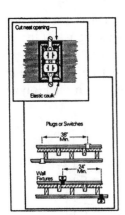

Fig. 3-47 *Light switches and receptacles. (Owens-Corning)*

Wood joist floors, Fig. 3-48, usually have a standard carpet and pad with ⅜-inch plywood or particle board subfloor. Single-layer ½-inch gypsum ceiling mounted on resilient channels spaced every 24 inches also improves the sound deadening quality of the ceil-

ing and floor. This can be improved more by adding a 3.5-inch-thick fiberglass acoustic batts.

Doors with a solid wood core will improve the noise control. See Fig. 3-49. Install a threshold closure at the bottom of the door to reduce sound transmission. Ducts can be insulated with fiberglass batts to aid in maintaining constant air temperatures while minimizing sound transmission. Duct work featuring a smooth acrylic coating on the inner surface for easy cleaning and maintenance also improves the sound qualities. See Fig. 3-50.

Caulking also can decrease the sound level. Caulking around the perimeter of drywall panels, plumbing fixtures, pipes, and wall plates can reduce

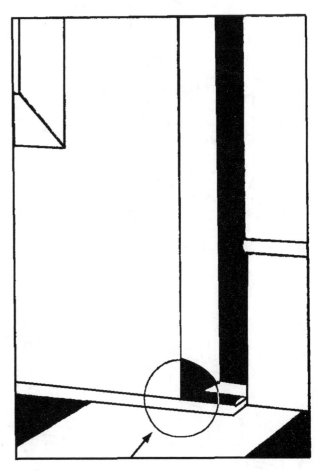

Fig. 3-49 *Doors need a threshold closure to keep out noise. (Owens-Corning)*

Fig. 3-50 *Installing fiberglass on ductwork.* (Owens-Corning)

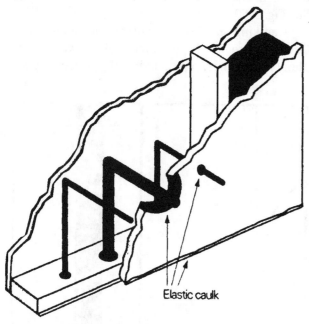

Elastic caulk

Fig. 3-51 *Caulking around plumbing helps seal out noise.* (Owens-Corning)

the incidence of unwanted noises considerably. See Fig. 3-51.

Small openings Small openings are often needed in walls. They are needed for air ducts, plumbing, and drains. Recessed cabinets, such as bathroom medicine cabinets, also need openings. These openings are framed for strength and support. See Fig. 3-52.

SHEATHING

Several types of sheathing are used. Sheathing is the first layer put on the outside of a wall. Sheathing makes the frame still and rigid and provides insulation. It will also keep out the weather until the building is finished.

There are five main types of sheathing used today: fiberboard, gypsum board, boards, plywood, and rigid foam. Corner bracing is needed for all types except plywood and boards.

Fiberboard Sheathing

The most common type of sheathing is treated fiberboard. It should be applied vertically for best bracing and strength characteristics. It is usually ½ inch thick. Plywood ½ inch thick can be used for corner bracing. The ½-inch fiber sheathing and plywood fit flush for a smooth surface. The exterior siding can easily be attached. Fiber sheathing should be fastened with roofing nails. Nails 1½ inches long are spaced 3 to 6 inches apart. Nails should always be driven at least ⅜ inch from the edge. If plywood is not used, diagonal corner braces are required. See Fig. 3-53A to C.

Gypsum Sheathing

Sheathing made from gypsum material is also used. See Fig. 3-54. This gypsum sheathing is not the same as the sheets used on interior walls. This gypsum sheathing is treated to be weather-resistant. The most common thickness is ½ inch. A ½-inch-thick plywood

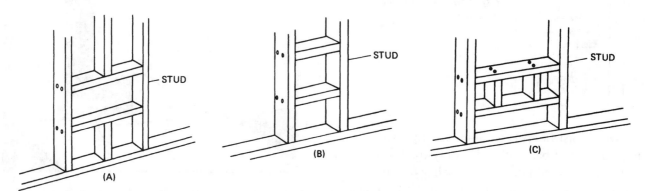

Fig. 3-52 *Framing for small openings in walls.*

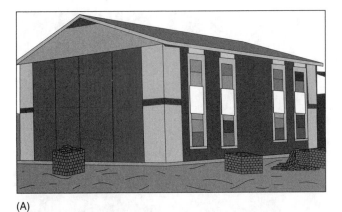
(A)

(B)

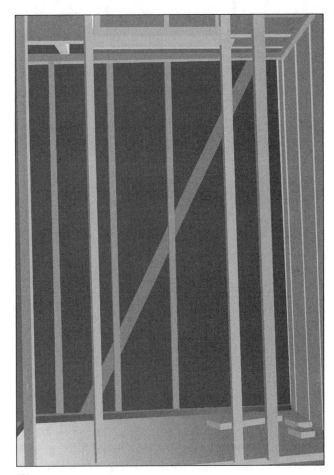
(C)

Fig. 3-53 *Corner bracing; (A) Fiberboard sheathing with plywood corner braces; (B) Fiberboard sheathing with inlet board corner braces—seen from the outside; (C) Fiberboard sheathing with inlet board corner braces—seen from the inside.*

Fig. 3-54 *Gypsum sheathing is inexpensive and fire resistive.*

corner brace may be used. The sheathing should be fastened with roofing nails. Nails 1¾ inches long should be used. The nails should be 4 inches apart.

Plywood Sheathing

Plywood is also used for sheathing. When it is used, no corner bracing is required. Plywood is both strong and fire-resistant. It can be nailed up quickly with little cutting. A moisture barrier must be added when plywood is used. However, plywood and board sheathing are both expensive.

When plywood is used, it should be at least ⁵⁄₁₆ inch thick. Half-inch thickness is recommended. Exterior siding can be nailed directly to ½-inch plywood. Plywood sheets can be applied vertically or horizontally. The sheets should be fastened with 6d or 8d nails. The nails should be 6 to 12 inches apart.

Energy Sheathing

Two types of energy sheathing are commonly used. The first is a special plastic foam called *rigid foam*. See Fig. 3-55. Its use can greatly reduce energy consumption. It

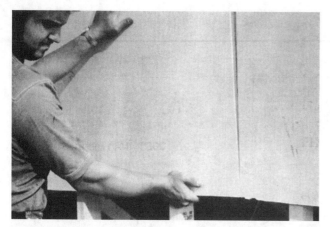

Fig. 3-55 *Rigid polystyrene foam sheathing provides more insulation.* (Dow Chemical)

is roughly equivalent to 3 more inches of regular wall insulation. It is grooved on the sides and ends. It is fitted together horizontally and is nailed in place with 1¼-inch nails spaced 9 to 12 inches apart. Joints may be made at any point. Rigid foam may also be covered with a shiny foil on one or both sides. See Fig. 3-56. The foil surface further reduces energy losses. It does so because the shiny surface reflects heat. Also, the foil prevents air from passing through the foam. However, foam burns easily. A gypsum board interior wall should be used with the foam. The gypsum wall reduces the fire hazard.

Fig. 3-56 *Rigid polystyrene foam may be coated with reflective foil to increase its effectiveness as an insulation sheathing.*

The second type is a special fiber. It is also backed on both sides with foil. Its fibers do not insulate as well as foam. It does prevent air movement better than plain foam. The foil surface is an effective reflector. Also, it is more fire-resistant and costs less. See Fig. 3-57.

Boards

Boards are still used for sheathing. However, diagonal boards are seldom used. Plain boards may be used. Boards with both side and end tongue-and-groove

Fig. 3-57 *Another style of energy efficient sheathing is similar to hardboard but is coated with reflective foil.* (Simplex)

joints are used. For grooved boards, joints need not occur over a stud. This saves installation time. Carpenters need not cut boards to make joints over studs. However, for plain boards joints should be made over studs. A special moisture barrier should be added on the outside. Builder's felt is used most often. It is nailed in place with 1-inch nails through metal disks. The bottom layers are applied first. See Fig. 3-58.

FACTORS IN WALL CONSTRUCTION

A carpenter may learn the procedure for making a wall. However, the carpenter may not know why walls are made as they are. Each part of a wall frame has a specific role. The corner pieces help tie the walls together. The double top plates also add strength to corners. The top plate is doubled to help support the weight of the rafters and ceiling. A rafter or a ceiling joist between studs is held up by the top plate. A single top plate could eventually sag and bend.

Standard Spacing

The spacing of wall members is also important. Standard construction materials are 4 feet wide and 8 feet long. Standard finish floor-to-ceiling heights are 8 feet ½ inch. The extra ½ inch lets pieces 8 feet long be used without binding.

Further, using even multiples of 4 or 8 means that less cutting is needed. Buildings are often designed in multiples of 4 feet. Two-foot roof overhangs can be used. The overhang at each end adds up to 4 feet. This

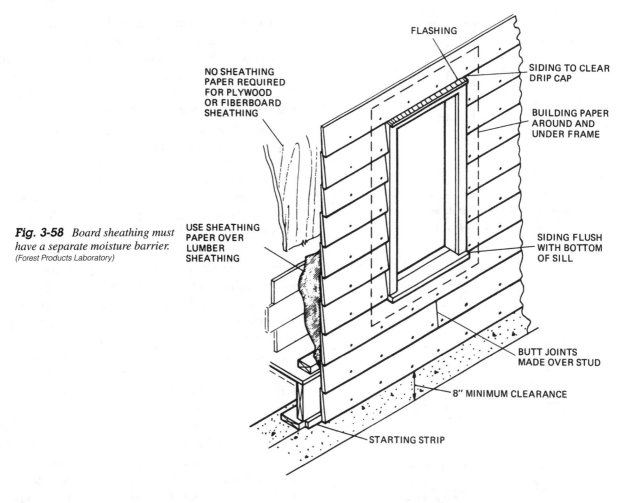

FLASHING

SIDING TO CLEAR
DRIP CAP

NO SHEATHING
PAPER REQUIRED
FOR PLYWOOD
OR FIBERBOARD
SHEATHING

BUILDING PAPER
AROUND AND
UNDER FRAME

Fig. 3-58 *Board sheathing must have a separate moisture barrier.* (Forest Products Laboratory)

USE SHEATHING
PAPER OVER
LUMBER
SHEATHING

SIDING FLUSH
WITH BOTTOM
OF SILL

BUTT JOINTS
MADE OVER STUD

8″ MINIMUM CLEARANCE

STARTING STRIP

reduces the cutting done on siding, floors, walls, and ceilings, thereby reducing time and costs.

Notching and Boring

Whenever a hole or notch is cut into a structural member, the structural capacity of the piece is weakened and a portion of the load supported by the cut member must be transferred properly to other joists. It is best to design and frame a project to accommodate mechanical systems from the outset, as notching and boring should be avoided whenever possible. However, unforeseen circumstances sometimes arise during construction.

If it is necessary to cut into a framing member, the following figures (3-59, 3-60, 3-61) provide a guide for doing so in the least destructive manner. These drawings comply with the requirements of the three major model building codes: Uniform (UBC), Standard (SBC), and National (BOCA), and the CABO one- and

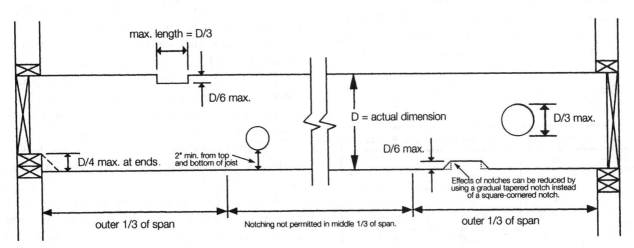

max. length = D/3

D/6 max.

D = actual dimension

D/3 max.

D/4 max. at ends.

2″ min. from top and bottom of joist

D/6 max.

Effects of notches can be reduced by using a gradual tapered notch instead of a square-cornered notch.

outer 1/3 of span

Notching not permitted in middle 1/3 of span.

outer 1/3 of span

Fig. 3-59 *Placement of cuts in floor joists.* (Western Wood Products Association)

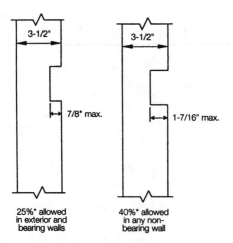

Fig. 3-60 *Notches in 2 × 4 studs.* (Western Wood Products Assocation)

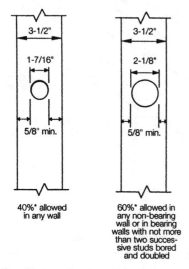

Fig. 3-61 *Bored holes in 2 × 4 studs.* (Western Wood Products Association)

of installing plumbing, electrical wiring, security systems, and sound systems. However, there are some very good reasons for not indiscriminately drilling a hole where you want. There can be serious results when the lumber being used for its strength is weakened by the placement of the hole or notch.

For instance, when structural wood members are used vertically to carry loads in compression, the same engineering procedure is used for both studs and columns. However, differences between studs and columns are recognized in the model building codes for conventional light-frame residential construction.

The term *column* describes an individual major structural member subjected to axial compression loads, such as columns in timber-frame or post-and-beam structures. The term *stud* describes one of the members in a wall assembly or wall system carrying axial compression loads, such as 2 × 4 studs in a stud wall that includes sheathing or wall board. The difference between columns and studs can be further described in terms of the potential sequences of failure.

Columns function as individual major structural members. Consequently, failure of a column is likely to result in partial collapse of a structure (or complete collapse in extreme cases due to the domino effect). However, studs function as members in a system. Due to the system effects (load sharing, partial composite action, redundancy, load distribution, etc.), studs are much less likely to fail and result in a total collapse than are columns.

Notching or boring into columns is not recommended and rarely acceptable; however, model codes established guidelines for allowable notching and boring into studs used in a stud-wall system.

Figures 3-60 and 3-61 illustrate the maximum allowable notching and boring of 2×4 studs under all model codes except BOCA. BOCA allows a hole one-third the width of the stud in all cases.

Bored holes shall not be located in the same cross section of a stud as a cut or notch.

It is important to recognize the point at which a notch becomes a rip, such as when floor joists at the entry of a home are ripped down to allow underlayment for a floor.

Ripping wide-dimension lumber lowers the grade of the material, and is unacceptable under all building codes.

When a sloped surface is necessary, a non-structural member can be ripped to the desired slope and fastened to the structural member in a position above the top edge. Do not rip the structural member.

two-family dwelling code. Figure 3-59 shows the placement of cuts in floor joists. Notches and holes in 2 × 4 studs are shown in Figs. 3-60 and 3-61. Table 3-2 presents maximum sizes for cuts in floor joists.

The Western Wood Products Association provides information on the woods harvested in their area and used for studs and floor joists. Notching and boring holes in wood are a common practice for use as means

Table 3-2 *Maximum Sizes for Cuts in Floor Joists (Western Wood Products Association)*

Joist Size	Max. Hole	Max Notch Depth	Max. End Notch
2x4	none	none	none
2x6	1-1/2"	7/8"	1-3/8"
2x8	2-3/8"	1-1/4"	1-7/8"
2x10	3"	1-1/2"	2-3/8"
2x12	3-3/4"	1-7/8"	2-7/8"

Modular Standards

Currently, a new modular system of construction is being used more. The modular system uses stud and rafter systems 24 inches O.C. The wider spacing does not provide as much support for wall sheathing. However, the ability to support the roof is not reduced much. A building with studs on 16-inch centers is very strong. Yet the difference in strength between 16- and 24-inch centers is very small. The advantage of using the 24-inch modular system is that it saves on costs. For a three-bedroom house, the cost of wall studs may be reduced about 25 percent. Moreover, the costs of labor are also reduced. This saves the time that would be used for cutting and nailing that many more studs.

The savings are even more in terms of money and wages paid. This saving is made because fewer resources are used with almost the same results. In the modular system, several ways of reducing material use are employed. Floor joists are butted and not lapped. Building and roof size are exact multiples of 4 or 8 feet. Also, single top plates are used on partitions.

Energy

Energy is also a matter of concern today. Insulated walls save energy used for heating and cooling. The amount of insulation helps determine the efficiency. A 6-inch-thick wall can hold more insulation than a 4-inch wall. It will reduce the energy used by about 20 percent. However, it takes more material to build such a wall. Canadian building codes often specify 6-inch walls.

The double-wall system of frame buildings is a better insulator than a solid wall. Wood is good insulation when compared with other materials. But the best insulation is a hollow space filled with material that does not conduct heat. A wall should have three layers. An outside, weatherproof wall is needed. This layer stops rain and wind. A thick layer of insulation is next. This helps keep heat energy from being conducted through the wall. The inside wall is the third layer. It helps reduce air movement. It also helps seal and hold the insulation in place.

More insulation effect can also be added by adding a shiny surface. See Figs. 3-56 and 3-57.

Energy sheathing adds insulation in two ways: it insulates in the same way as regular insulation, and it also covers the studs. This way, the stud does not conduct heat directly through the wall. See Fig. 3-63.

The use of headers also affects wall construction. Headers reduce time and construction costs; however,

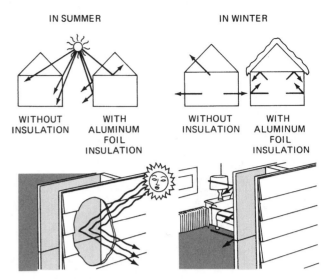

Fig. 3-62 *Reflected heat makes the house cooler in summer and warmer in winter.* (U.S. Gypsum)

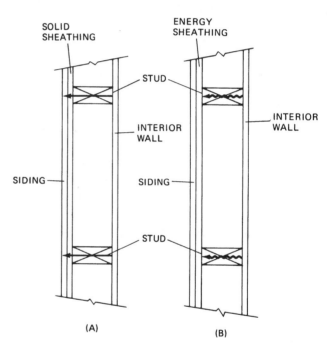

Fig. 3-63 *(A) Solid sheathing, stud, and wall are a solid path. This conducts energy loss straight through a wall. (B) Energy sheathing forms a barrier. There is no solid path for energy.*

they are also difficult to insulate. Truss headers or a single header is better. This allows more insulation to be used.

The builder and the carpenter can alter a wall frame to save energy. Walls can be made thicker so that more insulation can be used. Thicker walls also let insulation cover studs in the same way as for sound insulation. Energy sheathing can be used to cover the studs. It adds insulation without requiring thick walls. Reflective foil also makes it more efficient. See Fig. 3-62.

4
CHAPTER

Building
Roof Frames

THE FRAMING FOR A ROOF IS DETERMINED by the choice of roof style. There are a number of types of rooflines. Some of them are the gable, mansard, shed, hip, and gambrel. Each has identifying characteristics. Each style presents unique problems. Rafters and sheathing will be shaped according to the roofline desired.

In this unit you will learn how to build roof frames. You will learn how to make roofs of different types. You will also learn how to make openings for chimneys and soil pipes. Things you can learn to do are:

- Design a particular type of roof rafter
- Mark off and cut a roof rafter
- Put roof trusses in place
- Lay sheathing onto a roof frame
- Build special combinations of framing for roofs
- Identify the type of plywood needed for a particular roof

INTRODUCTION

The roof is an important part of any building. It is needed to keep out the weather and to control the heat and cooling provided for human comfort inside. There are many types of roofs. Each serves a purpose. Each one is designed to keep the inside of the house warm in the winter and cool in the summer. The roof is designed to keep the house free of moisture, whether rain, snow, or fog.

The roof is made of rafters and usually supported by ceiling joists. When all the braces and forms are put together, they form a roof. In some cases roofs have to be supported by more than the outside walls. This means some of the partitions must be weight-bearing. However, with truss roofs it is possible to have a huge open room without supports for the roof.

Figure 4-1 shows various roof shapes. The shape can affect the type of roofing materials used. The various shapes call for some special details. Roofs have to withstand high winds and ice and snow. The weight of snow can cause a roof to collapse unless it is properly designed. It is necessary to make sure the load on the roof can be supported. This calls for the proper size rafters and decking.

Shingles on a roof protect it from rain, wind, and ice. They have to withstand many years of exposure to all types of weather conditions.

The hip roof with its hip and hip jack rafters is of particular concern. It is one of the most popular types. The common rafter is the simplest in the gable roof. The

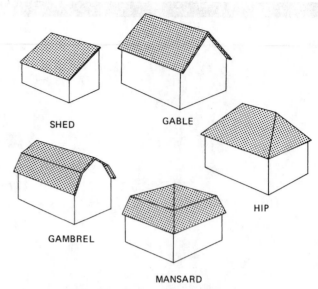

Fig. 4-1 *Various roof shapes or styles.*

mansard roof and the gambrel roof present some interesting problems with some very interesting solutions. All this will be explained in detail later in this chapter.

Various valleys, ridge boards, and cripple rafters will be described in detail here.

SEQUENCE

The carpenter should build the roof frame in this sequence:

1. Check the plans to see what type of roof is desired.
2. Select the ceiling joist style and spacing.
3. Lay out ceiling joists for openings.
4. Cut ceiling joists to length and shape.
5. Lay out regular rafter spacing.
6. Lay out the rafters and cut to size.
7. Set the rafters in place.
8. Nail the rafters to the ridge board.
9. Nail the rafters to the wall plate.
10. Figure out the sheathing needed for the roof.
11. Apply the sheathing according to specifications.
12. Attach the soffit.
13. Put in braces or lookouts where needed. See Fig. 4-2.
14. Cut special openings in the roof decking.

There may be a need for cutting dormer rafters after the rest of the roof is finished. Structural elements may be installed where needed. Truss roof rafters will need special attention as to spacing and nailing. Double-check to make sure they fit the manufacturer's specifications.

Fig. 4-2 *Braces (lookouts) are put in where they are needed.* (Duo-Fast)

ERECTING TRUSS ROOFS
Truss Construction

Trussed rafters are commonly used in light frame construction. They are used to support the roof and ceiling loads. Such trusses are designed, fabricated, and erected to meet the design and construction criteria of various building codes. They efficiently use the excellent structural properties of lumber. They are spaced either 16 inches on centers (O.C.) or in some cases 24 inches O.C. See Fig. 4-3.

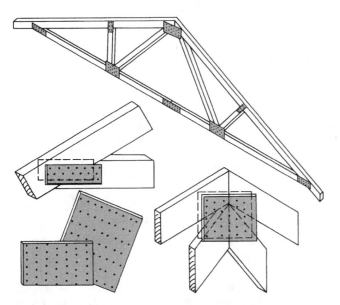

Fig. 4-3 *W-type truss roof and metal plates used to make the junction points secure.*

Truss Disadvantages

You should keep in mind that the truss type of construction does have some disadvantages. The attic space is limited by the supports that make up the truss.

Truss types of construction may need special equipment to construct them. In some instances it is necessary to use a crane to lift the trusses into position on the job site.

Roof framing The roof frame is made up of rafters, ridge board, collar beams, and cripple studs. See Fig. 4-4. In gable roof construction, all rafters are precut to the same length and pattern. Figure 4-5 shows a gable roof. Each pair of rafters is fastened at the top to a ridge board. The ridge board is commonly a 2 × 8 for 2 × 6 rafters. This board provides a nailing area for rafter ends. See Fig. 4-4.

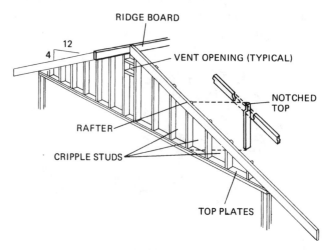

Fig. 4-4 *Gable-type rafter with cripple studs and ridge board. Note the notched top in the cripple stud.* (American Plywood Association)

Fig. 4-5 *Two-story house with gable roof. Note the bay windows with hip roof.*

Getting started Getting started with erection of the roof framing is the most complicated part of framing a house. Plan it carefully. Make sure you have all materials on hand. It is best to make a "dry run" at ground level. The erection procedure will be much easier if you have at least two helpers. A considerable amount of temporary bracing will be required if the job must be done with only one or two persons.

Steps in framing a roof Take a look at Fig. 4-6. It shows two persons tipping up trusses. They are tipped up and nailed in place one at a time. Two people, one working on each side, will get the job done quickly. This is one of the advantages of trussed-roof construction. The trusses are made at a lumber yard or some other location. They are usually hauled to the construction site on a truck. Here they are lifted to the roof of the building with a crane or by hand. In some cases the sheet metal bracket shown in Fig. 4-7 holds the truss to the wall plate. Figure 4-8 shows how the metal bracket is used to fasten the truss in place. In Fig. 4-9 toenailing is used to fasten the truss to the wall plate.

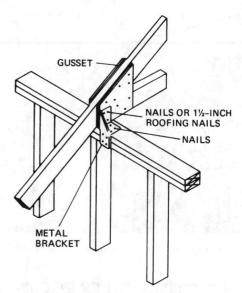

Fig. 4-8 *Using a nailing bracket to attach a truss to the wall plate and nailing into the gusset to attach the rafter.*

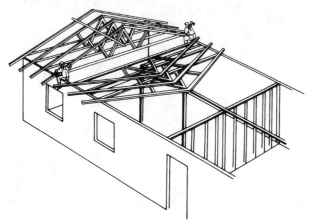

Fig. 4-6 *Roof trusses are placed on the walls, then tipped up and nailed in place.*

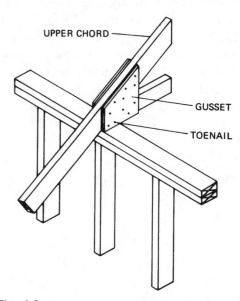

Fig. 4-9 *Toenailing the truss rafter to the wall plate.*

Fig. 4-7 *Rafter truss attached to the wall plate with a sheet-metal bracket.*

Advantages of trusses Manufacturers point out that the truss has many advantages. In Fig. 4-10 the conventional framing is used. Note how bearing walls are required inside the house. With the truss roof you can place the partitions anywhere. They are not weight-bearing walls. With conventional framing it is possible for ceilings to sag. This causes cracks. The truss has

supports to prevent sagging ceilings. See Fig. 4-11. Notice how the triangle shape is obvious in all the various parts of the truss. The triangle is a very strong structural form.

Details of trusses There are a number of truss designs. A W-type is shown in Fig. 4-12. Note how the 2 × 4s are brought together and fastened by plywood that is both nailed and glued. In some cases, as shown in Fig. 4-3, steel brackets are used. Whenever plywood is used for the gussets, make sure the glue fits the climate. In humid parts of the country, where the attic may be damp, the glue should be able to take the humidity. The manufacturer will inform you of the glue's use and how it will perform in humid climates. The

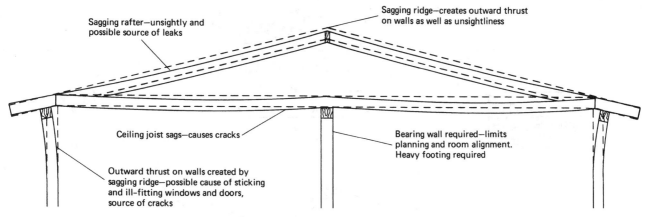

Fig. 4-10 *Disadvantages of conventional roof framing.*

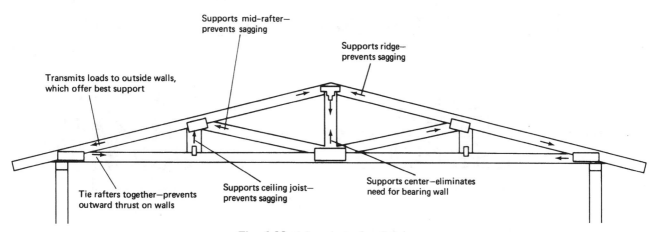

Fig. 4-11 *Advantages of roof trusses.*

glue should not lose its strength when the weather turns humid. Most glue containers have this information on them. If not, check with the manufacturer before making trusses. In most instances, the manufacturer of trusses is very much aware of the glue requirements for a particular location.

Figure 4-13 shows how the three suggested designs are different. The truss in Fig. 4-13A is known as the W type. Note the W in the middle of the truss. In Fig. 4-13B the king post is simple. It is used for a low-pitch roof. The scissors type is shown in Fig. 4-13C.

The W type is the most popular. It can be used on long spans. It can also be made of low-grade lumber. The scissor type is used on houses with sloping living room ceilings. It can be used for a cathedral ceiling. This truss is cheaper to make than the conventional type of construction for cathedral ceilings. King-post trusses are very simple. They are used for houses. This truss is limited to about a 26-foot span—that is, if 2 × 4s are used for the members of the truss. Compare this with a W type, which could be used for a span of 32 feet. King type is economical for use in medium spans. It is also useful in short spans. See Fig. 4-14.

The type of truss used depends upon the wind and snow. The weight applied to a roof is an important factor.

Make sure the local codes allow for truss roofs. In some cases, the FHA has to inspect them before insuring a mortgage with them in the house.

Lumber to use in trusses The lumber used in construction of trusses must be that which is described in Table 4-1. The moisture content should be from 7 to 16 percent. Glued surfaces have to be free of oil, dirt, or any foreign matter. Each piece must be machine-finished, but not sanded.

Lumber with roughness in the gusset area cannot be used. Twisted, cupped, or warped lumber should not be used either. This is especially true if the twist or cup is in the area of the gusset. Keep the intersecting surfaces of the lumber within $\frac{1}{32}$ inch.

Glue for trusses For dry or indoor locations use casein-type glue. It should meet Federal Specification MMM—A—125, Type 11. For wet conditions use resorcinol-type glue. Military specifications are MIL—A—46051 for wet locations. If the glue joint is

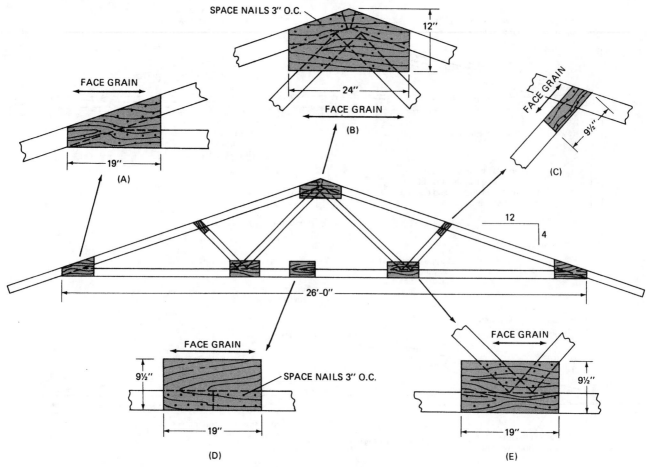

Fig. 4-12 *Construction of the W truss.*

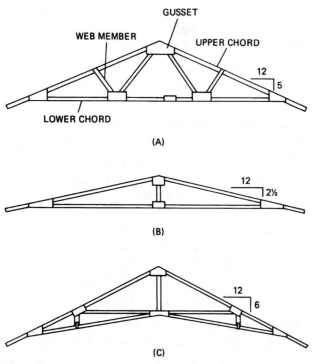

Fig. 4-13 *Wood trusses: (A) W type; (B) King post type; (C) Scissors type.*

exposed to the weather or used at the soffit, use the resorcinol-type glue.

Load conditions for trusses Table 4-2 shows the loading factors needed in designing trusses. Note the 30 pounds per square foot (psf) and 40 psf columns. Then look up the type of gusset—either beveled-heel or square-heel type. Standard sheathing or C–C Ext—APA grade plywood is the type used here for the gussets and sheathing (Table 4-3). C-D plywood is often used for roof sheathing.

Covering the trusses Once the trusses are in place, cover them with sheathing or plywood. This will make a better structure once the sheathing is applied. The underlayment is applied and the roofing attached properly. The sheathing or plywood makes an excellent nail base for the shingles. See Chapter 8 for applying shingles.

THE FRAMING SQUARE

In carpentry it is necessary to be able to use the framing square. This device has a great deal of informa-

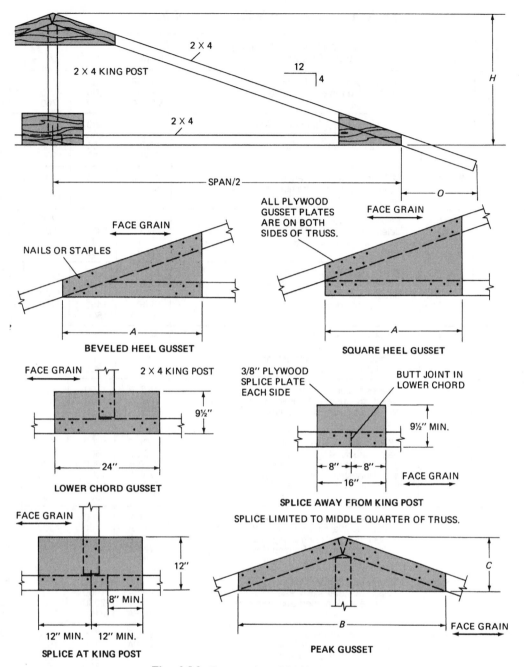

Fig. 4-14 *Construction of the king-post truss.*

tion stamped on its body and tongue. See Fig. 4-15 for what the square can do in terms of a right angle. Figure 4-16 shows the right angle made by a framing square. The steel square or carpenter's square is made in the form of a right angle. That is, two arms (the body and the tongue of the square) make an angle of 90°.

Note in Fig. 4-15 how a right triangle is made when points A, B, and C are connected. Figure 4-16 shows the right triangle. A right triangle has one angle which is 90°.

Parts of the Square

The steel square consists of two parts—the body and the tongue. The body is sometimes called the *blade*. See Fig. 4-17.

Body The body is the longer and wider part. The body of the Stanley standard steel square is 24 inches long and 2 inches wide.

Tongue The tongue is the shorter and narrower part and usually is 16 inches long and 1½ inches wide.

Table 4-1 *Chord Code Table for Roof Trusses*

Chord Code	Size	Grade and Species Meeting Stress Requirements	Grading Rules	f	t//	c//
1	2 × 4	Select structural light framing WCDF	WCLIB	1950	1700	1400
		No. 1 dense kiln-dried Southern pine	SPIB	2000	2000	1700
		1.8E	WWPA	2100	1700	1700
2	2 × 4	1500f industrial light framing WCDF	WCLIB	1500	1300	1200
		1500f industrial light framing WCH	WCLIB	1450	1250	1100
		No. 1 2-inch dimension Southern pine	SPIB	1450	1450	1350
		1.4E	WWPA	1500	1200	1200
3	2 × 4	1200f industrial light framing WCDF	WCLIB	1200	1100	1000
		1200f industrial light framing WCH	WCLIB	1150	1000	900
		No. 2 2-inch dimension Southern pine	SPIB	1200	1200	900
4	2 × 6	Select structural J&P WCDF	WCLIB	1950	1700	1600
		Select structural J&P Western larch	WWPA	1900	1600	1500
		No. 1 dense kiln-dried Southern pine	SPIB	2000	2000	1700
		1.8E	WWPA	2100	1700	1700
5	2 × 6	Construction grade J&P WCDF	WCLIB	1450	1300	1200
		Construction grade J&P WCH	WCLIB	1450	1250	1150
		Structural J&P Western larch	WWPA	1450	1300	1200
		No. 1 2-inch dimension Southern pine	SPIB	1450	1450	1350
		1.4E	WWPA	1500	1200	1200
6	2 × 6	Standard grade J&P WCDF	WCLIB	1200	1100	1050
		Standard grade J&P WCH	WCLIB	1150	1000	950
		Standard structural Western larch	WWPA	1200	1100	1050
		No. 2 2-inch dimension Southern pine	SPIB	1200	1200	900
7	2 × 4	Select structural light framing WCDF	WCLIB	1900	1900	1400
		Select structural light framing Western larch	WWPA	1900	1900	1400
		No. 1 dense kiln-dried Southern pine	SPIB	2050	2050	1750
		1.8E	WWPA	2100	1700	1700
8	2 × 4	1500f industrial light framing WCDF	WCLIB	1500	1500	1200
		Select structural light framing WCH	WCLIB	1600	1600	1100
		Select structural light framing WH	WWPA	1600	1600	1100
		1500f industrial light framing Western larch	WWPA	1500	1500	1200
		No. 1 2-inch dimension Southern pine	SPIB	1500	1500	1350
		1.4E	WWPA	1500	1200	1200
9	2 × 4	1200f industrial light framing WCDF	WCLIB	1200	1200	1000
		1200f industrial light framing Western larch	WWPA	1200	1200	1000
		1500f industrial light framing WCH	WCLIB	1500	1500	1000
		1500f industrial light framing WH	WWPA	1500	1500	1000
		No. 2 2-inch dimension Southern pine	SPIB	1200	1200	900
10	2 × 6	Select structural J&P WCDF	WCLIB	1900	1900	1500
		Select structural J&P Western larch	WWPA	1900	1900	1500
		No. 1 dense kiln-dried Southern pine	SPIB	2050	2050	1750
		1.8E	WWPA	2100	1700	1700
11	2 × 6	Construction grade J&P WCDF	WCLIB	1500	1500	1200
		Construction grade J&P Western larch	WWPA	1500	1500	1200
		Select structural J&P WCH	WCLIB	1600	1600	1200
		Select structural J&P WH	WWPA	1600	1600	1200
		No. 1 2-dimension Southern pine	SPIB	1500	1500	1350
		1.4E	WWPA	1500	1200	1200
12	2 × 6	Standard grade J&P WCDF	WCLIB	1200	1200	1000
		Standard grade J&P Western larch	WWPA	1200	1200	1000
		Standard grade J&P WCH	WCLIB	1200	1200	1000
		Standard grade J&P WH	WWPA	1200	1200	1000
		No. 2 2-inch dimension dense Southern pine	SPIB	1400	1400	1050

Heel The point at which the body and the tongue meet on the outside edge of the square is called the *heel*. The intersection of the inner edges of the body and tongue is sometimes also called the heel.

Face The face of the square is the side on which the manufacturer's name, Stanley in this case, is stamped, or the visible side when the body is held in the left hand and the tongue is held in the right hand. See Fig. 4-17.

Back The back is the side opposite the face. See Fig. 4-18.

The modern scale usually has two kinds of marking: scales and tables.

Table 4-2 Designs When Using Standard Sheathing as C-C EXT

Loading Condition, Total Roof Load, psf	Span	Beveled-Heel Gusset							Square-Heel Gusset						
		Dimensions, Inches					Chord code		Dimensions, Inches					Chord code	
		A	B	C	H	O	Upper	Lower	A	B	C	H	O	Upper	Lower
30 psf (20 psf live load, 10 psf dead load) on upper chord and 10 psf dead load on lower chord. Meets FHA requirements.	20'8"	32	48	12	45⅛	44	2	3	19	32	12	48¾	44	2	3
	22'8"	32	48	12	49⅛	48	1	2	19	32	12	52¾	48	1	3
	24'8"	48	60	16	53⅛	48	1	2	24	48	12	56¾	48	1	3
	26'8"	48	72	16	57⅛	48	1	2	32	60	16	60¾	48	1	2
40 psf (30 psf live load, 10 psf dead load) on upper chord and 10 psf dead load on lower chord.	20'8"	32	48	12	45⅛	43	8	9	19	32	12	48¾	48	7	9
	22'8"	32	60	16	49⅛	48	7	8	19	48	12	52¾	48	7	9
	24'8"								32	60	16	56¾	48	7	9

Table 4-3 Plywood Veneer Grades

Grade	Description
N	Special order "natural finish" veneer. Select all heartwood or all sapwood. Free of open defects. Allows some repairs.
A	Smooth and paintable. Neatly made repairs permissible. Also used for natural finish in less demanding applications.
B	Solid surface veneer. Circular repair plugs and tight knots permitted.
C	Knotholes to 1″. Occasional knotholes ½″ larger permitted providing total width of all knots and knotholes within a specified section does not exceed certain limits. Limited splits permitted. Minimum veneer permitted in Exterior type plywood.
C Plugged	Improved C veneer with splits limited to ⅛″ in width and knotholes and borer holes limited to ¼″ by ½″.
D	Permits knots and knotholes to 2½″ in width, and ½″ larger under certain specified limits. Limited splits permitted.

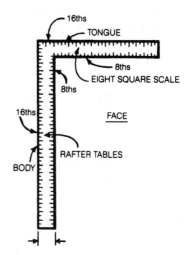

Fig. 4-17 *Parts of the steel square—face side.* (Stanley Tools)

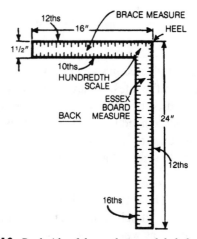

Fig. 4-18 *Back side of the steel square labeled.* (Stanley Tools)

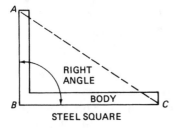

Fig. 4-15 *The steel square has a 90° angle between the tongue and the body.* (Stanley Tools)

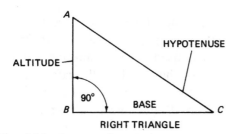

Fig. 4-16 *The parts of a right triangle.* (Stanley Tools)

Scales The scales are the inch divisions found on the outer and inner edges of the square. The inch graduations are in fractions of an inch. The Stanley square has the following graduations and scales:

Face of body— outside edge	Inches and sixteenths
Face of body— inside edge	Inches and eighths

Face of tongue—outside edge	Inches and sixteenths
Back of tongue—inside edge	Inches and eighths
Back of body—outside edge	Inches and twelfths
Back of body—inside edge	Inches and sixteenths
Back of tongue—outside edge	Inches and twelfths
Back of tongue—inside edge	Inches and tenths

Hundredths scale This scale is located on the back of the tongue, in the corner of the square, near the brace measure. The hundredths scale is 1 inch divided into 100 parts. The longer lines indicate 25 hundredths, while the next shorter lines indicate 5 hundredths. With the aid of a pair of dividers, fractions of an inch can be obtained. See Fig. 4-19 for the location of the hundredths scale.

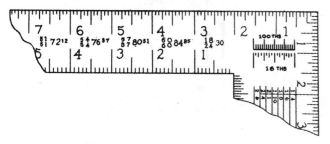

Fig. 4-19 *Location of the hundredths scale on a square.* (Stanley Tools)

One inch graduated in sixteenths is below the hundredths scale on the latest squares, so that the conversion from hundredths to sixteenths can be made at a glance without the need to use dividers. This comes in handy when you are determining rafter lengths using the figures of the rafter tables, where hundredths are given.

Rafter scales These tables will be found on the face of the body and will help you to determine rapidly the lengths of rafters as well as their cuts.

The rafter tables consist of six lines of figures. Their use is indicated on the left end of the body. The first line of figures gives the lengths of common rafters per foot of run. The second line gives the lengths of hip-and-valley rafters per foot of run. The third line gives the length of the first jack rafter and the differences in the length of the others centered at 16 inches. The fourth line gives the length of the first jack rafter and the differences in length of the others spaced at 24-inch

centers. The fifth line gives the side cuts of jacks. The sixth line gives the side cuts of hip-and-valley rafters.

Octagon scale The octagon or "eight-square" scale is found along the center of the face of the tongue. Using this scale a square timber may be shaped into one having eight sides, or an "octagon."

Brace scale This table is found along the center of the back of the tongue and gives the exact lengths of common braces.

Essex board measure This table is on the back of the body and gives the contents of any size lumber.

Steel Square Uses

The steel square has many applications. It can be used as a try square or to mark a 90° line along a piece of lumber. See Fig. 4-20.

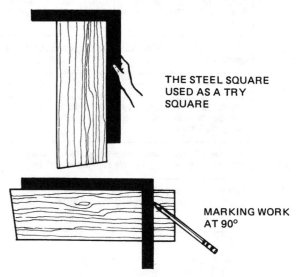

THE STEEL SQUARE USED AS A TRY SQUARE

MARKING WORK AT 90°

Fig. 4-20 *Uses of the steel square.* (Stanley Tools)

The steel square can also be used to mark 45° angles and 30–60° angles. See Fig. 4-21.

In some instances you may want to use the square for stepping off the length of rafters and braces. See Fig. 4-22. Another use for the square is the laying out of stairsteps. Figure 4-23 shows how this is done.

However, one of the most important roles the square plays in carpentry is the layout of roof framing. Here it is used to make sure the rafters fit the proper angles. The length of the rafters and the angles to be cut can be determined by the use of the framing square. Rafter cuts are shown in the following sections.

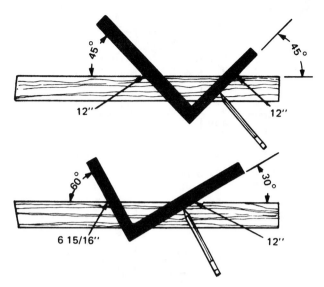

Fig. 4-21 *Using the steel square for marking angles.* (Stanley Tools)

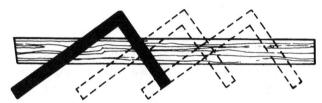

Fig. 4-22 *Using the steel square to step off the length of rafters and braces.* (Stanley Tools)

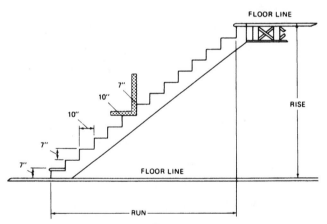

Fig. 4-23 *Using the square to lay out stairs.* (Stanley Tools)

ROOF FRAMING

There are a number of types of roofs. A great variety of shapes is prevalent, as you can see in any neighborhood. Some of the most common types will be identified and worked with here.

Shed roof The shed roof is the most common type. Easy to make, it is also sometimes called the lean-to roof. It has only a single slope. See Fig. 4-24.

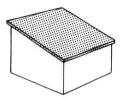

Fig. 4-24 *Lean-to or shed roof.* (Stanley Tools)

Gable or pitch roof This is another type of roof that is commonly used. It has two slopes meeting at the center or ridge, forming a gable. It is a very simple form of roof, and perhaps the easiest to construct. See Fig. 4-25.

Fig. 4-25 *Gable roof.* (Stanley Tools)

Gable-and-valley or hip-and-valley roof This is a combination of two intersecting gable or hip roofs. The valley is the place where two slopes of the roof meet. The roofs run in different directions. There are many modifications of this roof, and the intersections usually are at right angles. See Figs. 4-26 and 4-27.

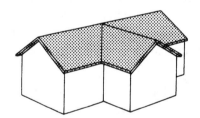

Fig. 4-26 *Gable-and-valley roof.* (Stanley Tools)

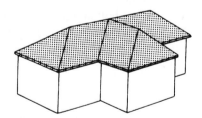

Fig. 4-27 *Hip-and-valley roof.* (Stanley Tools)

Hip roof This roof has four sides, all sloping toward the center of the building. The rafters run up diagonally to meet the ridge, into which the other rafters are framed. See Fig. 4-28.

Roof Terms

There are a number of terms you should be familiar with in order to work with roof framing. Each type of

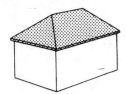

Fig. 4-28 *Hip roof.* (Stanley Tools)

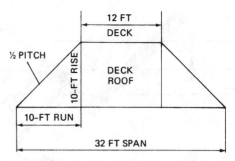

Fig. 4-30 *Location of the deck roof.* (Stanley Tools)

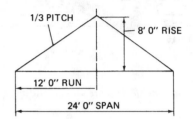

Fig. 4-31 *One-third pitch.* (Stanley Tools)

roof has its own particular terms; however, some of the common terms are:

Span The span of a roof is the distance over the wall plates. See Fig. 4-29.

Run The run of a roof is the shortest horizontal distance measured from a plumb line through the center of the ridge to the outer edge of the plate. See Fig. 4-29. In equally pitched roofs, the run is always equal to half the span or generally half the width of the building.

Rise The rise of a roof is the distance from the top of the ridge and of the rafter to the level of the foot. In figuring rafters, the rise is considered as the vertical distance from the top of the wall plate to the upper end of the measuring line. See Fig. 4-29.

Deck roof When rafters rise to a deck instead of a ridge, the width of the deck should be subtracted from the span. The remainder divided by 2 will equal the run. Thus, in Fig. 4-30 the span is 32 feet and the deck is 12 feet wide. The difference between 32 and 12 is 20 feet, which divided by 2 equals 10 feet. This is the run of the common rafters. Since the rise equals 10 feet, this is a ½-pitch roof.

Pitch The pitch of a roof is the slant or the slope from the ridge to the plate. It may be expressed in several ways:

1. The pitch may be described in terms of the ratio of the total width of the building to the total rise of the roof. Thus, the pitch of a roof having a 24-foot span with an 8-foot rise will be 8 divided by 24, which equals ⅓ pitch. See Fig. 4-31.

2. The pitch of a roof may also be expressed as so many inches of vertical rise to each foot of horizontal run. A roof with a 24-foot span and rising 8 inches to each foot of run will have a total rise of 8 × 12 = 96 inches or 8 feet. Eight divided by 24 equals ⅓. Therefore, the roof is ⅓ pitch. See Fig. 4-31.

Note that in Fig. 4-32 the building is 24 feet wide. It has a roof with a 6-foot rise. What is the pitch of the roof? The pitch equals 6 divided by 24, or ¼.

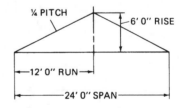

Fig. 4-32 *One-fourth pitch.* (Stanley Tools)

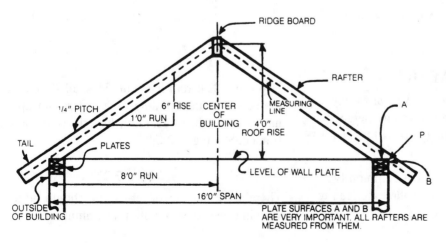

Fig. 4-29 *Span, run, rise, and pitch.* (Stanley Tools)

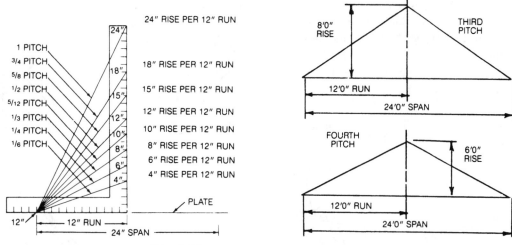

Fig. 4-33 *Principal roof pitches.* (Stanley Tools)

Principal roof pitches Figure 4-33 shows the principal roof pitches. They are called ½ pitch, ⅓ pitch, or ¼ pitch, as the case may be, because the height from the level of the wall plate to the ridge of the roof is one-half, one-third, or one-quarter of the total width of the building.

Keep in mind that roofs of the same width may have different pitches, depending upon the height of the roof.

Take a look at Fig. 4-34. This will help you interpret the various terms commonly used in roof construction.

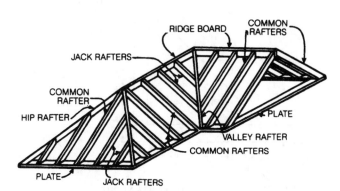

Fig. 4-34 *Different types of rafters used in a roof frame.*

Principal Roof Frame Members

The principal members of the roof frame are the plates at the bottom and the ridge board at the top. To them the various rafters are fastened. See Fig. 4-34.

Plate The plate is the roof member to which rafters are framed at their lower ends. The top, *A*, and the outside edge of the plate, *B*, are the important surfaces from which rafters are measured in Fig. 4-29.

Ridge board The ridge board is the horizontal member that connects the upper ends of the rafters on one side to the rafters on the opposite side. In cheap con-

struction the ridge board is usually omitted. The upper ends of the rafters are spiked together.

Common rafters A common rafter is a member that extends diagonally from the plate to the ridge.

Hip rafters A hip rafter is a member that extends diagonally from the corner of the plate to the ridge.

Valley rafters A valley rafter is one that extends diagonally from the plate to the ridge at the line of intersection of two roof surfaces.

Jack rafters Any rafter that does not extend from the plate to the ridge is called a jack rafter. There are different kinds of jacks. According to the position they occupy, they can be classified as hip jacks, valley jacks, or cripple jacks.

Hip jack A hip jack is a jack rafter with the upper end resting against a hip and the lower end against the plate.

Valley jack A valley jack is a jack rafter with the upper end resting against the ridge board and the lower end against the valley.

Cripple jack A cripple jack is a jack rafter with a cut that fits in between a hip-and-valley rafter. It touches neither the plate nor the ridge.

All rafters must be cut to proper angles so that they will fit at the points where they are framed. These different types of cuts are described below.

Top or plumb cut The cut of the rafter end which rests against the ridge board or against the opposite rafter is called the top or plumb cut.

Bottom or heel cut The cut of the rafter end which rests against the plate is called the bottom or heel cut. The bottom cut is also called the foot or seat cut.

Side cuts Hip-and-valley rafters and all jacks, besides having top and bottom cuts, must also have their sides at the end cut to a proper bevel so that they will fit into the other members to which they are to be framed. These are called side cuts or cheek cuts. All rafters and their cuts are shown in Fig. 4-35.

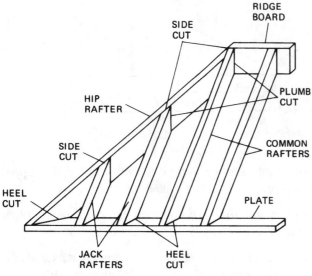

Fig. 4-35 *Rafter cuts. (Stanley Tools)*

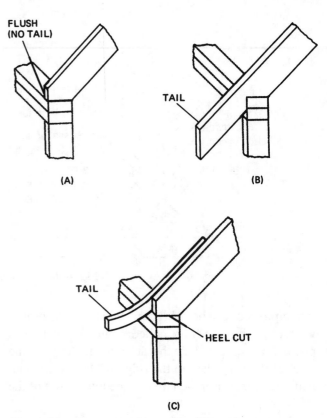

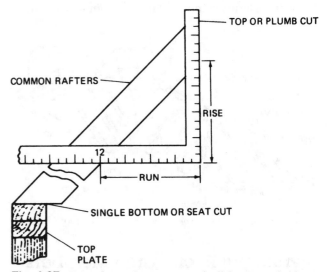

Fig. 4-36 *Tails or overhangs of rafters. (A) Flush tail (no tail); (B) Full tail; (C) Separate curved tail.*

RAFTERS
Layout of a Rafter

The *measuring line* of a rafter is a temporary line on which the length of the rafter is measured. This line runs parallel to the edge of the rafter and passes through the point *P* on the outer edge of the plate. Point *P* is where cut-lines *A* and *B* converged. This is the point from which all dimensions are determined. See Fig. 4-29.

Length The length of a rafter is the shortest distance between the outer edge of the plate and the center of the ridge line.

Tail That portion of the rafter extending beyond the outside edge of the plate is called the tail. In some cases it is referred to as the eave. The tail is figured separately and is not included in the length of the rafter as mentioned above. See Fig. 4-29.

Figure 4-36 shows the three variations of the rafter tail. Figure 4-36(A) shows the flush tail or no-tail. The rafter butts against the wall plate with no overhang. In Fig. 4-36(B), the full tail is shown. Note the overhang. Figure 4-36(C) shows the various shapes possible. This one indicates a separate tail that has been curved. It is nailed onto the no-tail or flush rafter.

All the cuts for the various types of common rafters are made at right angles to the sides of the rafter.

Fig. 4-37 *Using the steel square to mark off the top or plumb cut.*

Figure 4-37 shows how the framing square is used to lay out the angles. Find the 12-inch mark on the square. Note how the square is set at 12 for laying out the cut.

The distance 12 is the same as 1 foot of run. The other side of the square is set with the edge of the stock to the rise in inches per foot of run. In some cases the tail is not cut until after the rafter is in place. Then it is cut to match the others and aligns better for the fascia board.

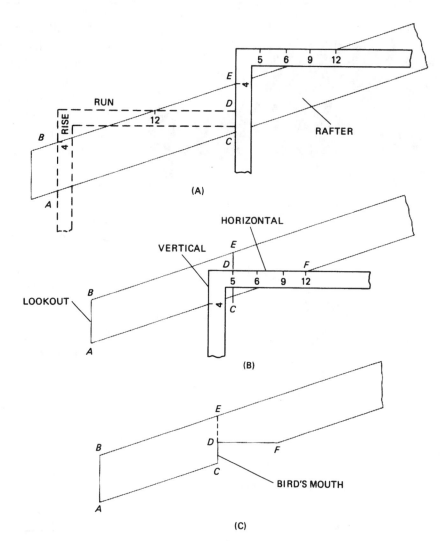

Fig. 4-38 *Using the steel square to lay out rafter lookout and bird's mouth.*

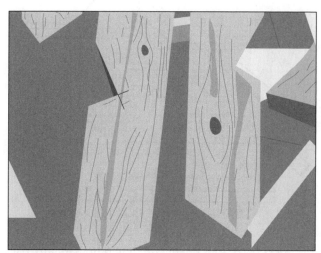

Fig. 4-39A *Notice the saw cuts past the bird's mouth.*

Figure 4-38 shows a method of using the square to lay out the bottom and lookout cuts. If there is a ridge board, you have to deduct one-half the thickness of the ridge from the rafter length. Figure 4-38 shows how to place the square for marking the rafter lookout. Scribe the cut line as shown in Fig. 4-38(A). The rise is 4 and the run is 12. Then move the square to the next position and mark from C to E. The distance from B to E is equal to the length of the lookout. Move the square up to E (with the same setting). Scribe line CE. On this line, lay off CD. This is the length of the vertical side of the bottom cut. Now apply the same setting to the bottom edge of the rafter. This is done so that the edge of the square cuts D. Scribe DF. See Fig. 4-38(B). This is the horizontal line of the bottom cut. In making this cut, the wood is cut out to the lines CD and DE. See Fig. 4-39(A) for an example of rafters cut this way. Note how the portable handsaw makes cuts beyond the marks.

In Figs. 4-39(B) and 4-40 you can see the rafters in place. Note the 90° angle. The rafter and the ridge should meet at 90°. The rafter in Fig. 4-40 has not had the lookout cut. The overhang is slightly different from the type just shown. Can you find the difference?

Length per foot of run The rafter tables on the Stanley squares are based on the rise per foot run, which means that the figures in the tables indicate the length

Fig. 4-39B *Rafters in place. They are nailed to the ridge board, the wall plate, and the ceiling joists. The other side of the roof is already covered with plywood sheathing.*

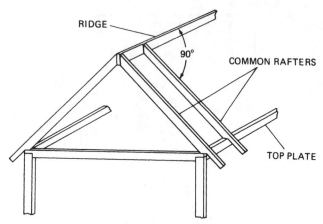

Fig. 4-40 *Common rafters in place.*

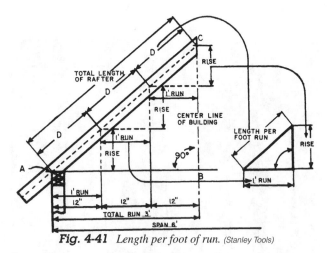

Fig. 4-41 *Length per foot of run.* (Stanley Tools)

of rafters per 1-foot run of common rafters for any rise of roof. This is shown in Fig. 4-41.

The roof has a 6-foot span and a certain rise per foot. The figure may be regarded as a right triangle *ABC*, having for its sides the run, the rise, and the rafter.

The run of the rafter has been divided into three equal parts, each representing 1-foot run.

It will be noted that by drawing vertical lines through each division point of the run, the rafter also will be divided into three equal parts *D*.

Since each part *D* represents the length of rafter per 1-foot run and the total run of the rafter equals 3 feet, it is evident that the total length of rafter will equal the length *D* multiplied by 3.

The reason for using this per-foot-run method is that the length of any rafter may be easily determined for any width of building. The length per foot run will be different for different pitches. Therefore, before you can find the length of a rafter, you must know the rise of roof in inches or the rise per foot of run.

> RULE: To find the rise per foot run, multiply the rise by 12 and divide by the length of the run.

The factor *12* is used to obtain a value in inches. The rise and run are expressed in feet. See Figs. 4-42 and 4-43.

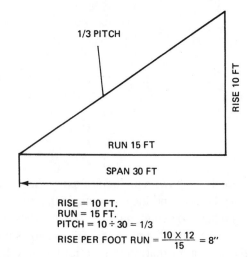

RISE = 10 FT.
RUN = 15 FT.
PITCH = 10 ÷ 30 = 1/3
RISE PER FOOT RUN = $\frac{10 \times 12}{15}$ = 8″

Fig. 4-42 *Finding the rise per foot of run.* (Stanley Tools)

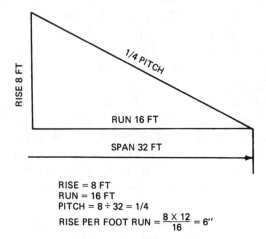

RISE = 8 FT
RUN = 16 FT
PITCH = 8 ÷ 32 = 1/4
RISE PER FOOT RUN = $\frac{8 \times 12}{16}$ = 6″

Fig. 4-43 *Finding the rise per foot of run.* (Stanley Tools)

The rise per foot run is always the same for a given pitch and can be easily remembered for all ordinary pitches.

PITCH	½	⅓	¼	⅙
Rise per foot run in inches	12	8	6	4

The members of a firmly constructed roof should fit snugly against each other. Rafters that are not properly cut make the roof shaky and the structure less stable. Therefore, it is very important that all rafters are the right lengths and that their ends are properly cut. This will provide a full bearing against the members to which they are connected.

Correct length, proper top and bottom cuts, and the right side or cheek cuts are the very important features to be observed when framing a roof.

Length of Rafters

The length of rafters may also be obtained in other ways. There are three in particular which can be used:

1. Mathematical calculations
2. Measuring across the square
3. Stepping off with the square

The first method, while absolutely correct, is very impractical on the job. The other two are rather unreliable and quite frequently result in costly mistakes.

The tables on the square have eliminated the need for using the three methods just mentioned. These tables let the carpenter find the exact length and cuts for any rafter quickly, and thus save time and avoid the possibility of errors.

Common Rafters

The common rafter extends from the plate to the ridge. Therefore, it is evident that the rise, the run, and the rafter itself form a right triangle. The length of a common rafter is the shortest distance between the outer edge of the plate and a point on the centerline of the ridge. This length is taken along the *measuring line*. The measuring line runs parallel to the edge of the rafter and is the hypotenuse or the longest side of a right triangle. The other sides of the triangle are the run and the rise. See Fig. 4-44.

The rafter tables on the face of the body of the square include the outside edge graduations on both

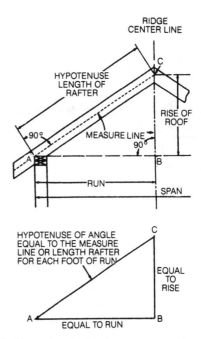

Fig. 4-44 *Measuring the line for a common rafter.* (Stanley Tools)

the body and the tongue, which are in inches and sixteenths of an inch.

The length of rafters The lengths of common rafters are found on the first line, indicated as *Length of main rafters per foot run*. There are seventeen of these tables, beginning at 2 inches and continuing to 18 inches. Figure 4-45 shows the square being used.

RULE: To find the length of a common rafter, multiply the length given in the table by the number of feet of run.

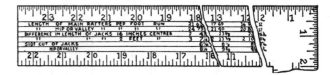

Fig. 4-45 *Note the labels for tables on the steel square.* (Stanley Tools)

For example, if you want to find the length of a common rafter where the rise of roof is 8 inches per foot run, or one-third pitch, and the building is 20 feet wide, you first find where the table is. Then on the inch line on the top edge of the body, find the figure that is equal to the rise of the roof, which in this case will be 8. On the first line under the figure 8 will be found 14.42. This is the length of the rafter in inches per foot run for this particular pitch. Examine Fig. 4-46.

The building is 20 feet wide. Therefore, the run of the rafter will be 20 divided by 2, which equals 10 feet.

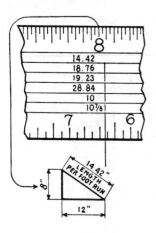

Fig. 4-46 Finding the proper number on the square. (Stanley Tools)

Since the length of the rafter per 1-foot run equals 14.42 inches, the total length of the rafter will be 14.42 multiplied by 10, which equals 144.20 inches, or 144.20 divided by 12, which equals 12.01 feet, or for all practical purposes, 12 feet. Check Fig. 4-46.

Top and bottom cuts of the common rafter The top or plumb cut is the cut at the upper end of the rafter where it rests against the opposite rafter or against the ridge board. The bottom cut or heel cut is the cut at the lower end which rests on the plate. See Fig. 4-47.

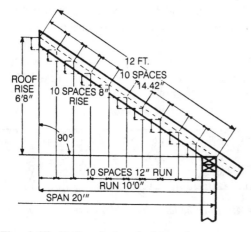

Fig. 4-47 Finding the length of the rafter. (Stanley Tools)

The top cut is parallel to the centerline of the roof, and the bottom cut is parallel to the horizontal plane of the plates. Therefore, the top and bottom cuts are at right angles to each other.

RULE: To obtain the top and bottom cuts of a common rafter, use 12 inches on the body and the rise per foot run on the tongue. Twelve inches on the body will give the horizontal cut, and the figure on the tongue will give the vertical cut.

To illustrate the rule, we will examine a large square placed alongside the rafter as shown in Fig. 4-48. Note that the edge of the tongue coincides with the top cut of the rafter. The edge of the blade coincides with the heel cut. If this square were marked in feet, it would show the run of the rafter on the body and the total rise on the tongue. Line *AB* would give the bottom cut and line *AC* would give the top cut.

However, the regular square is marked in inches. Since the relation of the rise to 1-foot run is the same as that the total rise bears to the total run, we use 12 inches on the blade and the rise per foot on the tongue to obtain the respective cuts. The distance 12 is used as a unit and is the 1-foot run, while the figure on the other arm of the square represents the rise per foot run. See Figs. 4-49 and 4-50.

Actual length of the rafter The rafter lengths obtained from the tables on the square are to the center-line of the ridge. Therefore, the thickness of half of the ridge board should always be deducted from the obtained total length before the top cut is made. See Fig. 4-51. This deduction of half the thickness of the ridge is measured at right angles to the plumb line and is marked parallel to this line.

Figure 4-52 shows the wrong and right ways of measuring the length of rafters. Diagram D shows the measuring line as the edge of the rafter, which is the case when there is no tail or eave.

Cutting the rafter After the total length of the rafter has been established, both ends should be marked and allowance made for a tail or eave. Don't forget to allow for half the thickness of the ridge.

Both cuts are obtained by applying the square so that the 12-inch mark on the body and the mark on the tongue that represents the rise are at the edge of the stock.

All cuts for common rafters are made at right angles to the side of the rafter.

For example, a common rafter is 12 feet 6 inches, and the rise per foot run is 9 inches. Obtain the top and bottom cuts. See Fig. 4-53.

Points *A* and *B* are the ends of the rafter. To obtain the bottom or seat cut, take 12 inches on the body of the square and 9 inches on the tongue. Lay the square on the rafter so that the body will coincide with point *A* or the lower end of the rafter. Mark along the body of the square and cut.

To obtain the top cut, move the square so that the tongue coincides with point *B*. This is the upper end of the rafter. Mark along the tongue of the square.

Deduction for the ridge The deduction for half the thickness of the ridge should now be measured. Half

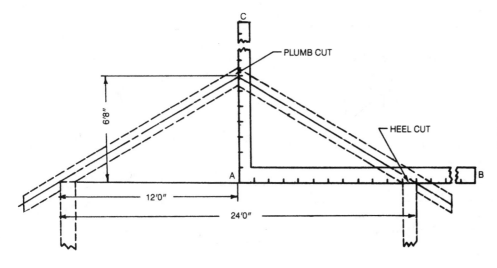

Fig. 4-48 *Using the steel square to check plumb and heel cuts.* (Stanley Tools)

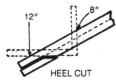

Fig. 4-49 *Using the steel square to lay out the heel cut.* (Stanley Tools)

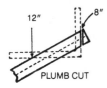

Fig. 4-50 *Using the steel square to lay out the plumb cut.* (Stanley Tools)

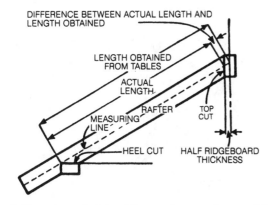

Fig. 4-51 *How to find the difference between the actual length and the length obtained.* (Stanley Tools)

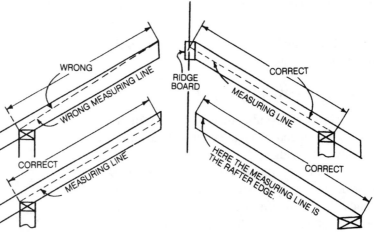

Fig. 4-52 *Right and wrong ways of measuring rafters.* (Stanley Tools)

the thickness of the ridge is 1 inch. One inch is deducted at right angles to the top cut mark or plumb line, which is point *C*. A line is then drawn parallel to the top cut mark, and the cut is made. You will notice that the allowance for half the ridge measured along the measuring line is 1¼ inches. This will vary according to the rise per foot run. It is therefore important to measure for this deduction at right angles to the top cut mark or plumb line.

Measuring rafters The length of rafters having a tail or eave can also be measured along the back or top edge instead of along the measuring line, as shown in Fig. 4-54. To do this it is necessary to carry a plumb line to the top edge from *P*, and the measurement is started from this point.

Occasionally in framing a roof, the run may have an odd number of inches; for example, a building might have a span of 24 feet 10 inches. This would mean a run of 12'5". The additional 5 inches can be added easily without mathematical division after the length obtained from the square for 12 feet of run is measured. The additional 5 inches is measured at right

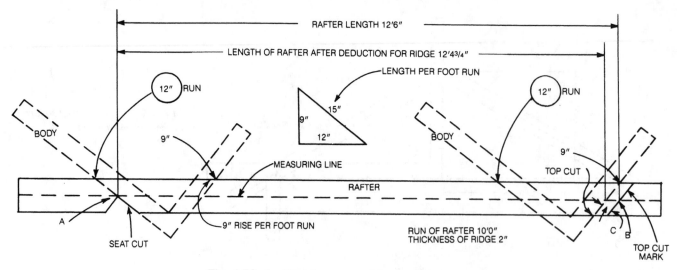

Fig. 4-53 *Applying the square to lay out cuts.* (Stanley Tools)

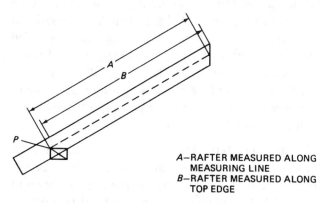

A—RAFTER MEASURED ALONG
MEASURING LINE
B—RAFTER MEASURED ALONG
TOP EDGE

Fig. 4-54 *Two places to measure a rafter.* (Stanley Tools)

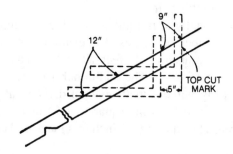

Fig. 4-55 *Adding extra inches to the length of a rafter.* (Stanley Tools)

angles to the last plumb line. See Fig. 4-55 for an illustration of the procedure.

Hip-and-Valley Rafters

The hip rafter is a roof member that forms a hip in the roof. This usually extends from the corner of the building diagonally to the ridge.

The valley rafter is similar to the hip, but it forms a valley or depression in the roof instead of a hip. It also extends diagonally from the plate to the ridge. Therefore, the total rise of the hip-and-valley rafters is the same as that of the common rafter. See Fig. 4-56.

The relation of hip-and-valley rafters to the common rafter is the same as the relation of the sides of a right triangle. Therefore, it will be well to explain here one of the main features of right triangles.

In a right triangle, if the sides forming the right angle are 12 inches each, the hypotenuse, or the side opposite the right angle, is equal to 16.97 inches. This is usually taken as 17 inches. See Fig. 4-57.

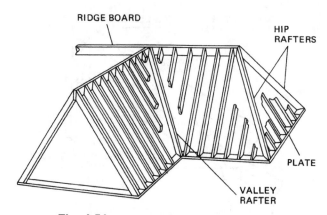

Fig. 4-56 *Hip-and-valley rafters.* (Stanley Tools)

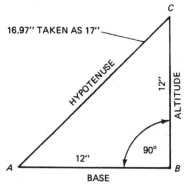

Fig. 4-57 *In a right triangle, 12-inch base and 12-inch altitude produces an isosceles triangle. This means the hypotenuse is 16.97 inches.* (Stanley Tools)

The position of the hip rafter and its relation to the common rafter is shown in Fig. 4-58, where the hip rafter is compared to the diagonal of a square prism. The prism (as shown in Fig. 4-58) has a base of 5 feet square, and its height is 3 feet 4 inches.

D   is the corner of the building
BC is the total rise of the roof
AC is the common rafter
DB is the run of the hip rafter
DC is the hip rafter

It should be noted that the figure *DAB* is a right triangle whose sides are the portion of the plate *DA*, the run of the common rafter *AB*, and the run of the hip rafter *DB*. The run of the hip rafter is opposite right angle *A*. The hypotenuse is the longest side of the right triangle.

If we should take only 1 foot of run of common rafter and a 1-foot length of plate, we would have a right triangle *H*. The triangle's sides are each 12 inches long. The hypotenuse is 17 inches, or more accurately 16.97 inches. See Figs. 4-57 and 4-59.

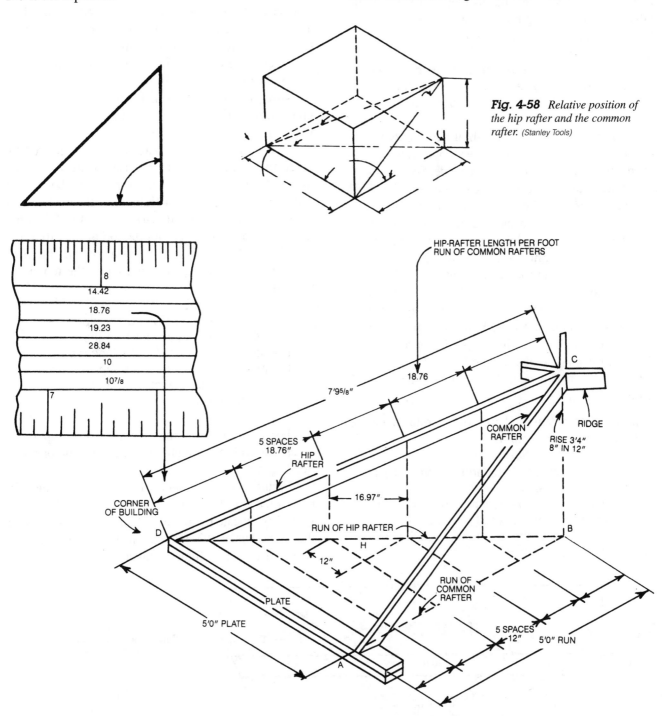

Fig. 4-58 *Relative position of the hip rafter and the common rafter.* (Stanley Tools)

Fig. 4-59 *Finding the length of the hip rafter. Note the location of the 18.76 on the square.* (Stanley Tools)

The hypotenuse in the small triangle *H* in Fig. 4-59 is a portion of the run of the hip rafter *DB*. It corresponds to a 1-foot run of common rafter. Therefore, the run of hip rafter is always 16.97 inches for every 12 inches of run of common rafter. The total run of the hip rafter will be 16.97 inches multiplied by the number of feet run of common rafter.

Length of hip-and-valley rafters Lengths of the hip-and-valley rafters are found on the second line of the rafter table. It is entitled *Length of hip or valley rafters per foot run*. This means that the figures in the table indicate the length of hip-and-valley rafters per foot run of common rafters. See Fig. 4-45.

> RULE: To find the length of a hip or valley rafter, multiply the length given in the table by the number of feet of the run of common rafter.

For example, find the length of a hip rafter where the rise of the roof is 8 inches per foot of run, or one-third pitch. The building is 10 feet wide. See Fig. 4-58.

Proceed as in the case of the common rafters. That is, find on the inch line of the body of the square the figure corresponding to the rise of roof—which is 8. On the second line under this figure is found 18.76. This is the length of hip rafter in inches for each foot of run of common rafter for one-third pitch. See Fig. 4-59.

The common rafter has a 5-foot run. Therefore, there are also five equal lengths for the hip rafter, as may be seen in Fig. 4-59. We have found the length of the hip rafter to be 18.76 inches per 1 foot run. Therefore, the total length of the hip rafter will be 18.76×5 = 93.80 inches. This is 7.81 feet, or for practical purposes 7'9¹³⁄₁₆" or 7'9⅝".

Top and bottom cuts The following rule should be followed for top and bottom cuts.

> RULE: To obtain the top and bottom cut for hip or valley rafters, use 17 inches on the body and the rise per foot run on the tongue. Seventeen on the body will give the seat cut, and the figure on the tongue will give the vertical or top cut. See Fig. 4-60.

Measuring hip-and-valley rafters The length of all hip-and-valley rafters must always be measured along the center of the top edge or back. Rafters with a tail or eave are treated like common rafters, except that the measurement or measuring line is at the center of the top edge.

Deduction from hip or valley rafter for ridge The deduction for the ridge is measured in the same way as for the common rafter (see Fig. 4-53), except that half the diagonal (45') thickness of the ridge must be used.

Side cuts In addition to the top and bottom cuts, hip-and-valley rafters must also have side or check cuts at the point where they meet the ridge.

These side cuts are found on the sixth or bottom line of the rafter tables. It is marked *Side cut hip or valley—use*. The figures given in this line refer to the graduation marks on the outside edge of the body. See Fig. 4-45.

The figures on the square have been derived by determining the figure to be used with 12 on the tongue for the side cuts of the various pitches by the following method. From a plumb line, the thickness of the rafter is measured and marked as at *A* in Fig. 4-61. A line is then squared across the top of the rafter and the diagonal points connected as at *B*. The line *B* or side cut is obtained by marking along the tongue of the square.

> RULE: To obtain the side cut for hip or valley rafters, take the figure given in the table on the body of the square and 12 inches on the tongue. Mark the side cut along the tongue where the tongue coincides with the point on the measuring line.

For example, find the side cut for a hip rafter where the roof has 8 inches rise per foot run or pitch. See Fig. 4-62.

Figure 4-62 represents the position of the hip rafter on the roof. The rise of the roof is 8 inches to the foot. First, locate the number 8 on the outside edge of the body. Under this number in the bottom line you will find 10⅞. This figure is taken on the body and 12

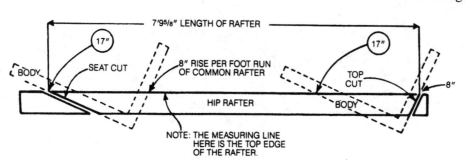

Fig. 4-60 *Using the square to lay out top and seat cuts on a hip rafter. (Stanley Tools)*

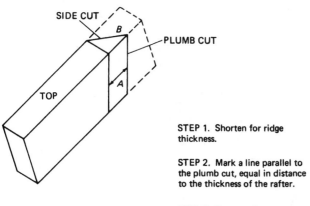

STEP 1. Shorten for ridge thickness.

STEP 2. Mark a line parallel to the plumb cut, equal in distance to the thickness of the rafter.

STEP 3. Square a line across top of hip rafter. The diagonal is the side cut.

Fig. 4-61 *Making side cuts so that the hip will fit into the intersection of rafters. (Stanley Tools)*

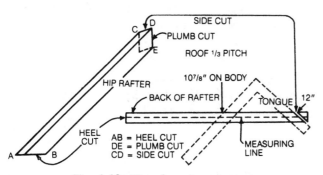

Fig. 4-62 *Hip rafter cuts. (Stanley Tools)*

inches is taken on the tongue. The square is applied to the edge of the back of the hip rafter. The side cut *CD* comes along the tongue.

In making the seat cut for the hip rafter, an allowance must be made for the top edges of the rafter. They would project above the line of the common and jack rafters if the corners of the hip rafter were not removed or backed. The hip rafter must be slightly lowered. Do this by cutting parallel to the seat cut. The distance varies with the thickness and pitch of the roof.

It should be noted that on the Stanley square the 12-inch mark on the tongue is always used in all angle cuts—top, bottom, and side. This leaves the worker with only one number to remember when laying out side or angle cuts. That is the figure taken from the fifth or sixth line in the table.

The side cuts always come on the right-hand or tongue side on rafters. When you are marking boards, these can be reversed for convenience at any time by taking the 12-inch mark on the body and using the body references on the tongue.

Obtain additional inches in run of hip or valley rafters by using the explanation given earlier for common rafters. However, use the diagonal (45°) of the ad-

ditional inches. This is approximately 7¹⁄₁₆ inches for 5 inches of run. This distance should be measured in a similar manner.

Jack Rafters

Jack rafters are *discontinued* common rafters. They are common rafters cut off by the intersection of a hip or valley before reaching the full length from plate to ridge.

Jack rafters lie in the same plane as common rafters. They usually are spaced the same and have the same pitch. Therefore, they also have the same length per foot run as common rafters have.

Jack rafters are usually spaced 16 inches or 24 inches apart. Because they rest against the hip or valley equally spaced, the second jack must be twice as long as the first one, the third three times as long as the first, and so on. See Fig. 4-63.

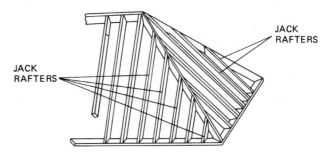

Fig. 4-63 *Location of jack rafters. (Stanley Tools)*

Length of jack rafters The lengths of jacks are given in the third and fourth lines of the rafter tables on the square. They are indicated:

> Third line: Difference in length of jacks—16 inch centers
> Fourth line: Difference in length of jacks—2 foot centers

The figures in the table indicate the length of the first or shortest jack, which is also the difference in length between the first and second jacks, between the second and third jacks, and so on.

> RULE: To find the length of a jack rafter, multiply the value given in the tables by the number indicating the position of the jack. From the obtained length, subtract half the diagonal (45°) thickness of the hip or valley rafter.

For example, find the length of the second jack rafter. The roof has a rise of 8 inches to 1 foot of run of common rafter. The spacing of the jacks is 16 inches.

On the outer edge of the body, find the number 8, which corresponds to the rise of the roof. On the third line under this figure find 19¼. This means that the first jack rafter will be 19¼ inches long. Since the length of the second jack is required, multiply 19¼ by 2, which equals 38½ inches. From this length half the diagonal (45°) thickness of the hip or valley rafter should be deducted. This is done in the same manner that the deduction for the ridge was made on the hip rafter.

Proceed in the same manner when the lengths of jacks spaced on 24-inch centers are required. It should be borne in mind that the second jack is twice as long as the first one. The third jack is three times as long as the first jack, and so on.

Top and bottom cuts for jacks Since jack rafters have the same rise per foot run as common rafters, the method of obtaining the top and bottom cuts is the same as for common rafters. That is, take 12 inches on the body and the rise per foot run on the tongue. Twelve inches will give the seat cut. The figure on the tongue will give the plumb cut.

Side cut for jacks At the end where the jack rafter frames to the hip or valley rafter, a side cut is required. The side cuts for jacks are found on the fifth line of the rafter tables on the square. It is marked: *Side cut of jacks—use*. See Fig. 4-45.

> RULE: To obtain the side cut for a jack rafter, take the figure shown in the table on the body of the square and 12 inches on the tongue. Mark along the tongue for side cut.

For example, find the side cut for jack rafters of a roof having 8 inches rise per foot run, or ⅓ pitch. See Figs. 4-64 and 4-65. Under the number 8 in the fifth line of the table find 10. This number taken on the outside edge of the body and 12 inches taken on the tongue will give the required side cut.

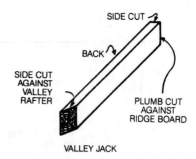

Fig. 4-65 *Valley jack rafter cuts.* (Stanley Tools)

BRACE MEASURE

In all construction there is the need for some braces to make sure certain elements are held securely. See Fig. 4-120(A). The brace measure table is found along the center of the back of the tongue of the carpenter's square. It gives the lengths of common braces. See Fig. 4-66.

Fig. 4-66 *Brace measure table on back side of steel square tongue.* (Stanley Tools)

For example, find the length of a brace whose run on post and beam equals 39 inches. See Fig. 4-67. In the brace table find the following expression:

$$\begin{array}{c} 39 \\ 55.15 \\ 39 \end{array}$$

This means that with a 39-inch run on the beam and a 39-inch run on the post, the length of the brace will be 55.15 inches. For practical purposes, use 55⅛ inches.

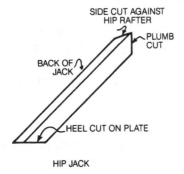

Fig. 4-64 *Hip jack rafter cuts.* (Stanley Tools)

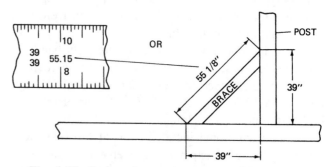

Fig. 4-67 *Cutting a brace using a square table.* (Stanley Tools)

Braces may be regarded as common rafters. Therefore, when the brace run on the post differs from the run on the beam, their lengths as well as top and bottom cuts may be determined from the figures given in the tables of the common rafters.

Rafter Layout Using a Speed Square

Since most blueprints label the angle of the roof using a slope index (see Fig. 4-68), a carpenter can easily lay out the top and bottom cuts of the rafter using the rise scale on the speed square. As illustrated in Fig. 4-69, just place the right angle of the speed square at the line marking the length of your rafter and then pivot it to the corresponding number labeled on the blueprints. In other words, if the rise of the roof is 8, just turn the speed square until the edge of the rafter is aligned with the number 8. Remember to use the scale marked "common" when laying out cuts for common rafters. Mark the edge as shown in Fig. 4-69. The outside and inside corner rafters for a hip-and-valley roof would use the scale labeled HIP-VAL, which is the upper scale shown in Fig. 4-70. Remember to tilt the saw 45° while cutting the top cut of a hip or valley rafter. This will give you the top plumb and side cuts with one sawing operation.

Speed squares do not have the length of rafter per foot of run stamped on them like a framing square. Instead, speed squares come with a compact 3" x 5" pocket reference booklet (Fig. 4-71), which contains charts listing rafter lengths for building widths from 3 to 40 feet. Common and hip rafter lengths are listed for roofs with a rise from 1 to 30 inches. Unlike the framing square, you do not have to multiply the number stamped on the square by half the span of your

Fig. 4-69 *Marking a plumb cut on a common rafter with an 8-inch rise.*

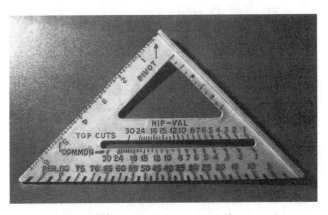

Fig. 4-70 *Rafter scales on a speed square.*

Fig. 4-71 *Mini 3" × 5" speed square pocket reference booklet.*

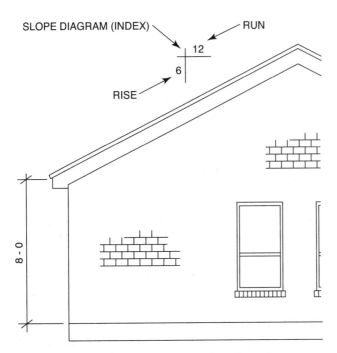

Fig. 4-68 *Slope diagram (index) of a roof.*

house. The entire length of the rafter is listed in the chart. Additional lengths can be determined by adding two lengths listed in the charts. For example, if you had a 47-foot roof span, then you would just add the 7- and 40-foot listed lengths together. Lengths do not include overhangs, so make sure you add the length of your overhang to the rafter length before you cut it. Further, remember to subtract half the width of your ridge board before cutting the rafter.

ERECTING THE ROOF WITH RAFTERS

Rafters are cut to fit the shape of the roof. Roofs are chosen by the builder or planner. The design of the rafter is determined by the pitch, span, and rise of the type of roof chosen. The gable roof is simple. It can be made easily with a minimum of difficult cuts. In this example we start with the gable type and then look at other variations of rooflines. See Fig. 4-72 for an example of the gable roof.

Rafter Layout

One of the most important tools used to lay out rafters is the carpenter's square. All rafters must be cut to the proper angle or bevel. They fasten to the wall plate or to the ridge board. In some cases there is an overhang. This overhang must be taken into consideration when the rafter is cut. Gable siding, soffits, and overhangs are built together (Fig. 4-73).

Raising Rafters

Mark rafter locations on the top plate of the side walls. The first rafter pair will be flush with the outside edge of the end wall. See Fig. 4-74. Note the placement of the gable end studs. The notch in the gable end stud is made to fit the 2 × 4, or whatever thickness of rafter you are using. Space the first interior rafter 24 inches, measured from the end of the building to the center of the rafter. In some cases 16 inches O.C. is used for spacing. All succeeding rafter locations are measured

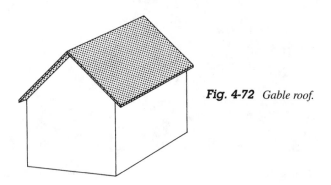

Fig. 4-72 Gable roof.

Fig. 4-73 *Gable siding, soffits, and gable overhangs can be built together.* (Fox and Jacobs)

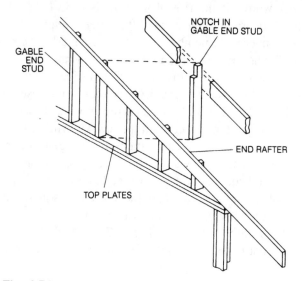

Fig. 4-74 *Placement of the end rafter.* (American Plywood Association)

24 inches center to center. They will be at the sides of ceiling joist ends. See Fig. 4-75.

Next, mark rafter locations on the ridge board. Allow for the specified gable overhang. To achieve the required total length of ridge board, you may have to splice it. See Fig. 4-76. Do not splice at this time. It is easier to erect it in shorter sections, then splice it after it is in place.

Check your house plan for roof slope. For example, a 4-inch rise in 12 inches of run is common. It is usually considered the minimum for asphalt or wood shingles.

Lay out one pair of rafters as previously shown. Mark the top and bottom angles and seat-cut location. Make the cuts and check the fit by setting them up at floor level. Mark this set of rafters for identification and use it as a pattern for the remainder.

Cut the remaining rafters. For a 48-foot house with rafters spaced 24 inches O.C. you will need 24 pairs

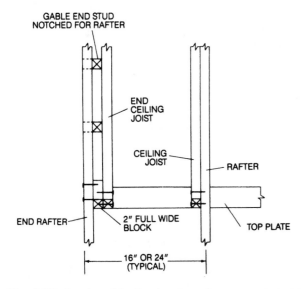

Fig. 4-75 *Spacing of the first interior rafter.* (American Plywood Association)

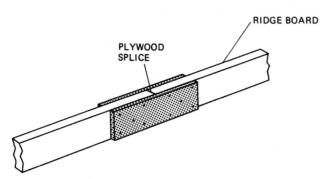

Fig. 4-76 *Method of splicing a ridge board.* (American Plywood Association)

cut to the pattern (25 pairs counting the pattern). In addition, you will need two pairs of fascia rafters for the ends of the gable overhang. See Fig. 4-77. Since they cover the end of the ridge board, they must be longer than the pattern rafters by half the width of the ridge

board. Fascia rafters have the same cuts at the top and bottom as the regular rafters. However, they do not have a seat cut.

Build temporary props of 2 × 4s to hold the rafters and ridge board in place during roof framing installation. The props should be long enough to reach from the top plate to the bottom of the ridge board. They should be fitted with a plywood gusset at the bottom. When the props are installed, the plywood gusset is nailed temporarily to the top plate or to a ceiling joist. The props are also diagonally braced from about midpoint in both directions to maintain true vertical (check with a plumb bob). See Fig. 4-78.

Move the ridge board sections and rafters onto the ceiling framing. Lay plywood panels over the ceiling joists for a safe footing. First erect the ridge board and the rafters nearest its ends. See Fig. 4-78. If the ridge of the house is longer than the individual pieces of ridge board, you'll find it easier to erect each piece separately, rather than splice the ridge board full-length first. Support the ridge board at both ends with the temporary props. Toenail the first rafter pair securely to the ridge board using at least two 8d nails on each side. Then nail it at the wall. Install the other end rafter pair in the same manner.

Make the ridge board joints using plywood gussets on each side of the joint. Nail them securely to the ridge board.

Check the ridge board for level. Also check the straightness over the centerline of the house.

After the full length of the ridge board is erected, put up the remaining rafters in pairs. Nail them securely in place. Check them occasionally to make sure the ridge board remains straight. If all rafters are cut and assembled accurately, the roof should be self-aligning.

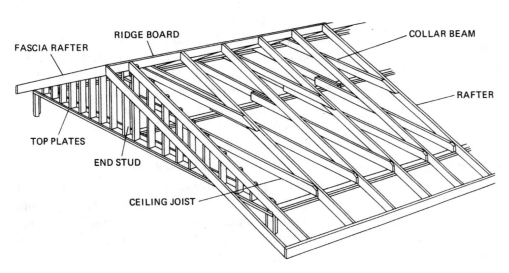

Fig. 4-77 *Placement of rafters, ridge board, and collar beam.* (American Plywood Association)

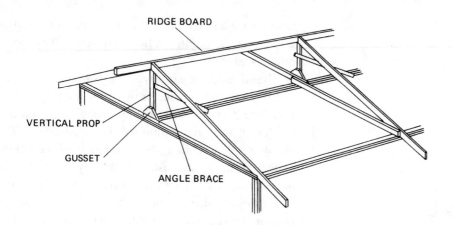

Fig. 4-78 *Placement of angle braces and vertical props.* (American Plywood Association)

Toenail the rafters to the wall plate with 10d nails. Use two per side. Also nail the ceiling joists to the rafters. For a 24-foot-wide house, you will need 16d nails at each lap. In high-wind areas, it is a good idea to add metal-strap fasteners for extra uplift resistance. See Fig. 4-79.

Cut and install 1 × 6 collar beams at every other pair of rafters (4 feet O.C.). See Fig. 4-77. Nail each end with four 8d nails. Collar beams should be in the upper third of the attic crawl space. Remove the temporary props.

Square a line across the end wall plate directly below the ridge board. If a vent is to be installed, measure half its width on each side of the center mark to locate the first stud on each side. Mark the positions for the remaining studs at 16 inches O.C. Then measure and cut the studs. Notch the top end to fit under the rafter so that the stud bottom will be flush with the top plate. Cut the cripple studs and headers to frame in the vent opening. See Fig. 4-80.

Cut and install fascia board to the correct length of the ridge board. Bevel the top edge to the roof slope. Nail the board to the rafter ends. Then, install fascia

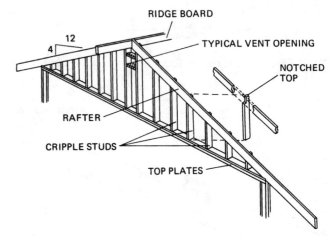

Fig. 4-80 *Vent openings should be blocked in.* (American Plywood Association)

rafters. Fascia rafters cover the end of the ridge board. See Fig. 4-77. Where the nails will be exposed to weather, use hot-dipped galvanized or other nonstaining nails.

SPECIAL RAFTERS

There are some rafters needed to make special roof shapes. The mansard roof and the hip roof both call for special rafters. Jack rafters are needed for the hip roof. This type of roof may also have valleys and have to be treated especially well. Dormers call for some special rafters, too. For bay windows and other protrusions, some attention may have to be given to rafter construction.

Dormers

Dormers are protrusions from the roof. They stick out from the roof. They may be added to allow light into an upstairs room. Or, they may be added for architectural effect. See Fig. 4-81. Dormers may be made in three types. They are:

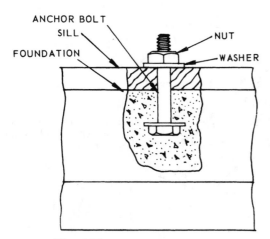

Fig. 4-79 *Metal framing anchors.*

Fig. 4-81 *A dormer.*

1. Dormers with flat, sloping roofs that have less slope than the roof in which they are located. This can be called a shed-type dormer (Fig. 4-82).

2. Dormers with roofs of the gable type at right angles to the roof (Fig. 4-83). No slope in this one.

3. The two types can be combined. This is called the hip-roof dormer.

Bay Windows

Bay windows are mostly for decoration. They add to the architectural qualities of a house. They stick out

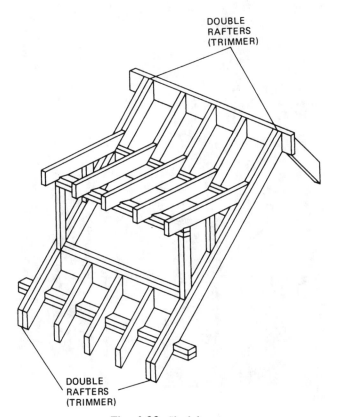

DOUBLE
RAFTERS
(TRIMMER)

DOUBLE
RAFTERS
(TRIMMER)

Fig. 4-82 *Shed dormer.*

from the main part of the house. This means they have to have special handling. The floor joists are extended out past the foundation the required amount. A band is then used to cap off the joist ends. See Fig. 4-84. Take a closer look at the ceiling joists and rafters for the bay. The rafters are cut according to the rise called for on the plans. Cuts and lengths have already been discussed. No special problems should be presented by this method of framing. In order to make it easier, it is best to lay out the rafter plan at first so that you know which are the common and which are the hip rafters. In some cases you may need a jack rafter or two depending upon the size of the bay. See Fig. 4-85.

CEILING JOISTS

Ceiling joists serve a number of purposes. They keep the wall from falling inward or outward. They anchor the rafters to the top plate. Ceiling joists also hold the finished ceiling in place. The run of the joist is important. The distance between supports for a joist determines its size. In some cases the ceiling joist has to be spliced. Figure 4-86 shows one method of splicing. Note how the splice is made on a supporting partition. Figure 4-87 shows how the joists fit on top of the plate. In some cases it is best to tie the joist down to the top plate by using a framing bracket. Figure 4-84 shows how one type of bracket is used to hold the joist. This helps in high-wind areas.

The ceiling joists in Fig. 4-89 A and B have been trimmed to take the angle of the rafter into consideration once the rafter is attached to the top plate and joist.

The size of the joist is determined by the local code. However, there are charts that will give you some idea of what size piece of dimensional lumber to use. Table 4-4 indicates some allowable spans for ceiling joists. These are given using non-stress graded lumber.

There are some special arrangements for ceiling joists. In some cases it is necessary to interrupt the free flow of lines represented by joists. For example, a chimney may have to be allowed for. An attic opening may be called for on the drawings. These openings have to be reinforced to make sure the joists maintain their ability to support the ceiling and some weight in the ceiling at a later time. See Fig. 4-90.

Figure 4-91 shows how framing anchors are used to secure the joists to the double header. This method can be used for both attic openings and fireplace openings.

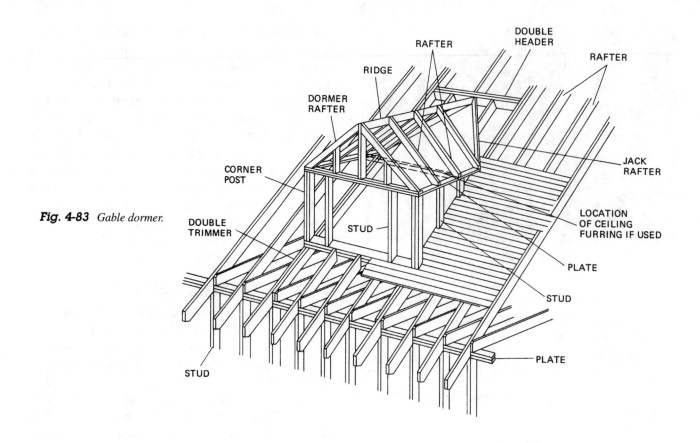

Fig. 4-83 *Gable dormer.*

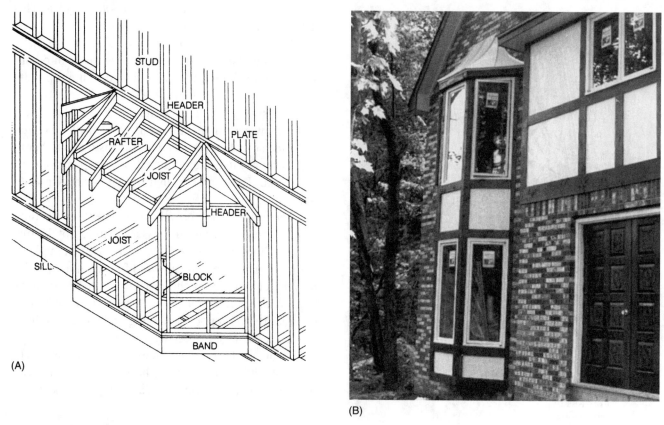

(A)

(B)

Fig. 4-84 *(A) Bay window framing; (B) Two bay windows stacked for a two-story house. The metal cover does not require rafters.*

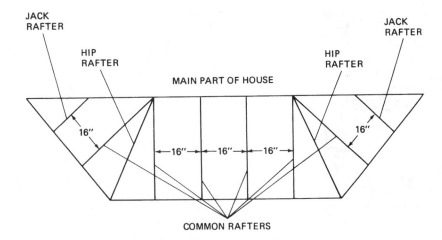

Fig. 4-85 *Rafter layout for a bay window.*

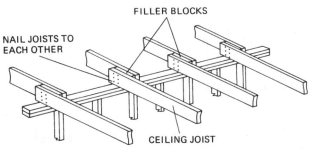

Fig. 4-86 *Ceiling joist splices are made on a supporting partition.*

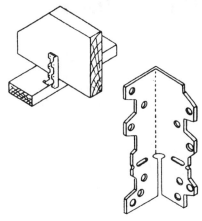

Fig. 4-88 *Steel bracket used to hold ceiling joist to the top plate. Note how it is bent to fit.*

Fig. 4-87 *Looking up toward the ceiling joists. Notice how these are supported on the wall plate and are not cut or tapered. Two nails are used to toenail the joists to the plate.*

OPENINGS

As mentioned before, fireplaces do come out through the roof. This must be allowed for in the construction process. The floor joists have to be reinforced. The area around the fireplace has to be strong enough to hold the hearth. However, what we're interested in here is how the fireplace opening comes through the rafters. See Fig. 4-92. The roofing here has been boxed off to allow the fireplace to come through. The chimney at the top has the flashing ready for installation as soon as the bricks are laid around the flue.

Other openings are for soil pipes. These are used for venting the plumbing system. A hole in the ply-

wood deck is usually sufficient to allow their exit from the inside of the house. See Fig. 4-93.

DECKING

A number of types of roof deckings are used. One is plywood. It is applied in 4-×-8-foot sheets. Plywood adds structural strength to the building. Plywood also saves time, since it can be placed on the rafters rather quickly.

Boards of the 1-×-6- or 1-×-8-inch size can be used for sheathing. This decking takes a longer time to apply. Each board has to be nailed and sawed to fit. This type of decking adds strength to the roof also.

Another type of decking is nothing more than furring strips applied to the rafters. This is used as a nail base for cedar shingles.

Plywood Decking

Roof sheathing, the covering over rafters or trusses, provides structural strength and rigidity. It makes a solid base for fastening the roofing material.

(A)

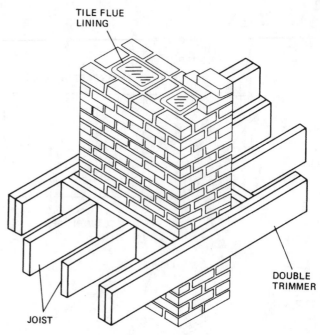

TILE FLUE LINING

JOIST

DOUBLE TRIMMER

Fig. 4-90 *Blocking in the joists to allow an opening for a chimney.*

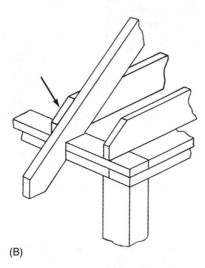

(B)

Fig. 4-89 *(A) Ceiling joists in place on the top plate. Note the cuts on the ends of the joists. (B) Set the first ceiling joist on the inside of the end of the wall. This will allow the ceiling drywall to be nailed to it later.*

A roof sheathing layout should be sketched out first to determine the amount of plywood needed to cover the rafters. See Fig. 4-94.

Draw your layout. It may be a freehand sketch, but it should be relatively close to scale. The easiest method is to draw a simple rectangle representing half of the roof. The long side will represent the length of the ridge board. Make the short side equal to the length of your rafters, including the overhangs. If you have open soffits, draw a line inside the ends and bottoms. Use a dotted line as shown in Fig. 4-94. This area is to be covered by *exterior* plywood. Remember that this is only half of the roof. Any cutting of panels on this side can be planned so that the cut-off portions can be used on the other side. If your eave overhang is less than 2 feet and you have an open soffit, you may wish to start with

Table 4-4 *Ceiling Joists*

Size of Ceiling Joists, Inches	Spacing of Ceiling Joists, Inches	Maximum Allowable Span			
		Group I	Group II	Group III	Group IV
2 × 4	12	11′6″	11′0″	9′6″	5′6″
	16	10′6″	10′0″	8′6″	5′0″
2 × 6	12	18′0″	16′6″	15′6″	12′6″
	16	16′0″	15′0″	14′6″	11′0″
2 × 8	12	24′0″	22′6″	21′0″	19′0″
	16	21′6″	20′6″	19′0″	16′6″

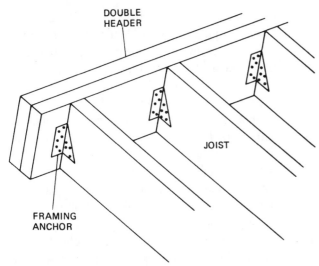

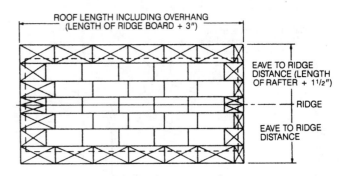

NOTE: For "open soffits" all panels marked with an "x" must be EXT-DFPA "soffit" panels.

Fig. 4-94 *Roof sheathing layout for a plywood deck.* (American Plywood Association)

Fig. 4-91 *Using framing anchors to hold the tail joists to the header.*

Fig. 4-92 *Arrows show the openings in the roof for the chimney.*

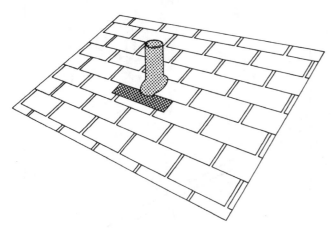

Fig. 4-93 *Soil pipe stack coming through the roof.*

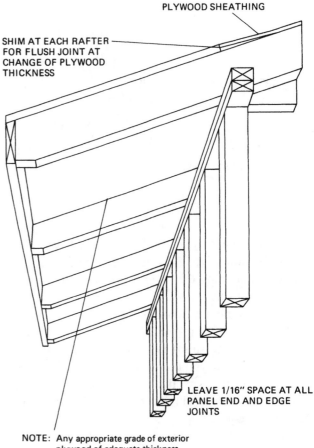

NOTE: Any appropriate grade of exterior plywood of adequate thickness (½" or more) prevents protrusion of roofing nails or staples at exposed underside, and carries design roof load.

Fig. 4-95 *Open soffit.* (American Plywood Association)

a half panel width of soffit plywood. Figure 4-95 shows the open soffit. Figure 4-96 shows the boxed soffit. Otherwise you will probably start with a full 4-×-8-foot sheet of plywood at the bottom of the roof and work

upward toward the ridge. This is where you may have to cut the last row of panels. Stagger panels in succeeding rows.

Complete your layout for the whole roof. The layout shows panel size and placement as well as the number of sheathing panels needed. This is shown in Fig. 4-94.

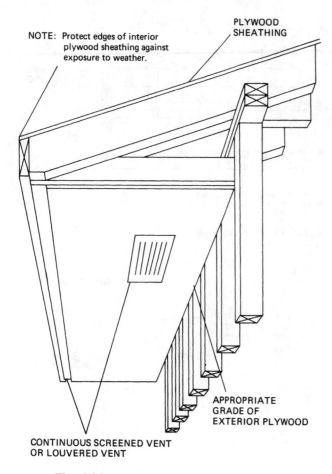

NOTE: Protect edges of interior plywood sheathing against exposure to weather.

PLYWOOD SHEATHING

APPROPRIATE GRADE OF EXTERIOR PLYWOOD

CONTINUOUS SCREENED VENT OR LOUVERED VENT

Fig. 4-96 *Boxed soffit. (American Plywood Association)*

If your diagram shows that you will have a lot of waste in cutting, you may be able to reduce scrap by slightly shortening the rafter overhang at the eave, or the gable overhang.

An example is shown in Fig. 4-94, where nearly half of the panels are "soffit" panels. In such a case, rather than using shims to level up soffit and interior sheathing panels, you may want to use interior sheathing panels of the same thickness as your soffit panels, even though they might then be a little thicker than the minimum required.

Cut panels as required, marking the cutting lines first to ensure square corners.

Begin panel placement at any corner of the roof. If you are using special soffit panels, remember to place them best or textured side down.

Fasten each panel in the first course (row), in turn, to the roof framing using 6d common smooth, ring-shank, or spiral-threaded nails. Space nails 6 inches O.C. along the panel ends and 12 inches O.C. at intermediate supports.

Leave a 1/16-inch space at panel ends and 1/8 inch at edge joints. In areas of high humidity, double the spacing.

Apply the second course, using a soffit half panel in the first (overhang) position. If the main sheathing panels are thinner than the soffit sheathing, install small shims to ease the joint transition. See Fig. 4-95 for location of the shims.

Apply the remaining courses as above.

Note that if your plans show closed soffits, the roof sheathing will all be the same grade thickness. To apply plywood to the underside of closed soffits, use nonstaining-type nails.

Figure 4-97 shows plywood decking being applied with a stapler. Figure 4-98A and B shows an H clip for plywood support along the long edges. This gives extra support for the entire length of the panels.

Figure 4-99 shows the erection of the sidewalls to a house. In Fig. 4-100 the plywood sheathing has been placed on one portion of the rafters. Note the rig (arrow) that allows a sheet of plywood to be passed up from the ground to roof level. The sheet is first placed on the rack. Then it is taken by the person on the roof and moved over to the needed area.

Figure 4-101 shows some boxed soffit. Note the louvers already in the board. The temporary supports hold the soffit in place until final nailing is done and it can be supported by the fascia board. The fascia has a groove along its entire length to allow the soffit to slide into it.

Fig. 4-97 *Using a stapler to fasten plywood decking to the rafters. (American Plywood Association)*

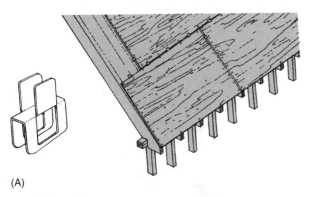

(A)

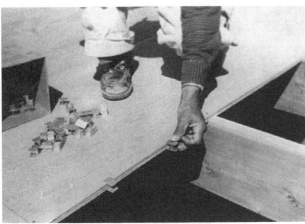

(B)

Fig. 4-98 *Plywood clips reinforce the surface area where the sheathing butts.*

Fig. 4-99 *Erection of sidewalls to a house.*

Fig. 4-100 *Plywood sheathing placed on one portion of rafters. Note the ladder made for plywood lifting.*

Boards for Decking

Lumber may be used for roof decking. In fact, it is necessary in some special-effects ceilings. It is needed where the ceiling is exposed and the underside of the decking is visible from below or inside the room.

Roof decking comes in a variety of sizes and shapes. See Fig. 4-102. A 2° angle is cut in the lumber decking ends to ensure a tight joint (Fig. 4-103). This type of decking is usually nailed, and so the nails must be concealed. This requires nailing as shown in Fig. 4-104. Eight-inch spikes are usually used for this type of nailing. Note the chimney opening in Fig. 4-105.

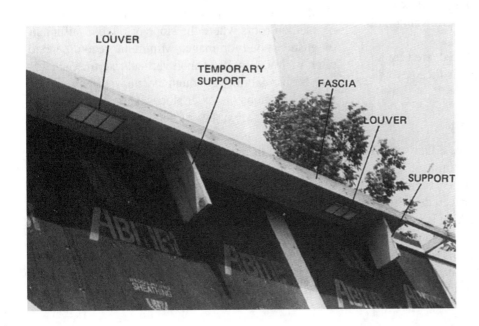

Fig. 4-101 *Louvered soffit held in place with temporary braces.*

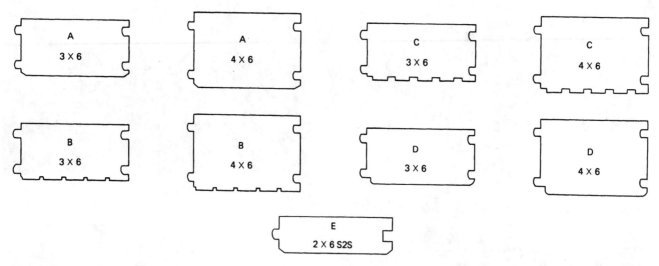

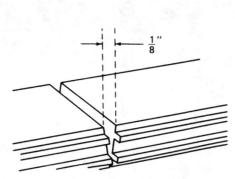

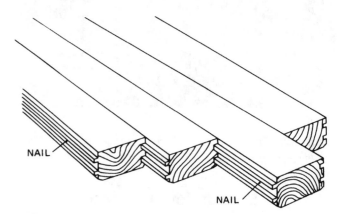

Fig. 4-102 *Different sizes and shapes of roof decking made of lumber. (A) Shows the regular V-jointed decking; (B) Indicates the straited decking; (C) Shows the grooved type; (D) Illustrates the eased joint or bullnosed type; (E) Indicates single tongue-and-groove with a V joint.*

Fig. 4-103 *A 2° angle is cut in the lumber decking ends. This ensures a tight face joint on the exposed ceiling below.*

Fig. 4-104 *Note the location of nails in the lumber decking.*

Figure 4-106 illustrates the application of 1-×-6 or 1-×-8 sheathing to the rafters. Note the two nails used to hold the boards down. The common nail is used here. In the concealed nailing it takes a finishing nail to be completely concealed.

Once the decking is in place, it is covered by an underlayment of felt paper. This paper is then covered by shingles as selected by the builder.

Shingle Stringers

For cedar shingles the roof deck may be either spaced or solid. The climatic conditions determine if it is a solid deck or one that is spaced. In areas where there are blizzards and high humidity, the spaced deck is not used. In snow-free areas, spaced sheathing is practical. Use 1 × 6s spaced on centers equal to the weather exposure at which the shakes are to be laid. However, the spacing should not be over 10 inches. In areas where wind-driven snow is encountered, solid sheathing is

recommended. See Fig. 4-106 for an example of spaced sheathing.

Roof pitch and exposure Handsplit shakes should be used on roofs where the slope or pitch is sufficient to ensure good drainage. Minimum recommended pitch is ⅙ or 4-in-12 (4-inch vertical rise for each 12-inch horizontal run), although there have been satisfactory installations on lesser slopes. Climatic conditions and skill and techniques of application are modifying factors.

Maximum recommended weather exposure is 10 inches for 24-inch shakes and 7½ inches for 18-inch shakes. A superior three-ply roof can be achieved at slight additional cost if these exposures are reduced to 7½ inches for 24-inch shakes and 5½ inches for 18-inch shakes.

Figure 4-107 shows the shakes in place on spaced sheathing. Note how the amount of exposure to the weather makes a difference in the spacing of the sheath-

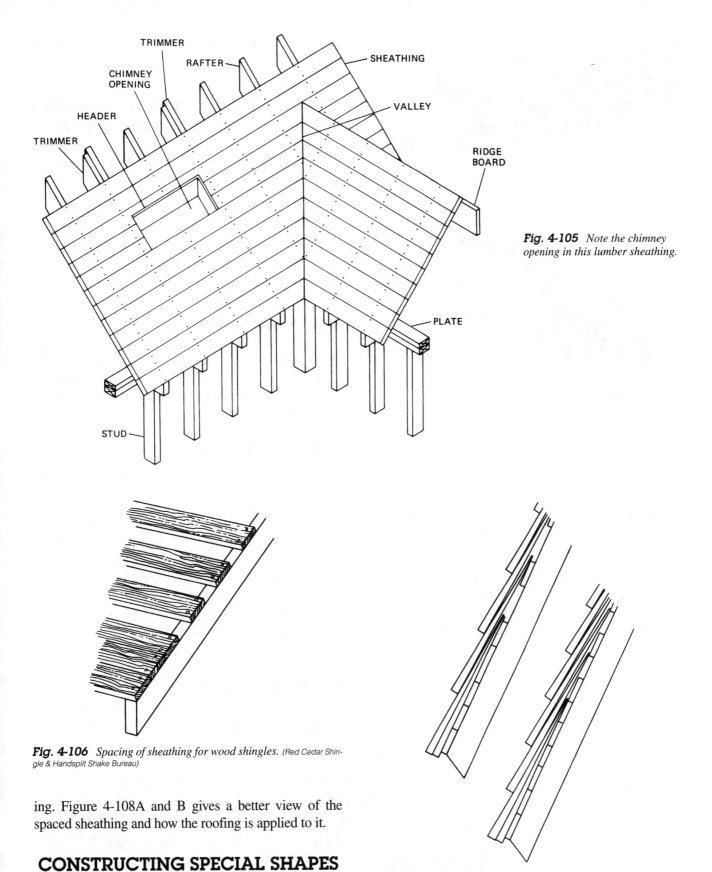

Fig. 4-105 *Note the chimney opening in this lumber sheathing.*

Fig. 4-106 *Spacing of sheathing for wood shingles. (Red Cedar Shingle & Handsplit Shake Bureau)*

ing. Figure 4-108A and B gives a better view of the spaced sheathing and how the roofing is applied to it.

CONSTRUCTING SPECIAL SHAPES

The gambrel shape is familiar to most people, since it is the favorite shape for barns. It consists of a double-slope roof. This allows for more space in the attic or

Fig. 4-107 *Handsplit shakes should be used on roofs where the slope is sufficient to ensure good drainage. Two different exposures to the weather are shown. Note the spacing of the sheathing under the shakes. (Red Cedar Shingle & Handsplit Shake Bureau)*

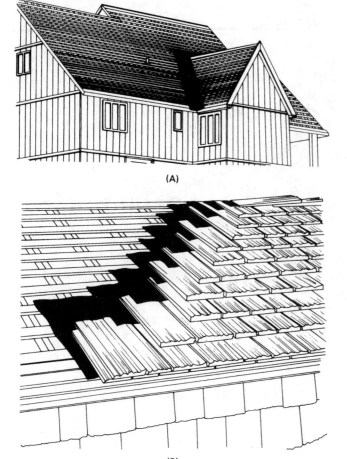

Fig. 4-108 *(A) Roof ready for application of shingles. (B) Cedar shingles being applied to prepared sheathing.* (Red Cedar Shingle & Handsplit Shake Bureau)

upper story. More can be stored there. In modern home designs, this type of roof has been used to advantage. It gives good headroom for an economical structure with two stories. This design was brought to

the United States by Germans in the early days of the country.

Gambrel-Shaped-Roof Storage Shed

A storage shed will give you an idea of the simplest way to utilize the gambrel-shaped roof. Examine the details and then obtain the equipment and supplies needed. The bill of materials lists the supplies that are needed. See Fig. 4-109. Now take a look at Figs. 4-110, 4-111, and 4-112. These show different ways of finishing the shed for different purposes. For instance, the structure can be covered with glass, Plexiglas, or polyvinyl as in Fig. 4-111 and made into a greenhouse. Then there is the rustic look shown in Fig. 4-110 and the contemporary look shown in Fig. 4-112.

Frame layout Note the dimensions of the shed. It is 7 feet high and 8 feet wide. A detail of the framing angle is shown in the frame-to-sill detail (Fig. 4-113). Note the spacing of the slopes of the roof. Figure 4-114 indicates how the vertical stud member and the roof member are attached with metal plates. Figure 4-115 indicates how an 18° angle is used for cutting the studs and roof members.

Frame-cutting instructions Mark off 18° angles on the 2 × 4s. See Fig. 4-115. Cut to length. Note the exact length required for the roof member and the vertical stud. When you have cut the required number of studs and roof members, place the sections on a hard, flat surface. A driveway or sidewalk can be used. Nail the metal plates equally into each member and flip the frame over. Then nail metal plates on this side. Make four frames for a shed with an 8-foot depth. You can change the number of frames to match your length or

Fig. 4-109 *Bill of materials for a storage shed.* (TECO Products and Testing Corporation, Washington, DC 20015)

BILL OF MATERIALS			
QUAN.	**DESCRIPTION**	**QUAN.**	**DESCRIPTION**
28	2″ x 4″ x 8 FT. LONG	40	TECO C-7 PLTS.
9	4′ x 8′ x ½″ PLYWD.	30	TECO JOIST HGR.
2	1″ x 4″ x 6 FT. LONG	12	TECO ANGLES
1 ROLL	ROOFING FELT	30	TECO A-5 PLTS.
1 GAL	ROOF. CEMENT	3	3 BUTT HINGES
1 GAL	BARN-RED PAINT	1	HASP & LOCK
5#	6d COM. NAILS	10 BG.	90# CONC. MIX
2#	12d COM. NAILS		OR
2#	½″ ROOF. NAILS	4	6″ x 8″ x 8′ RAIL TIE

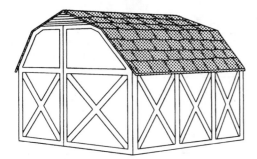

Fig. 4-110 *Rustic shed design. (TECO)*

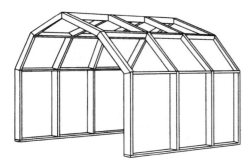

Fig. 4-111 *A greenhouse can be made by covering the frame with plastic. (TECO)*

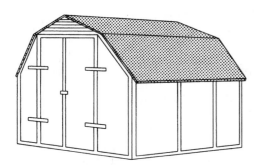

Fig. 4-112 *Contemporary shed design. (TECO)*

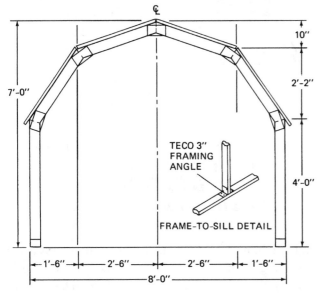

Fig. 4-113 *Frame layout for the shed. (TECO)*

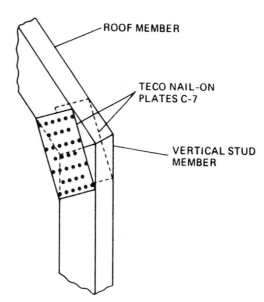

Fig. 4-114 *Detail of connection of vertical stud member and roof member. (TECO)*

depth requirements. Just add a frame for every 2 feet 8 inches of additional depth.

Roof framing plan Figure 4-116 shows the roof framing layout. Such a layout will eliminate any problems you might have later if you did not properly plan your project. The purlin is extended in length every time you extend the depth of the shed by 2 feet 6 inches. Note how the purlins are attached to the vertical stud member and the roof member. See Fig. 4-117. Once you have attached all the purlins, you have the standing frame. It is time to consider other details. Figure 4-118 shows how a vent is built into the rear elevation. It can be a standard window or constructed from 1 × 2s with a polyvinyl backing.

Door details You have to decide upon the design of the doors to be used. Figure 4-119 shows how the door is constructed for both rustic and contemporary styles. Cut the angles at 18° when making the door for the rustic style. Use the same template used for the studs and roof members.

The contemporary door is nothing more than squared-off corners. Figure 4-120 shows the placement of the hasps used on the door.

Mansard Roofs

The mansard roof has its origins in France. The mansard usually made in the United States is slightly different from the French style. Figure 4-121 shows the American style of mansard. It uses a roof rafter with a steep slope for the side portion and one with a very low slope for the top. Standard 2-×-4 or 2-×-6

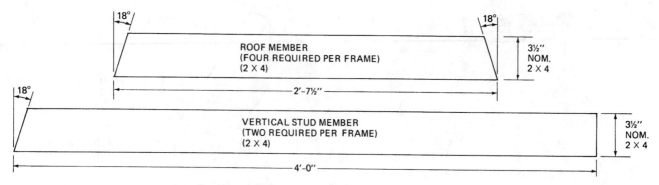

Fig. 4-115 *Frame cutting instructions.* (TECO)

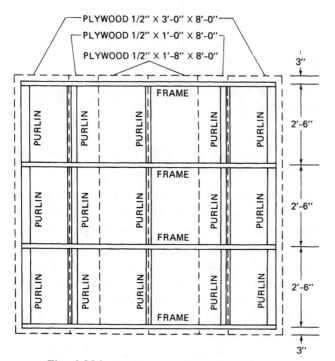

Fig. 4-116 *Roof-framing plan for the shed.* (TECO)

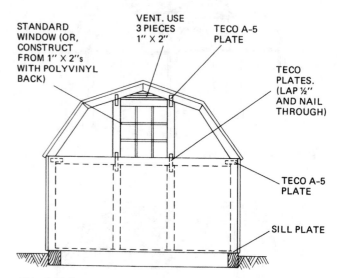

Fig. 4-118 *Rear elevation of the shed. Note how the vent is built in.* (TECO)

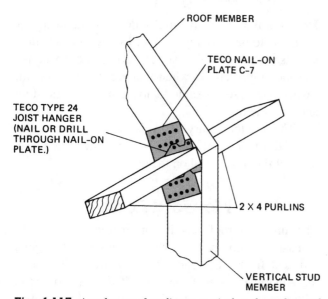

Fig. 4-117 *Attachment of purlins to vertical stud member and roof member.* (TECO)

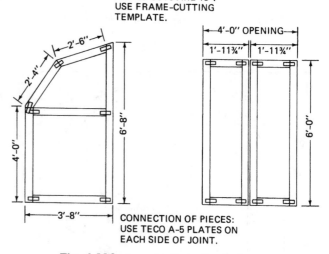

Fig. 4-119 *Door details for the shed.* (TECO)

framing lumber is used to make these rafters. Plywood sheathing is applied over the rafters, and roofing is usually applied over the entire surface. This is supposed to make the house look lower. It effectively low-

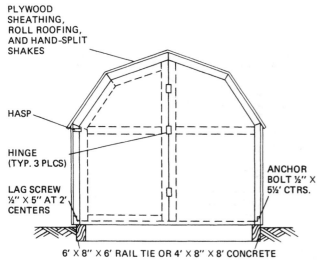

Fig. 4-120 *Front elevation of the shed.* (TECO)

PLYWOOD
SHEATHING,
ROLL ROOFING,
AND HAND-SPLIT
SHAKES

HASP

HINGE
(TYP. 3 PLCS)

LAG SCREW
½" X 5" AT 2'
CENTERS

ANCHOR
BOLT ½" X
5½' CTRS.

6' X 8" X 6' RAIL TIE OR 4' X 8" X 8' CONCRETE

Fig. 4-121 *Mansard roof.*

ers the "belt-line" and makes the roof look closer to the ground. Shingles have to be chosen for a steep slope so that they are not blown off by high winds.

Figure 4-122 shows a mansard roof with cedar shakes. The steep slope portion adds to the effect when covered by cedar shingles. It is best to use the wooden shakes for the low slope as well; however, in some cases the slope may be too low and other shingles may be needed to do the job properly.

Figure 4-123 shows how the French made the mansard roof truss. It was used on hotels and some homes. This style was popular during the nineteenth century. Note the elaborate framing used in those days to hold the various angles together. At that time they used wrought-iron straps instead of today's metal (steel) brackets and plates.

Fig. 4-122 *Handsplit shakes used on a mansard roof.* (Red Cedar Shingle & Handsplit Shake Bureau)

Post-and-Beam Roofs

The post-and-beam type of roof is used for flat or low-slope roofs. This type of construction can use the roof decking as the ceiling below. See Fig. 4-124A. The exposed ceiling or roof decking has to be finished. That means the wood used for the roof deck has to be surfaced and finished on the inside surface. Post-and-beam methods don't use regular rafters. See Fig. 4-124B. The rafters are spaced at a greater distance than in conventional framing. This calls for larger dimensional lumber in the rafters. These are usually exposed also and need to be finished according to plan.

ROOF LOAD FACTORS

Plywood roof decking offers builders some of their most attractive opportunities for saving time and money. Big panels really go down fast over large areas. They form a smooth, solid base with a minimum of joints. Waste is minimal, contributing to the low in-place cost. It is frequently possible to cut costs still further by using fewer rafters with a somewhat thicker panel for the decking. For example, use ¾-inch plywood over framing 4 feet O.C. Plywood and trusses are often combined in this manner. For recommended spans and plywood grades, see Table 4-5.

Plywood roof sheathing with conventional shingle roofing Plywood roof sheathing under shingles provides a tight deck with no wind, dust, or snow infiltration and high resistance to racking. Plywood has stood up for decades under asphalt shingles, and it has performed equally well under cedar shingles and shakes.

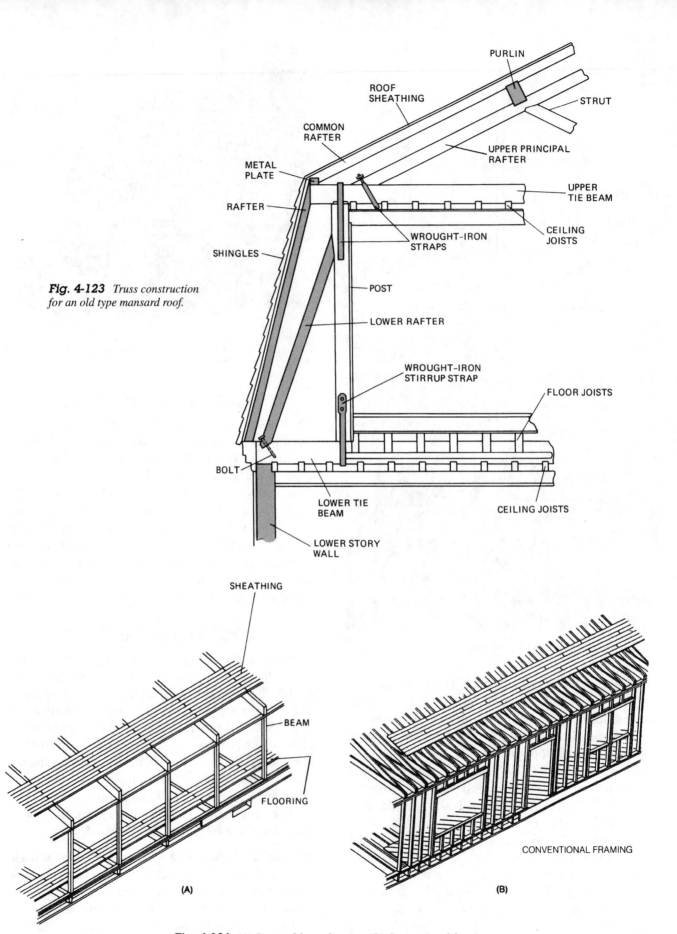

Fig. 4-123 *Truss construction for an old type mansard roof.*

PURLIN

ROOF SHEATHING

STRUT

COMMON RAFTER

UPPER PRINCIPAL RAFTER

METAL PLATE

UPPER TIE BEAM

RAFTER

CEILING JOISTS

SHINGLES

WROUGHT-IRON STRAPS

POST

LOWER RAFTER

WROUGHT-IRON STIRRUP STRAP

FLOOR JOISTS

BOLT

CEILING JOISTS

LOWER TIE BEAM

LOWER STORY WALL

SHEATHING

BEAM

FLOORING

(A)

CONVENTIONAL FRAMING

(B)

Fig. 4-124 *(A) Post-and-beam framing. (B) Conventional framing.*

Table 4-5 Plywood Roof Decking

| Identi-fication Index | Plywood Thickness, Inches | Maximum Span, Inches | Unsupported Edge—Max. Length, Inches | Allowable Live Loads, psf | | | | | | | | | |
| | | | | Spacing of Supports Center to Center, Inches | | | | | | | | | |
				12	16	20	24	30	32	36	42	48	60
12/0	5/16	12	12	150									
16/0	5/16, 3/8	16	16	160	75								
20/0	5/16, 3/8	20	20	190	105	65							
24/0	3/8, 1/2	24	24	250	140	95	50						
32/16	1/2, 5/8	32	28	385	215	150	95	50	40				
42/20	5/8, 3/4, 7/8	42	32		330	230	145	90	75	50	35		
48/24	3/4, 7/8	48	36			300	190	120	105	65	45	35	
2·4·1	1 1/8	72	48				390	245	215	135	100	75	45
1 1/8" Grp. 1 and 2	1 1/8	72	48				305	195	170	105	75	55	35
1 1/4" Grp. 3 and 4	1 1/4	72	48				355	225	195	125	90	85	40

Plywood sheathing over roof trusses spaced 24 inches O.C. is widely recognized as the most economical construction for residential roofs and has become the industry standard.

Design Plywood recommendations for plywood roof decking are given in Table 4-4. They apply for the following grades: C—D INT APA, C—C EXT APA, Structural II and II C—D INT APA, and Structural I and II C—C EXT APA. Values assume 5 pounds per square foot dead load. Uniform load deflection limit is $\frac{1}{180}$ of the span under live load plus dead load, or $\frac{1}{240}$ under live load only. Special conditions, such as heavy concentrated loads, may require constructions in excess of these minimums. Plywood is assumed continuous across two or more spans, and applied face grain across supports.

Application Provide adequate blocking, tongue-and-groove edges, or other edge support such as plyclips when spans exceed maximum length for unsupported edges. See Fig. 4-125 for installation of plyclips. Use two plyclips for 48-inch or greater spans and one for lesser spans.

Space panel ends $\frac{1}{16}$ inch apart and panel edges $\frac{1}{8}$ inch apart. Where wet or humid conditions prevail, double the spacings. Use 6d common smooth, ringshank,

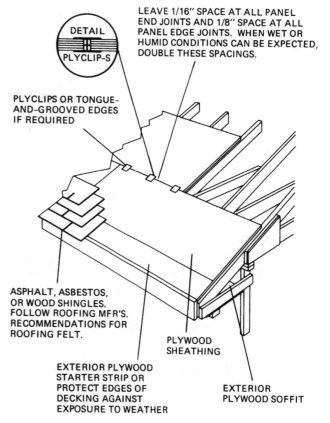

LEAVE 1/16" SPACE AT ALL PANEL END JOINTS AND 1/8" SPACE AT ALL PANEL EDGE JOINTS. WHEN WET OR HUMID CONDITIONS CAN BE EXPECTED, DOUBLE THESE SPACINGS.

DETAIL PLYCLIP-S

PLYCLIPS OR TONGUE-AND-GROOVED EDGES IF REQUIRED

ASPHALT, ASBESTOS, OR WOOD SHINGLES. FOLLOW ROOFING MFR'S. RECOMMENDATIONS FOR ROOFING FELT.

EXTERIOR PLYWOOD STARTER STRIP OR PROTECT EDGES OF DECKING AGAINST EXPOSURE TO WEATHER

PLYWOOD SHEATHING

EXTERIOR PLYWOOD SOFFIT

Fig. 4-125 Using a plyclip to reinforce plywood decking. (American Plywood Association)

or spiral-thread nails for plywood ½ inch thick or less. Use 8d nails for plywood to 1 inch thick. Use 8d ring-shank or spiral nails or 10d common smooth for 2 • 4 • 1, 1⅛ inch and 1¼ inch panels. Space nails 6 inches at panel edges and 12 inches at intermediate supports, except where spans are 48 inches or more. Then space nails 6 inches at all supports.

Plywood nail holding   Extensive laboratory and field tests, reinforced by more than 25 years experience, offer convincing proof that even ⁵⁄₁₆-inch plywood will hold shingle nails securely and permanently in place, even when the shingle cover is subjected to hurricane-force winds.

The maximum high wind pressure or suction is estimated at 25 psf except at the southern tip of Florida, where wind pressures may attain values of 40 to 50 psf. Because of shape and height factors, however, actual suction or lifting action even in Florida should not exceed 25 psf up to 30-foot heights. Thus any roof sheathing under shingles should develop at least that much withdrawal resistance in the nails used.

Plywood sheathing provides more than adequate withdrawal resistance. A normal wood-shingled room will average more than 6 nails per square foot. Each nail need carry no more than 11 pounds. Plywood sheathing only ⁵⁄₁₆ inch thick shows a withdrawal resistance averaging 50 pounds for a single 3d shingle nail in laboratory tests and in field tests of wood shingles after 5 to 8 years' exposure. In addition, field experience shows asphalt shingles consistently tear through at the nail before the nail pulls out of the plywood.

Figure 4-126 shows the markings found on plywood used for sheathing. Note the interior and exterior glue markings. APA stands for the American Plywood Association.

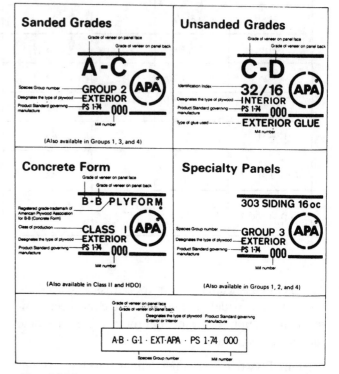

Fig. 4-126   *Plywood grades identified.* (American Plywood Association)

LAYING OUT A STAIR

So far you have used the framing square to lay out rafters. There is also another use for this type of instrument. It can be used to lay out the stairs going to the basement or going upstairs in a two-story house.

Much has been written about stairs. Here we only lay out the simplest and most useful of the types available. The fundamentals of stair layout are offered here.

1. Determine the height or rise. This is from the top of the floor from which the stairs start, to the top of the floor on which they are to end. See Fig. 4-127.
2. Determine the run or distance measured horizontally.

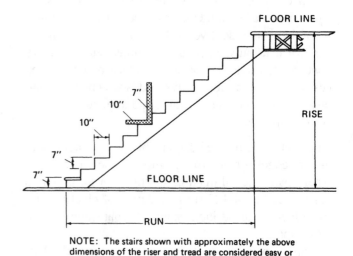

NOTE: The stairs shown with approximately the above dimensions of the riser and tread are considered easy or comfortable to climb.

Fig. 4-127   *Laying out stairs with the steel square.* (Stanley Tools)

3. Mark the total rise on a rod or a piece of 1-×-2-inch furring to make a so-called *story pole*. Divide the height or rise into the number of risers desired. A simple method is to lay out the number of risers wanted by spacing off the total rise with a pair of compasses. It is common to have this result in fractions of an inch. For example, a total rise of 8 feet 3 ¾ inches or 99¾ inches divided by 14 = 7.125 or 7⅛-inch riser. This procedure is not necessary in the next step because the horizontal distance, or run, is seldom limited to an exact space as is the case with the rise.

4. Lay out or space off the number of treads wanted in the horizontal distance or run. There is always one less tread than there are risers. If there are 14 risers in the stair, there are only 13 treads. For example, if the tread is 10 inches wide and the riser is 7 inches, the stair stringer would be laid out or "stepped off" with the square, ready for cutting as shown in Fig. 4-127. The thickness of the tread should be deducted from the first riser as shown. This is in order to have this first step the same height as all the others.

ALUMINUM SOFFIT

So far the soffit has been mentioned as the covering for the underside of the overhang. This has been shown to be covered with a plywood sheathing of ¼-inch thickness or with a cardboard substance about the thickness of the plywood suggested. The cardboard substitute is called Upson board. This is because the Charles A. Upson Company makes it. If properly installed and painted, it will last for years. However, it should not be used in some climates.

One of the better materials for soffit is aluminum. More and more homes are being fitted with this type of maintenance-free material. This particular manufacturer no longer makes this roll-type soffit.

Material Availability

Aluminum soffit can be obtained in 50-foot rolls with various widths. They can be obtained (Fig. 4-128) in widths of 12, 18, 24, 30, 36, and 48 inches. These are pushed or pulled into place as shown in Fig. 4-129. The hip roof with an overhang all around the house would require soffit material pulled in as shown in Fig. 4-130. The runners supporting the material are shown in Fig. 4-131. Covering the ends is important to ensure a neat job. Corner trim and fascia closure are available to help give the finished job a look not unlike that of an all-wood soffit.

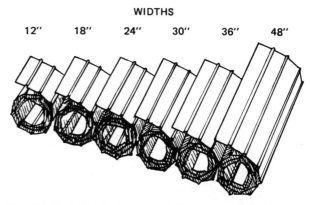

Fig. 4-128 *Rolls of soffit material. These are made of aluminum.*

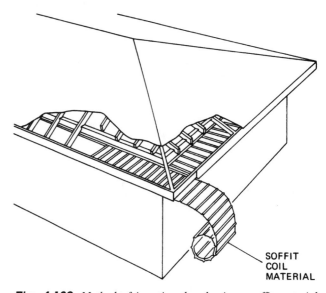

Fig. 4-129 *Method of inserting the aluminum soffit material.*
(Reynolds Metals Product)

Figure 4-132 shows how the fascia runner, the frieze runner, and corner trim are located for ease of installation of the soffit coil.

After the material has been put in place and the end cuts have been made, the last step is to insert a plastic liner to hold the aluminum in place. This prevents rattling when the wind blows. The material can be obtained with a series of holes prepunched. This will serve as ventilation for the attic. See Fig. 4-133.

Figure 4-134 gives more details on the installation of the runners that support the soffit material. The fascia runner is notched at points *b* for about 1½ inches maximum. Then the tab is bent upward and against the inside of the fascia board. Here it is nailed to the board for support. Take a look at *c* in Fig. 4-134 to see how the tab is bent up. Note how the width of the channels is the soffit coil width plus at least ⅜ inch and not more than ⅞ inch. This allows for expansion by the aluminum. Aluminum will expand in hot weather.

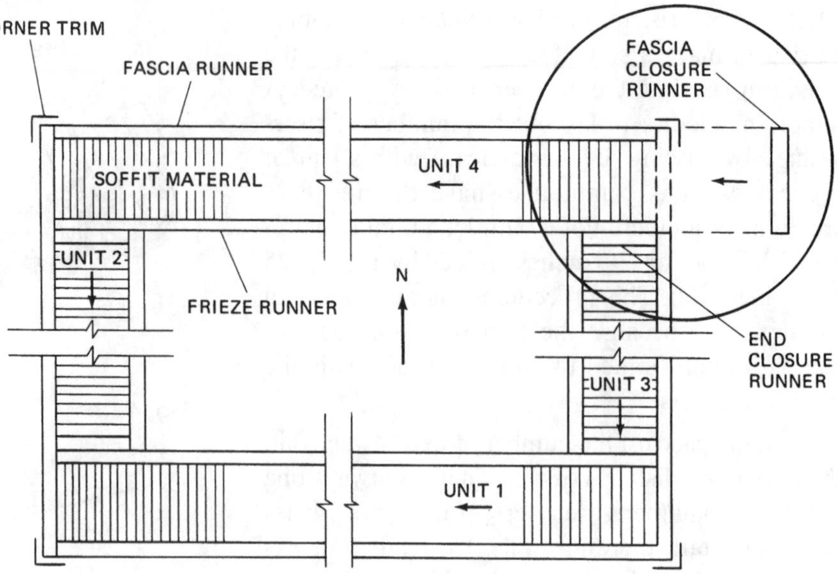

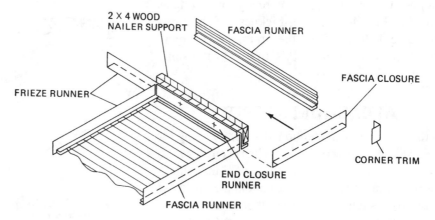

Fig. 4-130 *Steps in installing aluminum soffit in a hip roof with an overhang all around.* (Reynolds Metals Products)

Fig. 4-131 *Closing off the ends of the soffit with aluminum.* (Reynolds Metal Products)

Cutting the runner to desired lengths can be done by cutting the channel at *a* and *b* of Fig. 4-135. Then bend the metal back and forth along a line such as at *c* until it breaks. Of course you can use a pair of tin snips to make a clean cut.

Figure 4-136 shows how the soffit is installed with a brick veneer house. Part (A) shows how the frieze runner is located along the board. Insert (B) shows how the runner is nailed to the board.

When the fascia board is more than 1 inch wide, it is necessary to place 1-inch aluminum strips as shown in Fig. 4-137. The tabs are hooked in between the fold in the runner. Then the runner is brought under the fascia board and bent back and nailed. This will allow the runner to expand when it is hot. Do not nail the overlapping runners to one another.

In some instances it is necessary to use a double-channel runner. This is done so that there will be no sagging of the soffit material. See Fig. 4-138. Note how the frieze runner and double-channel runner are

located. Note the gravel stop on this flat roof. In some parts of the country more overhang is needed because it gives more protection from the sun.

Figure 4-139 shows how the H-molding joint works to support the two soffit materials as they are unrolled into the channel molding. Note the location of the vent strip, when needed.

As was mentioned before, the aluminum soffit makes for a practically maintenance-free installation. More calls will be made for this type of finish. The carpenter will probably have to install it, since the carpenter is responsible for the exterior finish of the building and the sealing of all the openings. With the advent of aluminum siding, it is only natural that the soffit be aluminum.

METAL CONNECTORS

Every year many houses are destroyed when the force of high winds cause roofs to fly off and walls to col-

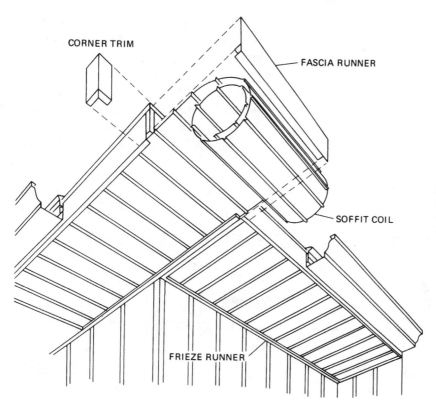

Fig. 4-132 *Installing soffit coil in the overhang space. Note the corner trim to give a finished appearance.* (Reynolds Metals Products)

CORNER TRIM

FASCIA RUNNER

SOFFIT COIL

FRIEZE RUNNER

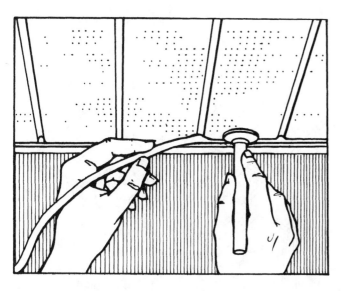

Fig. 4-133 *Finishing up the job with a vinyl insert to hold the aluminum in place.* (Reynolds Metals Products)

1" THICKNESS

FASCIA BOARD

FASCIA RUNNER

2' MAX.

c

b

b

1-1/2" MAX.

W

FRIEZE RUNNER

W = SOFFIT COIL WIDTH + 3/8" TO 7/8"
(3/8" MINIMUM) (7/8" MAXIMUM)

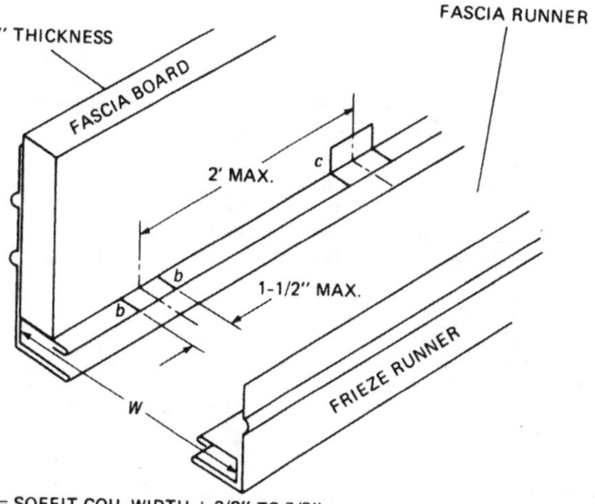

Fig. 4-134 *Method used to support the runner on the fascia.* (Reynolds Metals Products)

lapse. One of the methods used by builders located in high-wind areas such as along the shores of lakes, bays, oceans, and gulfs is the metal connector. Various fasteners are designed to increase the structural strength of homes built to withstand hurricanes, and in some instances, tornadoes and earthquakes.

Figure 4-140 shows how the use of metal connectors can increase the chances for a house to escape a hurricane's full force. The connectors are illustrated in the next few pages. Take a close look at the encircled number and then refer to the following pages for an illustration on where and how the connector is utilized to its best advantage.

High winds can be combatted by building a house that will withstand the force of Mother Nature. The only successful way to combat these forces is to employ proper construction techniques ahead of time to ensure the integrity of the structure. The best method is

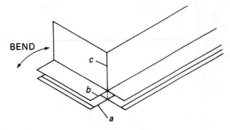

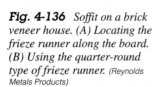

Fig. 4-135 *Bending and breaking the runner material.* (Reynolds Metals Products)

to use an uninterrupted load path from the roof members to the foundation. Metal connectors are engineered to satisfy the necessary wind uplift load requirements. See Fig. 4-141.

Figure 4-142 shows how foundations are prepared with metal connectors to cause the sole plate and studs to be permanently and solidly anchored to prevent damage by high winds. Figure 4-143 illustrates how

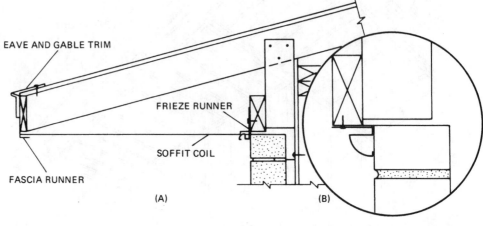

Fig. 4-136 *Soffit on a brick veneer house. (A) Locating the frieze runner along the board. (B) Using the quarter-round type of frieze runner.* (Reynolds Metals Products)

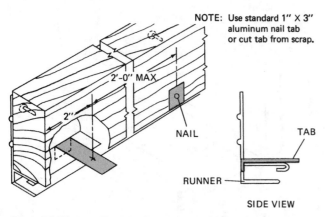

Fig. 4-137 *Installing a tab to keep the runner free to move as aluminum expands on hot days.* (Reynolds Metals Products)

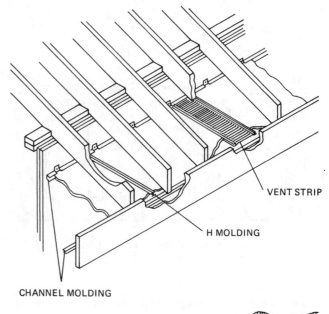

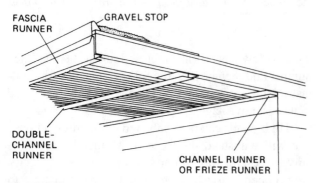

Fig. 4-138 *Using a double-channel runner and a double row of aluminum soffit material.* (Reynolds Metals Products)

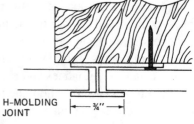

Fig. 4-139 *Installing the H-molding joint.* (Reynolds Metals Products)

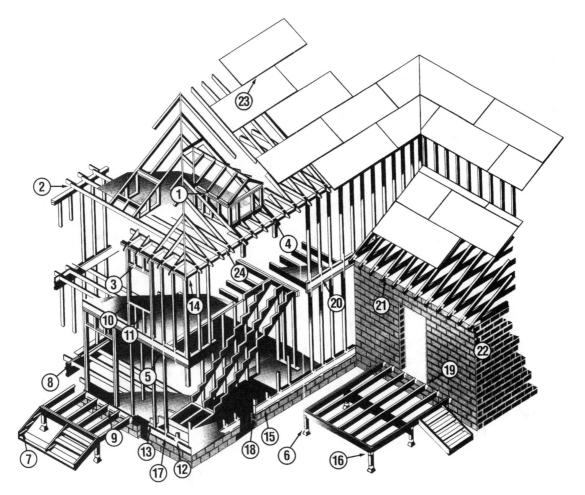

Fig. 4-140 *Location of metal connectors. (SEMCO)*

connectors are used for a Stem Wall System. A wood-to-wood type of construction enhanced by metal connectors is shown in Fig. 4-144. Note the allowable loads tables that give the nail size and uplift in pounds per square inch (psi).

A poured masonry header has the rafters attached by hangers anchored in the concrete. See Fig. 4-145.

Second floor problems can be solved by tying the first and second floors together with metal connectors as shown in Fig. 4-146. A variety of fasteners are illustrated. Regular truss-to-top plate construction is shown reinforced in Fig. 4-147 while

the top plate-to-stud reinforcement is illustrated by Fig. 4-148. These metal fasteners or connectors are standard requirements in areas buffeted by high winds, hurricanes, and earthquakes. The additional costs can often be offset by lower insurance rates on the finished house.

There are charts that show the expected winds in a given area. Figure 4-149 is a map of the United States with the wind speeds indicated. The wind load calculations placed on a structure are determined by a number of factors. Chief among these is the wind speed ratings for the location of the structure.

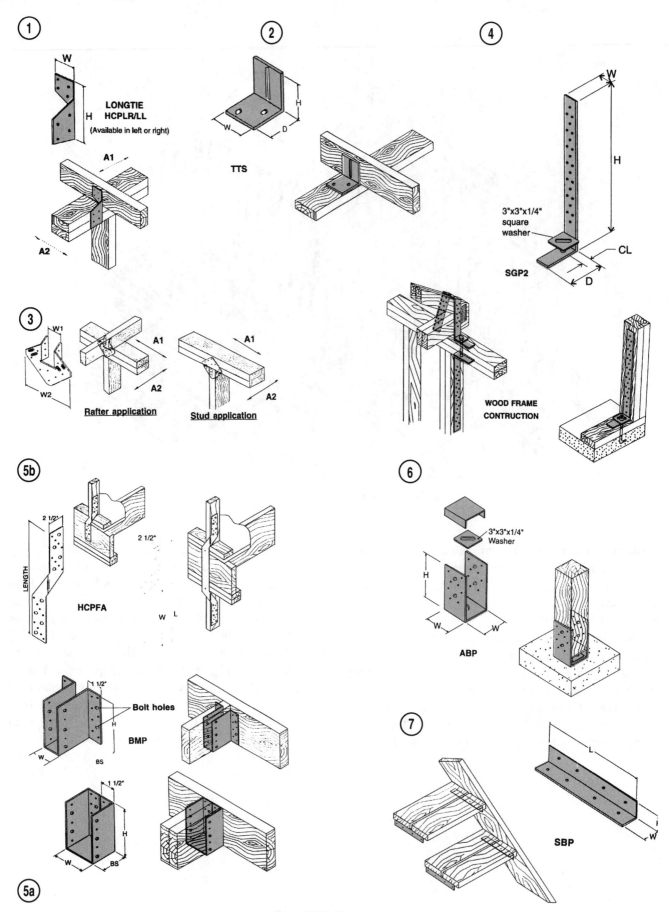

① LONGTIE HCPLR/LL (Available in left or right)
W
H
A1
A2

② TTS
W
H
D

③ W1
W2
A1
A2
Rafter application
A1
A2
Stud application

④ W
H
3"x3"x1/4" square washer
CL
D
SGP2
WOOD FRAME CONTRUCTION

⑤b 2 1/2"
LENGTH
HCPFA
2 1/2"
W   L

⑤a 1 1/2"
Bolt holes
H
BMP
W
BS
1 1/2"
H
W
BS

⑥ 3"x3"x1/4" Washer
H
W   W
ABP

⑦ L
W
SBP

Fig. 4-140 *Continued.*

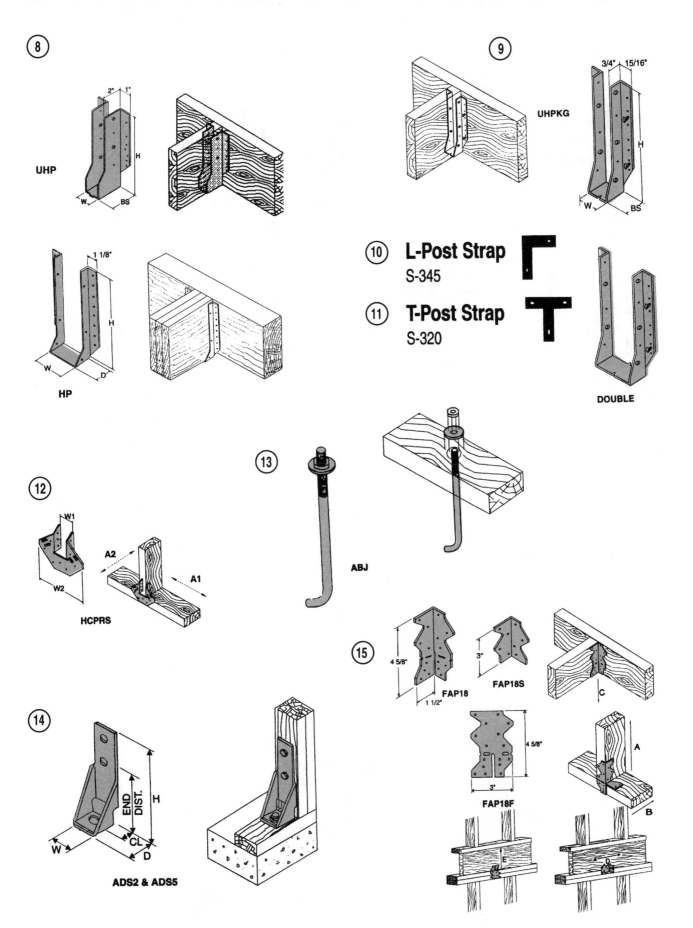

8

UHP

HP

9

UHPKG

3/4" 15/16"

10 **L-Post Strap**
S-345

11 **T-Post Strap**
S-320

DOUBLE

12

HCPRS

13

ABJ

14

ADS2 & ADS5

15

FAP18

4 5/8"

1 1/2"

FAP18S

3"

C

FAP18F

4 5/8"

3"

A

B

Fig. 4-140 Continued.

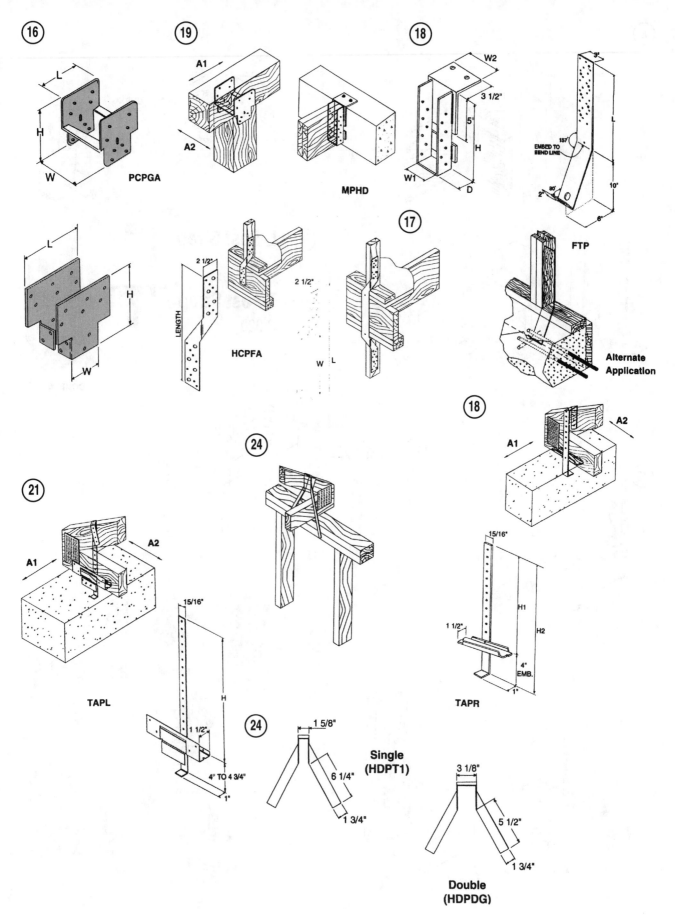

Fig. 4-140 *Continued.*

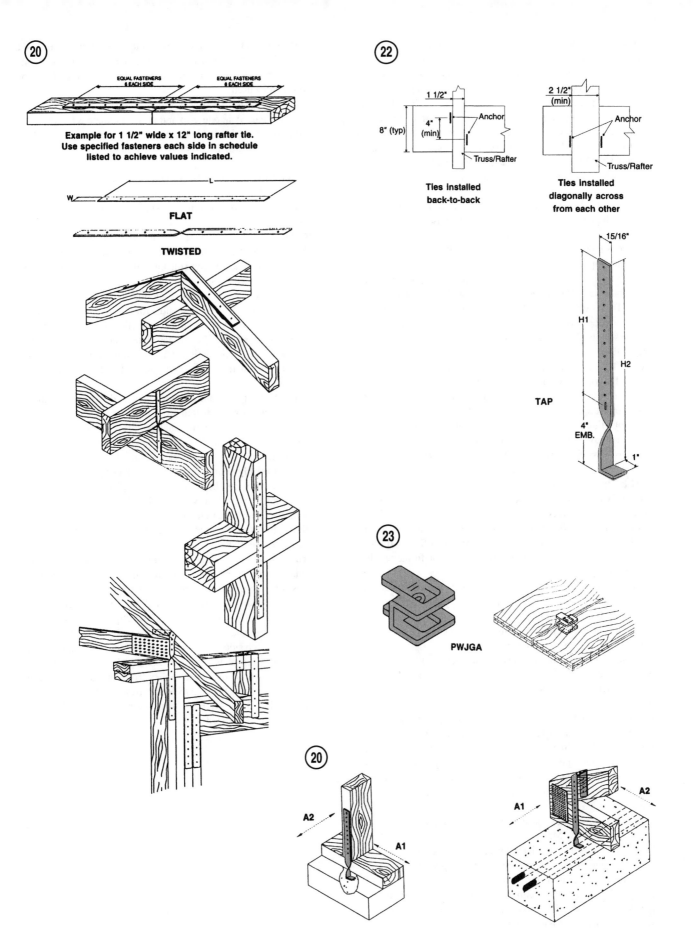

㉒ Ties installed back-to-back

Ties installed diagonally across from each other

㉓ PWJGA

Example for 1 1/2" wide x 12" long rafter tie.
Use specified fasteners each side in schedule listed to achieve values indicated.

FLAT

TWISTED

TAP

Fig. 4-140 Continued.

① **Hurricane Anchor** This tie adds increased resistance to wind uplift. The tie eliminates toe-nailing utilizing correctly located nail holes for fast, easy, and strong attachment of rafters and trusses to plates and studs. They are made of 20-18 gage galvanized steel. They can be installed on each side of the rafter for twice the loads when the rafter thickness is a minimum of 2.5 inches or diagonally when rafter is 1.5 inches.

② **90° Bracket** Used for tying trusses to non-load bearing walls.

③ **Rafter Clip/Stud Plate** A fast, economical tie to secure rafters or trusses to wall studs and top plates, and from studs to sill plate.

④ **Girder Truss Strap** Ideal for girder truss connections when there is a high uplift load requirement. Can be used for wood-to-wood application or concrete-to-wood application. A washer plate adds increased resistance to wind uplift.

⑤ **A. Beam Support** These face mount supports are designed to provide full support of the top chord, preventing joist rotation. Bolt holes are also provided for additional load capacity. **B. Floor Tie Anchor** These ties are designed especially for use with floors constructed above grade, as a connection between first or second floor level to studs. They are designed for engineered floor systems with larger clear spans of 21 inches and 24 inches. They can be fastened with bolts or nails.

⑥ **Heavy Duty Post Anchor** Post base is to elevate post above concrete to allow for ventilation. Heavy-duty design permits higher uplift loads and simplified installation with 0.5-inch anchor bolts that fit through pre-punched holes, slotted to permit adjustment to align for off-position anchor bolts.

⑦ **Staircase Bracket** These brackets are designed to simplify and reinforce stair construction.

⑧ **Joist Support** The regular joist support is made of 18 gage and the Heavy-Duty Joist Hanger is made to support headers, joists, and trusses. It is made from 14-gage galvanized steel.

⑨ **Kwik Grip Joist Support** These supports are designed for quick installation in place for easy nailing. The offset nail hole placement is for secure positive nailing. Precision formed high-strength 18-gage galvanized is used for long life.

⑩ **L-Post Strap** Can be used to tie the window framing together.

⑪ **T-Post Strap** Can be used at perpendicular junction points for cripple studs and the rough sill.

⑫ **Stud Plate/Rafter Clip** Used for tying studs to sill plate and top plates, or rafters to top plates.

⑬ **Anchor Bolt** 6-inch minimum embedment with 3000 psi concrete will resist 1,635 lbs. Wind unlift loads are based on the shear capacity of No. 2 Southern Pine. Compression perpendicular to grain 565 psi.

⑭ **Hold Down Anchor** Ideal for shear walls and vertical posts.

⑮ **Universal Framing Anchor** This is a multi-purpose anchor for almost any wood connection task. It anchors rafters and roof trusses to plates, and it anchors floor and ceiling joists to headers and solid blocking to plates. The 90° framing angles can be used to join posts to beams and make other right-angle connections.

⑯ **Post Cap** Simplifies installing 4×4 wood posts to wood beams and trusses. Makes a full-strength, positive connection between the post top and lateral beams or trusses. It is considerably stronger than random toe-nailing and spiking, and is less time-consuming than drilling and lag bolting. Post caps are of the split design to offer maximum flexibility in application. Direct-load path-side plates maximizes load capacities. Toe-nail slots speed plum and level adjustments.

⑰ **Floor Tie Anchor** As previously noted, these ties are designed especially for use with floors constructed above grade, as a connection between first or second floor level to studs. Nails or bolts can be used.

⑱ **Foundation Tie** There is a prepunched hole in the foot of the tie to increase concrete grip and allow alternate rebar rod support. Prepunched holes in the bend are there for nails to hold it in the form board. Ties should have a minimum of 3 inches in the concrete.

⑲ **Heavy-Duty Masonry Beam Hanger** Designed to work with standard block wall or concrete tie beam construction. Eliminates the need of constructing special seats to support floor joists.

⑳ **Ratter Tie** Tie straps meet a variety of application and design load conditions and specifications. Use rafter ties when tying rafters to plate, anchoring studs to sill, or framing over girders and bearing partitions.

㉑ **Lateral Truss Anchor** This anchor is designed to meet the lateral and uplift load demands for hurricane-resistant construction. It provides a custom connection to wood for trusses or rafters. An attached beam seat-plate eliminates treated sill or moisture-barrier installation. The riveted plate on the seat-plate design prevents truss movement parallel and perpendicular to the wall.

㉒ **Truss Anchor** Accommodates diverse design requirements for concrete-to-wood installation, allows a 4-inch embedment in concrete. You can use two anchors installed, one on each side of the rafter, for twice the load per single rafter thickness. Minimum edge distance is 2 inches.

㉓ **Galvanized Plywood Clip** This clip is designed for easy and fast application to sheathing edges. It gives a snug self-lock that keeps the clip firmly in position throughout panel placement. It eliminates unreliable wood edge blocking. This type of clip is also available in aluminum. The clip gives an automatic spacing to the plywood or sheathing.

㉔ **Truss Tie Down Strap (Gun Tie)** This strap provides additional increased resistance to wind uplift to secure rafters or trusses to top plates. It eliminates fastening through the truss nail plates and requires no truss nailing. It is designed to allow gun nailing for quick installation. The clip is made of 20-gage steel and the installer should wear eye protection. Fasteners are placed on each side of the truss into the top and bottom layer of the double top-plate members in equal quantity. Fasteners should be no less than 0.25 inches from the edge of the strap and placed no less than 0.375 inches from the edge of the framing member. **True Tie** is the same share as the **Gun Tie**, but it has 18-gage galvanized metal.

Fig. 4-140 Continued.

PRODUCT CODE	GAUGE	FASTENER SCHEDULE		ALLOWABLE LOADS	
		HEADER / PLATE	STUD	WIND / EARTHQUAKE	
				UPLIFT 133%	UPLIFT 160%
ABJBL10W	---	---	---	1635	1635
FOP41	12	12-16d	---	2190	2465
HCPFA	16	8-16d	8-16d	1200	1415
HCPSA	18	---	16-16d	1200	1415
HCPRS	18	5-8d	6-8d	540	540
CLP5W	18	11-8d	6-8d	530	540
SRP121630F	12	---	18-16d	2815	3380
RS150	16	---	11-10d	1645	1645
FAP18F	18	6-8d	6-8d	765	915
ADS2	12	(1) 5/8"	(2) 5/8"	2775	3330

PRODUCT CODE	GAUGE	FASTENER SCHEDULE		ALLOWABLE LOADS	
		STUD	PLATE	WIND / EARTHQUAKE	
				UPLIFT 133%	UPLIFT 160%
HCPLR	18	4-8d	4-8d	510	520
FAP18	18	6-8d	6-8d	745	745
HCPRS	18	6-10d	5-10d	540	540
CLP5W	18	6-10d	11-10d	540	540
TPP4	20	8-10d	8-10d	1335	1335

PRODUCT CODE	GAUGE	FASTENER SCHEDULE		ALLOWABLE LOADS			
		TRUSS / RAFTER	SEAT PLATE OR BEAM	LATERAL		WIND / EARTHQUAKE	
				PERP. TO WALL	PARAL. TO WALL	UPLIFT 133%	UPLIFT 160%
SGP2	14	14-16d	---	---	---	1455	1455
TAPL12	14&20	11-16d	4-10dx1 1/2"	1405	1405	1950	1950
TAP16	14	11-16d	---	595	210	1950	1950
TAPR216	14&20	11-16d	---	595	210	1950	1950
HDA6	1/4"	(2) 3/4"	(2) 3/4"	---	---	4256	4256

PRODUCT CODE	GAUGE	FASTENER SCHEDULE			ALLOWABLE LOADS			
		TRUSS / RAFTER	PLATE	STUD	LATERAL		WIND / EARTHQUAKE	
					PERP. TO WALL	PARAL. TO WALL	UPLIFT 133%	UPLIFT 160%
HDPT2	18	---	12-16d	---	450	450	1915	2300
RT10	20	5-8dx1 1/2"	8-8d	5-8dx1 1/2"	95	115	555	555
HCPLR	18	4-8d	4-8d	4-8d	95	145	510	520
HCPRF	18	6-10d	6-10d	6-10d	395	235	540	540
RTPGA818T	14	9-16d	---	9-16d	---	---	1360	1635
HCPFA	16	---	8-16d	8-16d	---	---	1200	1415
TPP4	20	---	---	8-10d	---	---	1290	1335

PRODUCT CODE	GAUGE	FASTENER SCHEDULE		ALLOWABLE LOADS	
		STUD / TOP	SILL	WIND / EARTHQUAKE	
				UPLIFT 133%	UPLIFT 160%
TAP18	14	(12) 16d		1950	1950
ADS2	12	(2) 5/8"	(1) 5/8"	2775	3330
FA3	16	(4) 8dX1 1/2"	(2) 8dX1 1/2"	1155	1155
FTP42*	12	(22) 16d		4050	4050
FAS118	18	(4) 8dX1 1/2"	(4) 8dX1 1/2"	755	755
SGP2	14	(14) 16d		1455	1455
ABJBL10W	---			1635	1635

Fig. 4-141 *Characteristics of metal connectors.* (SEMCO)

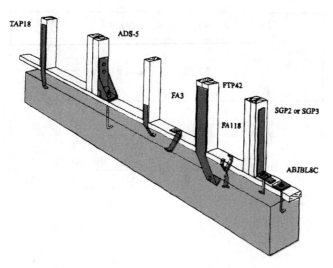

Fig. 4-142 *Wood-to-concrete foundation connections.* (SEMCO)

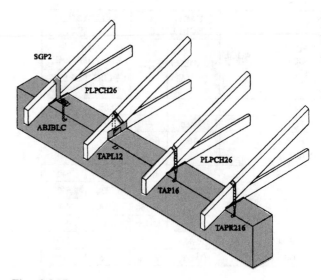

Fig. 4-145 *Poured masonry header with connectors holding the trusses in place.* (SEMCO)

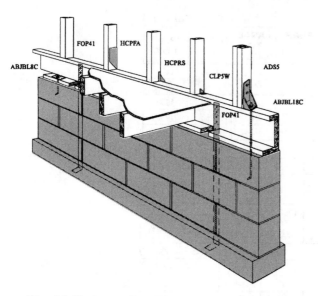

Fig. 4-143 *Stem wall system metal connectors.* (SEMCO)

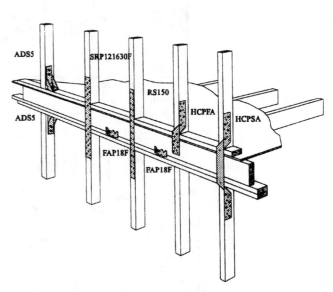

Fig. 4-146 *Wood frame second floor connections.* (SEMCO)

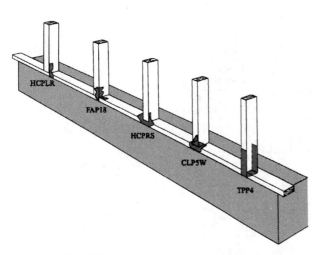

Fig. 4-144 *Wood-to-wood connectors.* (SEMCO)

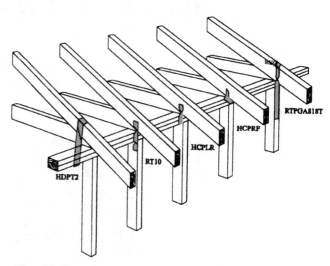

Fig. 4-147 *Top plate, truss or rafter connections with connectors making the truss to top plate a little more secure.* (SEMCO)

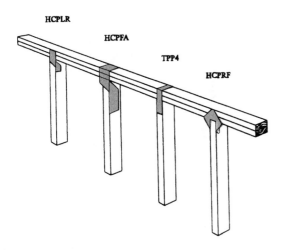

Fig. 4-148 *Various connectors used in the top plate-to-stud construction.* (SEMCO)

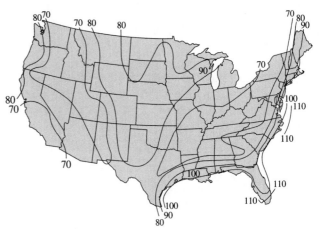

Fig. 4-149 *Basic Wind Speed Map for Continental USA.*

5
CHAPTER

Covering Roofs

IN THIS UNIT YOU WILL LEARN HOW THE carpenter covers roofs, how the roof is prepared for shingles, and how to place the shingles on the prepared surface. In addition, details for applying an asphalt shingle roof are given. Skills covered include how to:

- Prepare a roof deck for shingles
- Apply shingles to a roof deck
- Estimate shingles needed for a job
- Figure slope of a roof

INTRODUCTION

Roofing shingles are made in different sizes and shapes. Roofs have different angles and shapes. Pipes stick up through the roof. Roofing has to be fitted. There are crickets also to be fitted. (A cricket fits between a chimney and the roof.) A number of fine details are presented by roof shapes. All these have to be considered by the roofer.

Some industrial and commercial buildings use roll shingles. This material requires a slightly different approach. Asphalt shingles are safer than wooden shingles. The asphalt shingles will resist fire longer. Because of fire regulations, wooden shingles are not allowed in some sections of the country.

BASIC SEQUENCE

The carpenter should apply a roof in this order:

1. Check the deck for proper installation.
2. Decide which shingles to use for the job.
3. Estimate the amount needed for the job.
4. Apply drip strips.
5. Place the underlayment.
6. Nail the underlayment.
7. Start the first course of shingles.
8. Continue other courses of shingles.
9. Cut and install flashings:
   a. valleys
      (1) open
      (2) closed-cut
      (3) woven
   b. soil pipe flashing
   c. chimney flashing
   d. other flashings
10. Cover ridges.
11. Cover all nailheads with cement.
12. Glue down tabs, if needed.

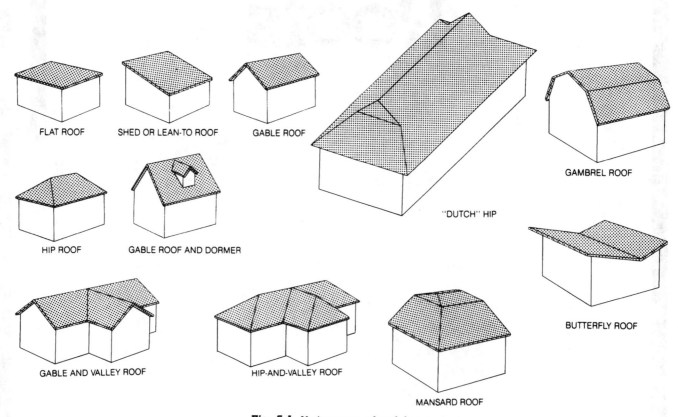

Fig. 5-1 *Various types of roof shapes.*

Types of Roofs

There are a number of roof types; each is classified according to its shape. Figure 5-1 shows the different types. Each type presents roofing problems, and different methods are used to cover the decking, ridges, and drip areas.

The *mansard roof* presents some unique problems. Figure 5-2 shows a mansard roof. Note that the dark area is covered with shingles. The angles presented by the various vertical and sloping sections need special bracing. Attaching the shingles also requires attention to details on the vertical sections. (Refer back to Fig. 4-123 for an illustration showing truss construction on a mansard style roof.) The attached garage in Fig. 5-2 has a hip roof. Figure 5-3 also shows a hip roof. Note how the entrance is also a hip, but shorter.

Fig. 5-2 *Mansard roof with hip on garage.*

Fig. 5-3 *Hip-and-valley roof.*

The *gable roof* is a common type. See Fig. 5-4. It is a simple roof that is easy to build. Figure 5-5 shows a variety of gable roofs. Each is a complete unit. The garage shows the angles of this type of roof very well.

Drainage Factors

The main purpose of a roof is to protect the inside of a building. This is done by draining the water from the

Fig. 5-4 *Gable roof.*

Fig. 5-5 *Gable roof with add-ons.*

roof. The water goes onto the ground or into the storm sewer. Some parts of the country allow the water to be dropped onto the earth below. Other sections require the collected water to be moved to a storm sewer. The main idea is to prevent water seepage. Roof water should not seep back into a basement.

Ice is frequently a problem in colder climates. Ice forms and makes a dam for melting snow. See Fig. 5-6A. Water backs up under the shingles and leaks into the ceiling below. This problem can be caused by insufficient insulation. The lack of soffit ventilation will also cause leaks. Heating cables can be installed to prevent ice dams. Leaking can be prevented by adding ventilation. If insulation can be added, this too should be done. See Fig. 5-6B.

The point where rooflines come together is called a *valley*. See Fig. 5-7A, B, and C. Valleys direct water to the drain. This keeps it out of the house. They need special attention during roofing.

Eave troughs and downspouts drain water from the roof. It is drained into gutters. Downspouts carry the water to the ground. Downspouts connect to other pipes. That piping sometimes goes to the street storm

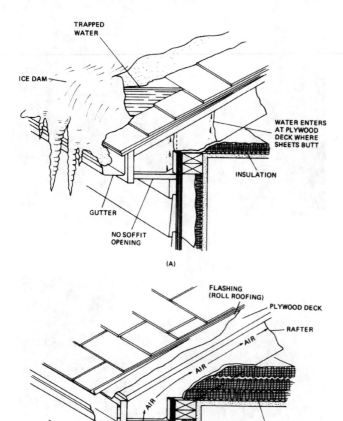

Fig. 5-6 *(A) Water leakage caused by ice dam. (B) Using ventilation and insulation to prevent leaks.*

sewer. This eliminates seepage into the basement or under the slab. See Fig. 5-8A. In most cases a splash block is located under the downspout to disperse the water. See Fig. 5-8B.

Roofing Terms

There are a number of roofing terms used by roofers and carpenters. You should become familiar with the terms; you will then be able to talk with roofing salespersons.

Square Shingles needed to cover 100 square feet of roof surface. That means 10 feet *square*, or 10 feet by 10 feet.

Exposure Distance between exposed edges of overlapping shingles. Exposure is given in inches. See Fig. 5-9. Note the 5-inch and 3-inch exposures.

Head lap Distance between the top of the bottom shingle and the bottom edge of the one covering it. See Fig. 5-10.

Top lap Distance between the lower edge of an overlapping shingle and the upper edge of the lapping shingle. See Fig. 5-10. Top lap is measured in inches.

Side lap Distance between adjacent shingles that overlap. Measured in inches.

Valley Angle formed by two roofs meeting. The internal part of the angle is the valley.

Rake On a gable roof, the inclined edge of the surface to be covered.

Flashing Metal used to cover chimneys and around things projecting through the roofing. Used to keep the weather out.

Underlayment Usually No. 15 or No. 30 felt paper applied to a roof deck. It goes between the wood and the shingle.

Ridge The horizontal line formed by the two rafters of a sloping roof being nailed together.

Hip The external angle formed by two sides of a roof meeting.

Roofing is part of the exterior building. The carpenter is called upon to place the covering over a frame. This frame is usually covered by plywood. Plywood comes in 4-×-8-foot sheets and can be quickly installed onto the rafters. Sheathing may be 1-×-6-inch or 1-×-10-inch boards. Sheathing takes longer to install than plywood. The frame is covered with a tar or felt paper. This paper goes on over the sheathing. The paper allows moisture to move from the wood upward. Moisture then escapes under the shingles. This prevents a buildup of moisture. If the weather is bad, moisture can freeze. This forms a frost under the shingles.

Shingles of wood, asphalt, asbestos, and Fiberglas are used for roofing. Tile and slate were once commonly used; however, they are rather expensive to install. Copper, galvanized iron, and tin are also used as roof coverings.

Commercial buildings may use a built-up roof, which has a number of layers. This type uses a gravel topping or cap sheet. Asphalt-saturated felt is mopped down with hot asphalt or tar. See Fig. 5-8C. Choice of roofing is determined by three things: cost, slope, and life expected. In some local climates (wind, rain, snow), flat roofs may have to be rejected.

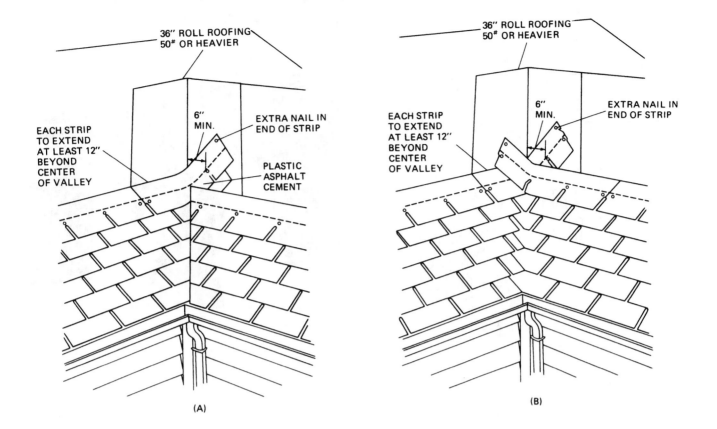

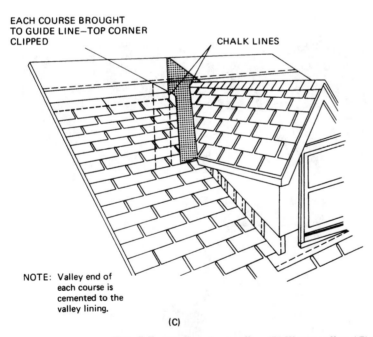

Fig. 5-7 *(A) Closed-cut valley; (B) Woven valley; (C) Open valley.*

In certain applications, such as homes, appearance is another important consideration. Shingles are used most frequently for homes. See Fig. 5-8D. In some areas, cedar shingles are not permitted because wood burns too easily. Once aged, however, it becomes more fire-resistant.

Pitch

Drainage of water from a roof surface is essential. This means that pitch should be considered. The pitch or slope of a roof deck determines the choice of shingle. It also determines drainage.

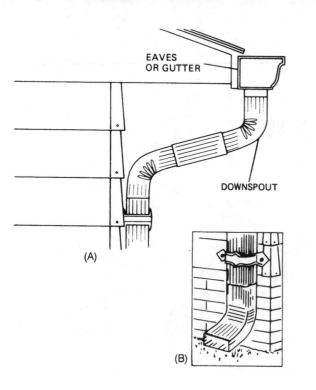

Fig. 5-8 *(A) Eave trough and downspout. (B) Downspout elbow turns water away from the basement. (C) Preparing a flat roof. Asphalt-saturated felt is mopped down with hot asphalt or tar. (D) Applying a shingle roof. The shingles are packaged so that they can be placed in a location convenient for the roofer.*

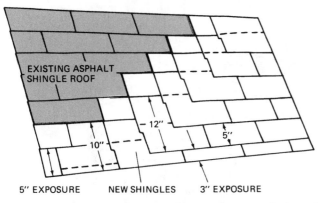

EXISTING ASPHALT
SHINGLE ROOF

12"

10"

5"

5" EXPOSURE   NEW SHINGLES   3" EXPOSURE

Fig. 5-9 *Exposure is the distance between the exposed edges of overlapping shingles.* (Bird and Son)

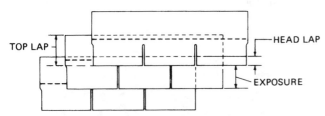

TOP LAP

HEAD LAP

EXPOSURE

Fig. 5-10 *Head lap and top lap.* (Bird and Son)

Pitch limitations are shown in Fig. 5-11. Any shingle may safely be used on roofs with normal slopes. Normal is 4 inches rise or more per horizontal foot. An exception exists for square-butt strip shingles. They may be used on slopes included in the shaded area in Fig. 5-11.

When the pitch is less than 4 inches per foot, it is best to use roll roofings. In the range of 4 inches down to 1 inch per foot, the following rules apply:

- Roll roofing may be applied by the exposed nail method if the pitch is not lower than 2 inches per foot.

- Roll roofings applied by the concealed nail method may be used on pitches down to, but not below, 1 inch per foot. This is true if (1) they have at least 3 inches of top lap, and (2) they have double coverage roofing with a top lap of 19 inches.

Any of the above may be applied on a deck with a pitch steeper than the stated minimum. Pitch is given as a fraction. For example, a roof has a rise of 8 feet and a run of 12 feet. Then its pitch is

$$\frac{8}{2 \times 12} \quad \text{or} \quad \frac{8}{24} \quad \text{or } \frac{1}{3}$$

Pitch is equal to the rise divided by 2 times the run. Or

$$\text{Pitch} = \frac{\text{rise}}{2 \times \text{run}}$$

Slope

Slope is how fast the roof rises from the horizontal. See Fig. 5-12. Slope is equal to the rise divided by the run. Or

$$\text{Slope} = \frac{\text{rise}}{\text{run}}$$

Slope and *pitch* are often used interchangeably. However, you can see there is a difference. Some roofers' manuals use them as if they were the same.

Before a roof can be applied, you have to know how many shingles are needed. This calls for estimating the area to be covered. First determine the number of square feet. Then divide the number of square feet by 100 to produce the number of squares needed.

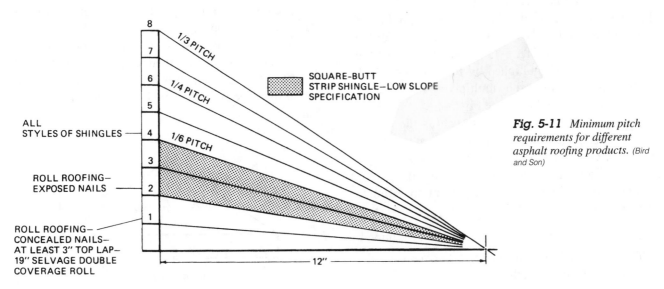

8
7
6
5
4
3
2
1

1/3 PITCH
1/4 PITCH
1/6 PITCH

SQUARE-BUTT
STRIP SHINGLE—LOW SLOPE
SPECIFICATION

ALL STYLES OF SHINGLES

ROLL ROOFING—EXPOSED NAILS

ROLL ROOFING—CONCEALED NAILS—AT LEAST 3" TOP LAP—19" SELVAGE DOUBLE COVERAGE ROLL

12"

Fig. 5-11 *Minimum pitch requirements for different asphalt roofing products.* (Bird and Son)

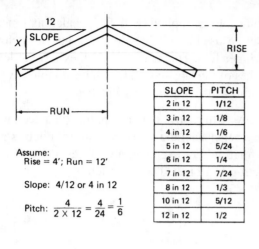

SLOPE	PITCH
2 in 12	1/12
3 in 12	1/8
4 in 12	1/6
5 in 12	5/24
6 in 12	1/4
7 in 12	7/24
8 in 12	1/3
10 in 12	5/12
12 in 12	1/2

Assume:
Rise = 4'; Run = 12'

Slope: 4/12 or 4 in 12

Pitch: $\dfrac{4}{2 \times 12} = \dfrac{4}{24} = \dfrac{1}{6}$

$$\text{Slope} = \frac{\text{rise}}{\text{run}} \qquad\qquad \text{Pitch} = \frac{\text{rise}}{2 \times \text{run}}$$

Fig. 5-12 *Slope, pitch, and run of a roof.*

ESTIMATING ROOFING QUANTITIES

Roofing is estimated and sold in squares. *A square of roofing is the amount required to cover 100 square feet.* To estimate the required amount, you have to compute the total area to be covered. This should be done in square feet. Then divide the amount by 100. This determines the number of squares needed. Some allowance should be made for cutting and waste. This allowance is usually 10 percent. If you use 10 percent for waste and cutting, you will have the correct number of shingles. A simple roof with no dormers will require less than 10 percent.

Complicated roofs will require more than 10 percent for cutting and fitting.

Estimating Area

The areas of simple surfaces can be computed easily. The area of the shed roof in Fig. 5-13 is the product of the eave line and the rake line ($A \times B$). The area of the simple gable roof in Fig. 5-13 equals the sum of the two rakes B and C multiplied by the eave line A. A gambrel roof is estimated by multiplying rake lines A, B, C, and D by eave line E. See Fig. 5-13.

Complications arise in roofs such as the one in Fig. 5-14. Ells, gables, or dormers can cause special problems. Obtain the lengths of eaves, rakes, valleys, and ridges from drawings or sketches. Measuring calls for dangerous climbing. You may want to estimate without climbing. To do this:

1. The pitch of the roof must be known or determined.
2. The horizontal area in square feet covered by the roof must be computed.

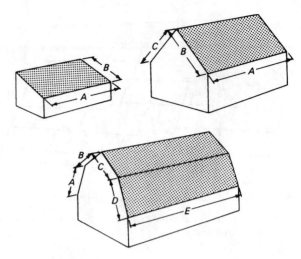

Fig. 5-13 *Simple roof types and their dimensions: Shed, gable, and gambrel.* (Bird and Son)

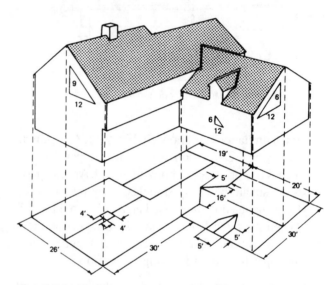

Fig. 5-14 *Complicated dwelling roof shown in perspective and plan views.* (Bird and Son)

Pitch is shown in Fig. 5-15. The pitch of a roof is stated as a relationship between rise and span. If the span is 24'0" and the rise is 8'0", the pitch will be 8/24 or ⅓. If the rise were 6'0", then the pitch would be 6/24 or ¼. The ⅓-pitch roof rises 8 inches per foot of horizontal run. The ¼-pitch roof rises 6 inches per foot of run.

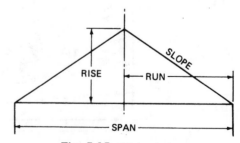

Fig. 5-15 *Pitch relations.*

Fig. 5-16 *Using the carpenter's rule to find the roof pitch.* (Bird and Son)

You can determine the pitch of any roof without leaving the ground. Use a carpenter's folding rule in the following manner.

Form a triangle with the rule. Stand across the street or road from the building. Hold the rule at arm's length. Align the roof slope with the sides of the rule. Be sure that the base of the triangle is held horizontal. It will appear within the triangle as shown in Fig. 5-16. Take a reading on the base section of the rule. Note the reading point shown in Fig. 5-17. Locate in the top line, headed *Rule Reading* in Fig. 5-18, the point nearest your reading. Below this point is the pitch and the rise per foot of run. Here the reading on the rule is 22. Under 22 in Fig. 5-18, the pitch is designated as a rise of 8 inches per foot of horizontal run.

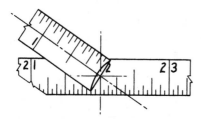

READING POINT

Fig. 5-17 *Reading the carpenter's rule to find the point needed for the pitch figures.* (Bird and Son)

Horizontal Area

Figure 5-14 is a typical dwelling. The roof has valleys, dormers, and variable-height ridges. Below the perspective the total ground area is covered by the roof. All measurements needed can be made from the ground. Or they can be made within the attic space of the house. No climbing on the roof is needed.

Computation of Roof Areas

Make all measurements. Draw a roof plan. Determine the pitches of the various elements of the roof. Use the carpenter's rule. The horizontal areas can now be quickly worked out.

Include in the estimate only those areas having the same pitch. The rise of the main roof is 9 inches per foot. That of the ell and dormers is 6 inches per foot.

The horizontal area under the 8-inch-slope roof will be

$$26 \times 30 = 780 \text{ square feet}$$
$$19 \times 30 = 570 \text{ square feet}$$
$$\text{Total} \quad 1350 \text{ square feet}$$

Less

$$8 \times 5 = 40 \text{ (triangular area under ell roof)}$$
$$4 \times 4 = 16 \text{ (chimney)}$$
$$56 \text{ square feet}$$

$$1350 - 56 = 1294 \text{ square feet total}$$

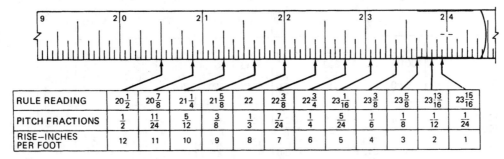

RULE READING	$20\frac{1}{2}$	$20\frac{7}{8}$	$21\frac{1}{4}$	$21\frac{5}{8}$	22	$22\frac{3}{8}$	$22\frac{3}{4}$	$23\frac{1}{16}$	$23\frac{3}{8}$	$23\frac{5}{8}$	$23\frac{13}{16}$	$23\frac{15}{16}$
PITCH FRACTIONS	$\frac{1}{2}$	$\frac{11}{24}$	$\frac{5}{12}$	$\frac{3}{8}$	$\frac{1}{3}$	$\frac{7}{24}$	$\frac{1}{4}$	$\frac{5}{24}$	$\frac{1}{6}$	$\frac{1}{8}$	$\frac{1}{12}$	$\frac{1}{24}$
RISE—INCHES PER FOOT	12	11	10	9	8	7	6	5	4	3	2	1

Fig. 5-18 *Reading point converted to pitch.* (Bird and Son)

The area under the 6-inch-rise roof will be

$20 \times 30 = 600$ square feet
$\underline{8 \times 5 = 40}$ (triangular area projecting over the
$\phantom{8 \times 5 = }640$ main house)

Duplications

Sometimes one element of a roof projects over another. Add duplicated areas to the total horizontal area. If the eaves in Fig. 5-14 project only 4 inches, there will be

1. A duplication of 2 (7 × ⅓) or 4⅔ square feet under the eaves of the main house. This is where they overhang the rake of the ell section.

2. A duplication under the dormer eaves of 2 (5 × ⅓) or 3⅓ square feet.

3. A duplication of 9½ × ⅓ or 3¹⁄₁₆ square feet under the eaves of the main house. This is where they overhang the rake of the ell section.

The total is 11⅙ or 12 square feet. Item 2 should be added to the area of the 6-inch-pitch roof. Items 1 and 3 should be added to the 9-inch-pitch roof. The new totals will be 640 + 4 or 644 for the 6-inch pitch and 1294 + 8 or 1302 for the 9-inch pitch.

Converting Horizontal to Slope Areas

Now convert horizontal areas to slope areas. Do this with the aid of the *conversion table*, Table 5-1. Horizontal areas are given in the first column. Corresponding slope areas are given in columns 2 to 12.

The total area under the 9-inch rise is 1302 square

Table 5-1 Conversion Table

Rise, Inches per Foot of Horizontal Run	1	2	3	4	5	6	7	8	9	10	11	12
Pitch, Fractions	1/24	1/12	1/8	1/6	5/24	1/4	7/24	1/3	3/8	5/12	11/24	1/2
Conversion Factor	1.004	1.014	1.031	1.054	1.083	1.118	1.157	1.202	1.250	1.302	1.356	1.414
Horizontal, area in square feet or length in feet												
1	1.0	1.0	1.0	1.1	1.1	1.1	1.2	1.2	1.3	1.3	1.4	1.4
2	2.0	2.0	2.1	2.1	2.2	2.2	2.3	2.4	2.5	2.6	2.7	2.8
3	3.0	3.0	3.1	3.2	3.2	3.2	3.5	3.6	3.8	3.9	4.1	4.2
4	4.0	4.1	4.1	4.2	4.3	4.5	4.6	4.8	5.0	5.2	5.4	5.7
5	5.0	5.1	5.2	5.3	5.4	5.6	5.8	6.0	6.3	6.5	6.8	7.1
6	6.0	6.1	6.2	6.3	6.5	6.7	6.9	7.2	7.5	7.8	8.1	8.5
7	7.0	7.1	7.2	7.4	7.6	7.8	8.1	8.4	8.8	9.1	9.5	9.9
8	8.0	8.1	8.3	8.4	8.7	8.9	9.3	9.6	10.0	10.4	10.8	11.3
9	9.0	9.1	9.3	9.5	9.7	10.1	10.4	10.8	11.3	11.7	12.2	12.7
10	10.0	10.1	10.3	10.5	10.8	11.2	11.6	12.0	12.5	13.0	13.6	14.1
20	20.1	20.3	20.6	21.1	21.7	22.4	23.1	24.0	25.0	26.0	27.1	28.3
30	30.1	30.4	31.0	31.6	32.5	33.5	34.7	36.1	37.5	39.1	40.7	42.4
40	40.2	40.6	41.2	42.2	43.3	44.7	46.3	48.1	50.0	52.1	54.2	56.6
50	50.2	50.7	51.6	52.7	54.2	55.9	57.8	60.1	62.5	65.1	67.8	70.7
60	60.2	60.8	61.9	63.2	65.0	67.1	69.4	72.1	75.0	78.1	81.4	84.8
70	70.3	71.0	72.2	73.8	75.8	78.3	81.0	84.1	87.5	91.1	94.9	99.0
80	80.3	81.1	82.5	84.3	86.6	89.4	92.6	96.2	100.0	104.2	108.5	113.1
90	90.4	91.3	92.8	94.9	97.5	100.6	104.1	108.2	112.5	117.2	122.0	127.3
100	100.4	101.4	103.1	105.4	108.3	111.8	115.7	120.2	125.0	130.2	135.6	141.4
200	200.8	202.8	206.2	210.8	216.6	223.6	231.4	240.4	250.0	260.4	271.2	282.8
300	301.2	304.2	309.3	316.2	324.9	335.4	347.1	360.6	375.0	390.6	406.8	424.2
400	401.6	405.6	412.4	421.6	433.2	447.2	462.8	480.8	500.0	520.8	542.4	565.6
500	502.0	507.0	515.5	527.0	541.5	559.0	578.5	601.0	625.0	651.0	678.0	707.0
600	602.4	608.4	618.6	632.4	649.8	670.8	694.2	721.2	750.0	781.2	813.6	848.4
700	702.8	709.8	721.7	737.8	758.1	782.6	809.9	841.4	875.0	911.4	949.2	989.8
800	803.2	811.2	824.8	843.2	864.4	894.4	925.6	961.6	1000.0	1041.6	1084.8	1131.2
900	903.6	912.6	927.9	948.6	974.7	1006.2	1041.3	1081.8	1125.0	1171.8	1220.4	1272.6
1000	1004.0	1014.0	1031.0	1054.0	1083.0	1118.0	1157.0	1202.0	1250.0	1302.0	1356.0	1414.0

feet. Under the column headed 9 (for 9-inch rise) on the conversion table is found:

	Horizontal Area		Slope Area
Opposite	1000	is	1250.0
Opposite	300	is	375.0
Opposite	00	is	00.0
Opposite	2	is	2.5
Totals	1302		1627.5

The total area under the 6-inch rise is 644 square feet.

	Horizontal Area		Slope Area
Opposite	600	is	670.8
Opposite	40	is	44.7
Opposite	4	is	4.5
Totals	644		720.0

The total area for both pitches is 1627.5 + 720 = 2347.5 square feet.

Now, add a percentage for waste. This should be 10 percent. That brings the total area of roofing required to 2582 square feet. Divide 2582 by 100 and get 25.82 or, rounded, *26 squares*.

One point about this method should be emphasized. The method is possible because of one fact. Over any given horizontal area, at a given pitch, a roof will always contain the same number of square feet regardless of its design. A shed, a gable, and a hip roof, with or without dormers, will each require exactly the same square footage of roofing—that is, if each is placed over the same horizontal area with the same pitch.

Accessories

Quantities of starter strips, edging strips, ridge shingles, and valley strips all depend upon linear measurements. These measurements are taken along the eaves, rake ridge, and valley. Eaves and ridge are horizontal. The rakes and valleys run on a slope. Quantities for the horizontal elements can be taken off the roof plan. True length of rakes and valleys must be taken from conversion tables.

LENGTH OF RAKE

Determine the length of the rake of the roof. Measure the horizontal distance over which it extends. In this case the rakes on the ends of the main house span dis-

tances are 26 and 19 feet. More rake footage is 26 + 19 + 13 + 3½ = 61½ feet.

Refer to Table 5-1 under the 9-inch-rise column. Opposite the figures in column 1 find the length of the rake.

Horizontal Run	Length of Rake
60	75.00
1	1.3
0.5	0.6
Totals 61.5	76.9 (actual length of the rake)

Use the same method and apply it to the rake of the ell. This will indicate its length, including the dormer, to be 39.1 inches. Add these amounts to the total length of eaves. The figure obtained can be used for an estimate of the amount of edging needed.

Hips and Valleys

Hip-and-valley lengths can be determined. Use the run off the common rafter. Then refer to the hip-and-valley table, Table 5-2.

Common rafter run is one-half the horizontal distance that the roof spans. This determines the length of a valley. The run of the common rafter should be taken at the lower end of the valley.

Figure 5-14 shows the portion of the ell roof that projects over the main roof. It has a span of 16 feet at the lower end of the valley. Therefore, the common rafter at this point has a run of 8'0".

There are two valleys at this roof intersection. Total run of the common rafter is 16'0". Refer to Table 5-2. Opposite the figures in the column headed *Horizontal*, find the linear feet of valleys. Then check the column under the pitch involved.

One of the intersecting roofs has a rise of 6 inches. The other has a rise of 9 inches. Length for each rise must be found. The average of the two is then taken. This gives a close approximation of the true length of the valley.

Thus,

Horizontal	6-inch rise	9-inch rise
10	15	16
6	9	9.6
16	24	25.6

24 + 25.6 = 49.6
49.6 ÷ 2 = 24.8 length of valleys

Table 5-2 Hip-and-Valley Conversions

Rise, Inches per Foot of Horizontal Run		4	5	6	7	8	9	10	11	12	14	16	18
Pitch	Degrees	18°26'	22°37'	26°34'	30°16'	33°41'	36°52'	39°48'	42°31'	45°	49°24'	53°8'	56°19'
	Fractions	1/6	5/24	1/4	7/24	1/3	3/8	5/12	11/24	1/2	7/12	2/3	3/4
Conversion Factor		1.452	1.474	1.500	1.524	1.564	1.600	1.642	1.684	1.732	1.814	1.944	2.062
Horizontal Length in Feet													
1		1.5	1.5	1.5	1.5	1.6	1.6	1.6	1.7	1.7	1.8	1.9	2.1
2		2.9	2.9	3.0	3.0	3.1	3.2	3.3	3.4	3.5	3.6	3.9	4.1
3		4.4	4.4	4.5	4.6	4.7	4.8	4.9	5.1	5.2	5.4	5.8	6.2
4		5.8	5.9	6.0	6.1	6.3	6.4	6.6	6.7	6.9	7.3	7.8	8.2
5		7.3	7.4	7.5	7.6	7.8	8.0	8.2	8.4	8.7	9.1	9.7	10.3
6		8.7	8.8	9.0	9.1	9.4	9.6	9.9	10.1	10.4	10.9	11.7	12.4
7		10.2	10.3	10.5	10.7	10.9	11.2	11.5	11.8	12.1	12.7	13.6	14.4
8		11.6	11.8	12.0	12.2	12.5	12.8	13.1	13.5	13.9	14.5	15.6	16.5
9		13.1	13.3	13.5	13.7	14.1	14.4	14.8	15.2	15.6	16.3	17.5	18.6
10		14.5	14.7	15.0	15.2	15.6	16.0	16.4	16.8	17.3	18.1	19.4	20.6
20		29.0	29.5	30.0	30.5	31.3	32.0	32.8	33.7	34.6	36.3	38.9	41.2
30		43.6	44.2	45.0	45.7	46.9	48.0	49.3	50.5	52.0	54.4	58.3	61.9
40		58.1	59.0	60.0	61.0	62.6	64.0	65.7	67.4	69.3	72.6	77.8	82.5
50		72.6	73.7	75.0	76.2	78.2	80.0	82.1	84.2	86.6	90.7	97.2	103.1
60		87.1	88.4	90.0	91.4	93.8	96.0	98.5	101.0	103.9	108.8	116.6	123.7
70		101.6	103.2	105.0	106.7	109.5	112.0	114.9	117.9	121.2	127.0	136.1	144.3
80		116.2	117.9	120.9	121.9	125.1	128.0	131.4	134.7	138.6	145.1	155.5	165.0
90		130.7	132.7	135.0	137.2	140.8	144.0	147.8	151.6	155.9	163.3	175.0	185.6
100		145.2	147.4	150.0	152.4	156.4	160.0	164.2	168.4	173.2	181.4	194.4	206.2

Dormer Valleys

The run of the common rafter at the dormer is 2.5 feet. Check Table 5-2. It is found that:

Horizontal	6-inch rise
2.0	3.0
0.5	0.75
2.5	3.75 (length of valley)

Two such valleys will total 7.5 feet.

The total length of valley will be 24.8 + 7.5 = 32.3 feet. Use these figures to estimate the flashing material required.

ROOFING TOOLS

Most roofing tools are already in the carpenter's toolbox. Tools needed for roofing are shown in Fig. 5-19.

Roof brackets Can be used to clamp onto a ladder.

Ladders A pair of sturdy ladders with ladder jacks are needed. Shingles are placed on the roof using a hoist on the delivery truck. These ladders come in handy for side roofing.

Staging These are planks for the ladder jacks. They hold the roofer or shingles. They are very useful on mansard roof jobs.

Apron The carpenter's apron is very necessary. It holds the nails and hammer. Other small tools can fit into it. It saves time in many ways. It keeps needed tools handy.

Hammer The hammer is a necessary device for roofing. It should be a balanced hammer for less wrist fatigue.

Chalk and line This combination is needed to draw guidelines. Shingles need alignment. The chalk marks are needed to make sure the shingles line up.

Tin snips Heavy-duty tin snips are needed for trimming flashing. They can also be used for trimming shingles.

Kerosene A cleaner is needed to remove tar from tools. Asphalt from shingles can be removed from tools with kerosene.

Tape measure A roofer has to make many measurements. This is a necessary tool.

Utility knife A general-purpose knife is needed for close trimming of shingles.

Putty knife This is used to spread roofing cement.

Carpenter's rule This makes measurements and also serves to determine the pitch of a roof. See Fig. 5-16.

Stapler Some new construction roofing can use a stapler. This device replaces the hammer and nails.

Fig. 5-19 *Carpenter's tools needed for roofing:* (Bird and Son) *(A) Planking support; (B) Ladder with planking; (C) Claw hammer; (D) Carpenter's apron; (E) Chalk and cord; (F) Snips; (G) Tape measure; (H) Kerosene; (I) Stapler; (J) Carpenter's folding rule; (K) Utility knife; (L) Putty knife.*

SAFETY

Working on a roof can be dangerous. Here are a few helpful hints. They may save you broken bones or pulled muscles.

1. Wear sneakers or rubber-soled shoes.
2. Secure ladders and staging firmly.
3. Stay off wet roofs.
4. Keep away from power lines.
5. Don't let debris accumulate underfoot.
6. Use roofing brackets, planks if the roof slopes 4 inches or more for every 12 inches of horizontal run.

APPEARANCE

How the finished job looks is important. Here are a few precautions to improve roof appearance.

1. Avoid shingling in extremely hot weather. Soft asphalt shingles are easily marred by shoes and tools.

2. Avoid shingling when the temperature is below 40°F. Cold shingles are stiff and may crack.

3. Measure carefully and snap the chalk line frequently. Roof surfaces aren't always square. You'll want to know about problems to come so that you can correct them.

4. Start at the rear of the structure. If you've never shingled before, this will give you a chance to gain experience before you reach the front.

APPLYING AN ASPHALT ROOF

Asphalt roofing products will serve well when they are correctly applied. Certain fundamentals must be considered. These have to do with the deck, flashing, and application of materials.

Roof Problems

A number of roof problems are caused by defects in the deck. A nonrigid deck may affect the lay of the roofing. Poorly seasoned deck lumber may warp. This can cause cocking of the shingle tabs. It can also cause wrinkling and buckling of roll roofing.

Improper ventilation can have an effect similar to that of green lumber. The attic area should be ventilated. This area is located directly under the roof deck. It should be free of moisture. In cold weather, be sure the interiors are well ventilated. This applies when plaster is used in the building. A positive ventilation of air is required through the building during roofing. This can usually be provided by opening one or two windows. Windows in the basement or on the first floor can be opened. This can create a positive draft through the house. Open windows at opposite ends of the building. This will also create a flow of air. The moving air has a tendency to dry out the roof deck. It helps to eliminate excess moisture. Condensation under the roofing can cause problems.

Deck construction Wood decks should be made from well-seasoned tongue-and-groove lumber that is more than 6 inches wide. Wider sheathing boards are more likely to swell or shrink, producing a buckling of the roof material. Sheathing should be tightly matched. It should be secured to the rafter with at least two 8d nails. One should be driven through the edge of the board. The other should be driven through the board face. Boards containing too many resinous areas should be rejected. Boards with loose knots should not be used. Do not use badly warped boards.

Figure 5-20 shows how a wood roof deck is constructed. In most cases today, 4-×-8-foot sheets of plywood are used as sheathing. The plywood goes over the rafters. C—D grade plywood is used.

Underlayment Apply one layer of no. 15 asphalt-saturated felt over the deck as an underlayment if the deck has a pitch of 4 inches per foot or greater. The felt should be laid horizontally. See Fig. 5-20. Do not use no. 30 asphalt felt. Do not use any tar-saturated felt. Laminated waterproof papers should not be used either. Do not use any vapor barrier-type material. Lay each course of felt over the lower course. Lap the courses 4 inches. Overlap should be at least 2 inches where ends join. Lap the felt 6 inches from both sides over all hips and ridges.

Apply underlayment as specified for low-slope roofs, where the roof slope is less than 4 inches per

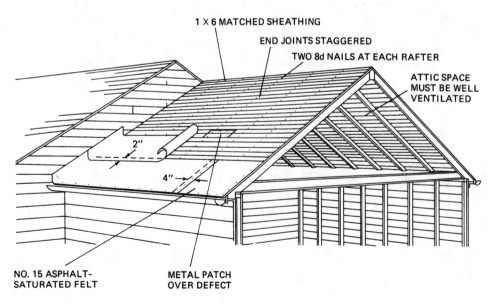

1 X 6 MATCHED SHEATHING

END JOINTS STAGGERED

TWO 8d NAILS AT EACH RAFTER

ATTIC SPACE MUST BE WELL VENTILATED

2"

4"

NO. 15 ASPHALT-SATURATED FELT

METAL PATCH OVER DEFECT

Fig. 5-20 *Features of a good wood roof deck.* (Bird and Son)

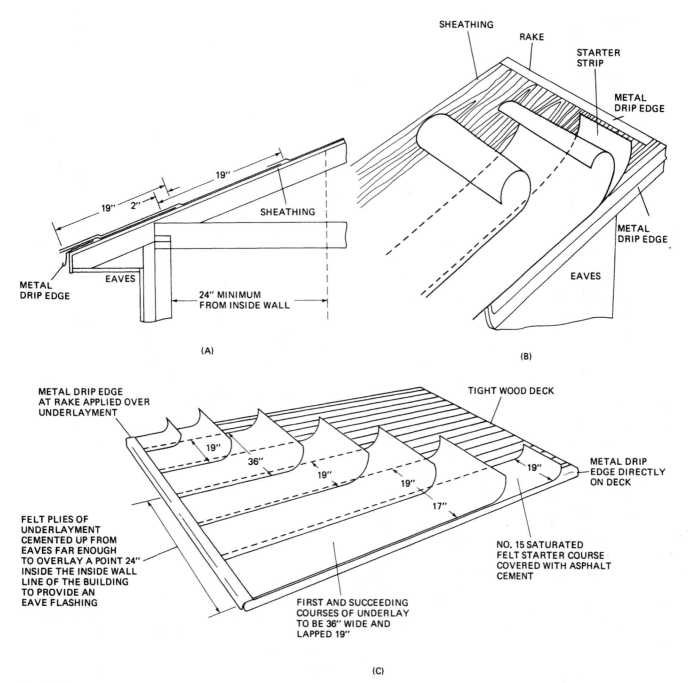

Fig. 5-21 (A) Eaves flashing for a low-slope roof. (B) Placing of sheathing and drip edge. (C) Placement of the underlayment for a shingle roof. (Bird and Son)

foot and not below 2 inches. Check the maker's suggestions. See Fig. 5-21. Felt underlayment performs three functions:

1. It ensures a dry roof for shingles. This avoids buckling and distortion of shingles. Buckling may be caused by shingles being placed over wet roof boards.

2. Felt underlay prevents the entrance of wind driven rain onto the wood deck. This may happen when shingles are lifted up.

3. Underlay prevents any direct contact between shingles and resinous areas. Resins may cause chemical reactions. These could damage the shingles.

Plywood decks Plywood decks should meet the Underwriters' Laboratories standards. Standards are set according to grade and thickness. Design the eaves, rake, and ridge to prevent problems. Openings through the deck should be made in such a way that the plywood will not be exposed to the weather. See Fig. 5-22.

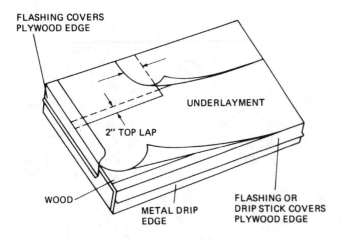

Fig. 5-22 *Preparing the roof deck for shingling.* (Bird and Son)

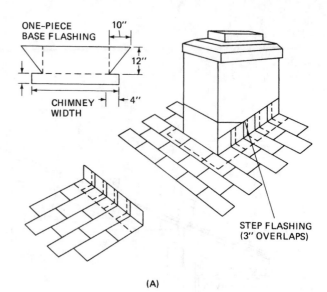

(A)

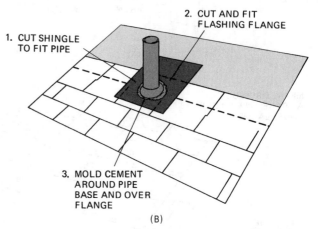

(B)

Fig. 5-23 *(A) Flashing patterns and in place around a chimney* (Bird and Son). *(B) Flashing around a soil pipe.* (Bird and Son)

Nonwood deck materials Nonwood materials are sometimes used in decks. Such things as fiberboard, gypsum, concrete plank, and tile are nonwoods used for decks. These materials have their own standards. Check with the manufacturer for suggestions.

Flashings Roofs are often complicated by intersections with other roofs. Some adjoining walls have projections through the deck. Chimneys and soil stacks create leakage problems. Special attention must be given to protecting against the weather here. Such precautions are commonly called *flashing*. Careful attention to flashing is critical. It helps provide good roof performance. See Fig. 5-23.

Valleys Valleys exist where two sloping roofs meet at an angle. This causes water runoff toward and along the joint. Drainage concentrates at the joint. This makes the joint an easy place for water to enter. Smooth, unobstructed drainage must be provided. It should have enough capacity to carry away the collected water.

There are three types of valleys: open, woven, and closed-cut. See Fig. 5-24.

Each type of valley calls for its own treatment. Figure 5-25 shows felt being applied to a valley. A 36-inch-wide strip of 15# asphalt-saturated felt is centered in the valley. It is secured with nails. They hold it in place until shingles are applied. Courses of felt are cut to overlay the valley strip. The lap should not be less than 6 inches. Eave flashing is then applied.

PUTTING DOWN SHINGLES

Before you put the shingles down, you need an underlayment. See Fig. 5-26. Covering the underlayment is a sheet of saturated felt or tar paper. Table 5-3 shows characteristics of typical asphalt rolls. It is best not to put down the first shingle until you know what is available. Study Table 5-4 to check the characteristics and sizes of typical asphalt shingles. Remember the # symbol means pound or lb., as in 285# or 285 pounds per square of shingles. A *square* covers an area of 100 square feet.

Nails

Nails used in applying asphalt roofings are large-headed and sharp-pointed. Some are hot-galvanized steel. Others are made of aluminum. They may be barbed or otherwise deformed on the shanks. Figure 5-27 shows three types of asphalt nails.

Roofing nails should be long enough to penetrate through the roofing material. They should go at least ¾ inch into the wood deck. This requires that they be of the lengths indicated in Table 5-5.

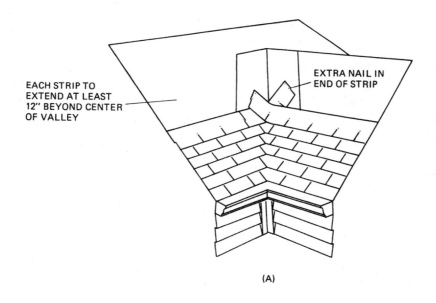

(A)

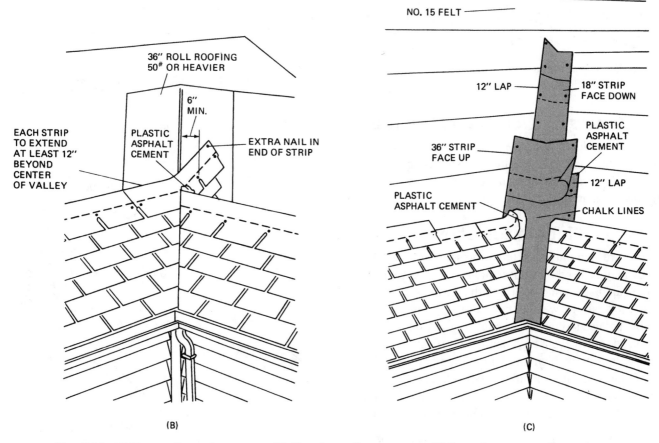

(B)   (C)

Fig. 5-24 *(A) Woven valley roof* (Bird and Son); *(B) Closed-cut valley* (Bird and Son); *(C) Preparing an open valley.* (Bird and Son)

Number of Nails The number of nails required for each shingle type is given by its maker. Manufacturer's recommendations come with each bundle.

Use 2-inch centers in applying roll roofing. This means 252 nails are needed per square. If 3-inch centers are used, then 168 nails are needed per square.

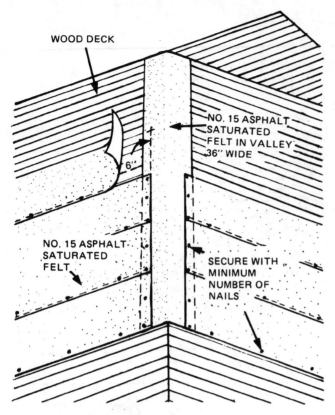

Fig. 5-25 *Felt underlay centered in the valley before valley linings are applied.* (Bird and Son)

Fasteners for Nonwood Materials

Gypsum products, concrete plank and tile, fiberboard, or unusual materials require special fasteners. This type of deck varies with its manufacturer. In such cases follow the manufacturer's suggestions.

Shingle Selection

There are a number of types of shingles available. They may be used for almost any type of roof. Various colors are used to harmonize with buildings. Some are made for various weather conditions. White and light colors are used to reflect the sun's rays. Pastel shingles are used to achieve a high degree of reflectivity. They still permit color blending with siding and trim. Fire and wind resistance should be considered. Simplicity of application makes asphalt roofings the most popular in new housing. They are also rated high for reroofing.

Farm buildings There is no one kind of asphalt roofing for every job. Building types are numerous. The style of roof on the farmhouse may affect the choice of roof on other buildings. They probably should have the same color roofing. This would make the group harmonize. A poultry laying house or machine storage shed near the house calls for a roof like the farm house.

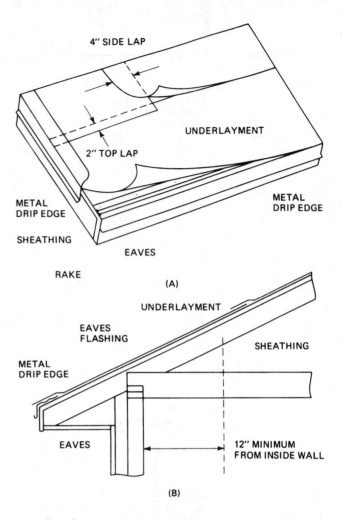

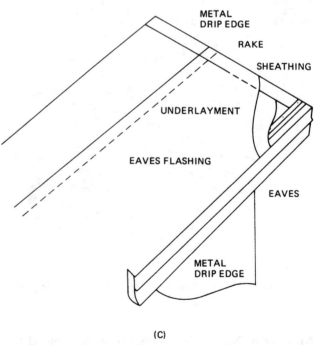

Fig. 5-26 *(A) Underlayment and drip edge; (B) Eaves flashing for a roof; (C) Underlayment, eaves flashing, and metal drip edge.* (Bird and Son)

Table 5-3 *Typical Asphalt Rolls*

Product	Approximate Shipping Weight		Squares Per Package	Length, Feet	Width, Inches	Side or End Lap, Inches	Top Lap, Inches	Exposure, Inches	Underwriters' Listing
	Per Roll	Per Square							
Mineral surface roll double coverage	75# to 90#	75# to 90#	1	36 / 38	36 / 38	6	2 / 4	34 / 32	C
	Available in some areas in 9/10 or 3/4 square rolls.								
Mineral surface roll	55# to 70#	55# to 70#	1/2	36	36	6	19	17	C
Coated roll	50# to 65#	50# to 65#	1	36	36	6	2	34	None
Saturated felt	60# / 60# / 60#	15# / 20# / 30#	4 / 3 / 2	144 / 108 / 72	36 / 36 / 36	4 to 6	2	34	None

An inexpensive roll roofing might be used for an isolated building.

The main idea is to select the right product for the building. The main reason for selecting a roofing material is protection of the contents of a building. The second reason for selecting a roofing is low maintenance cost.

Staples Staples may be used as an alternative to nails. This is only for new buildings. Staples must be zinc-coated. They should be no less than 16 gage. A semi-flattened elliptical cross section is preferred. They should be long enough to penetrate ¾ inch into the wood deck. They must be driven with *pneumatic* (air-driven) staplers. The staple crown must bear tightly against the shingle. However, it must not cut the shingle surface. Use four staples per shingle. See Fig. 5-28. The crown of the staple must be parallel to the tab edge. Position it as shown in Fig. 5-28. Figure 5-29 shows how shingles are nailed. Figure 5-30 shows how the shingles are overlapped.

Cements

Six types of asphalt coatings and cements are:

1. Plastic asphalt cements
2. Lap cements
3. Quick-setting asphalt adhesives
4. Asphalt water emulsions
5. Roof coatings
6. Asphalt primers

Methods of softening The materials are flammable. Cement should be applied to a dry, clean surface. It

Table 5-4 *Typical Asphalt Shingles*

1	2	3		4		5	6
		Per Square		**Size**			
Product*	Configuration	Approximate Shipping Weight	Shingles	Bundles	Width, Inches	Length Inches	Exposure, Inches	Underwriters' Listing
Wood appearance strip shingle more than one thickness per strip Laminated or job-applied	Various edge, surface texture, and application treatments	285# to 390#	67 to 90	4 or 5	11 1/2 to 15	36 or 40	4 to 6	A or C, many wind-resistant
Wood appearance strip shingle single thickness per strip	Various edge, surface texture, and application treatments	Various, 250# to 350#	78 to 90	3 or 4	12 or 12 1/4	36 or 40	4 to 5 1/8	A or C, many wind-resistant
Self-sealing strip shingle	Conventional three-tab	205# to 240#	78 or 80	3	12 or 12 1/4	36	5 or 5 1/8	A or C, all wind-resistant
	Two- or four-tab	Various, 215# to 325#	78 or 80	3 or 4	12 or 12 1/4	36	5 or 5 1/8	
Self-sealing strip shingle No cutout	Various edge and texture treatments	Various, 215# to 290#	78 to 81	3 or 4	12 or 12 1/4	36 or 36 1/4	5	A or C, all wind-resistant
Individual lock-down Basic design	Several design variations	180# to 250#	72 to 120	3 or 4	18 to 22 1/4	20 to 22 1/2		C, many wind-resistant

*Other types available from manufacturers in certain areas of the country. Consult your regional Asphalt Roofing Manufacturers Association manufacturer.

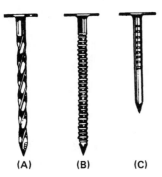

Fig. 5-27 *(A) Screw-threaded nail; (B) Annular threaded nail; (C) Asphalt shingle nail—smooth.*

Table 5-5 *Recommended Nail Length*

Purpose	Nail Length, Inches
Roll roofing on new deck	1
Strip or individual shingles—new deck	1 1/4
Reroofing over old asphalt roofing	1 1/4 to 1 1/2
Reroofing over old wood shingles	1 3/4

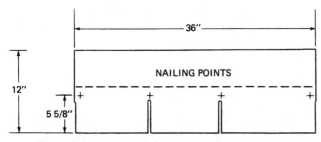

Fig. 5-28 *Nailing or stapling a strip asphalt shingle.* (Certain-Teed)

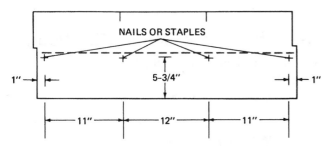

Fig. 5-29 *Nailing points on a strip shingle.* (Bird and Son)

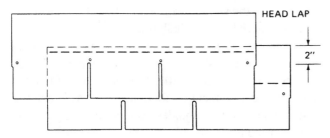

Fig. 5-30 *Overlap of shingles.* (Bird and Son)

should be troweled or brushed vigorously to remove air bubbles. The material should flow freely. It should be forced into all cracks and openings. An emulsion may be applied to damp or wet surfaces. It should not be applied in an exposed location. It should not be rained on for at least 24 hours. Emulsions are water soluble.

Uses Plastic asphalt cements are used for flashing cements. They are so processed that they will not flow at summer temperatures. They are elastic after setting. This compensates for normal expansion and contraction of a roof deck. They will not become brittle at low temperatures.

Lap cements come in various thicknesses. Follow the manufacturer's suggestions. Lap cement is not as thick as plastic cement. It is used to make a watertight bond. The bond is between lapping elements of roll roofing. It should be spread over the entire lapped area. Nails used to secure the roofing should pass through the cement. The shank of the nail should be sealed where it penetrates the deck material.

Seal down the free tabs of strip shingles with quick-setting asphalt adhesive. It can also be used for sealing laps of roll roofing.

Quick-setting asphalt adhesive is about the same thickness as plastic-asphalt cement. However, it is very adhesive. It is mixed with a solvent that evaporates quickly. This permits the cement to set up rapidly.

Roof coatings are used in spray or brush thickness. They are used to coat the entire roof. They can be used to resurface old built-up roofs. Old roll roofing or metal roofs can also be coated.

Asphalt water emulsions are a special type of roof coating made with asphalt and sometimes mixed with other materials. Because they are emulsified with water, they can freeze. Be sure to store them in a warm location. They should not be rained on for at least 24 hours.

Masonry primer is very fluid. Apply it with a brush or by spray. It must be thin enough to penetrate rapidly into the surface pores of masonry. It should not leave a continuous surface film. Thin, if necessary, by following instructions on the can.

Asphalt primer is used to prepare the masonry surface. It should bond well with other asphalt products. These are found on built-up roofs. Other products are plastic-asphalt cement or asphalt coatings.

Starter Course

Putting down the shingles isn't too hard—that is, if you have the roof deck in place. It should be covered

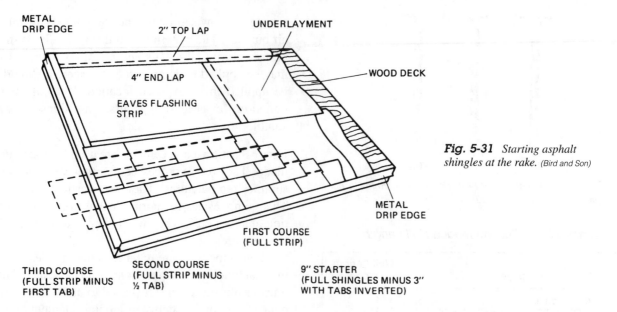

Fig. 5-31 *Starting asphalt shingles at the rake.* (Bird and Son)

METAL
DRIP EDGE

2" TOP LAP

UNDERLAYMENT

4" END LAP

EAVES FLASHING
STRIP

WOOD DECK

METAL
DRIP EDGE

FIRST COURSE
(FULL STRIP)

THIRD COURSE
(FULL STRIP MINUS
FIRST TAB)

SECOND COURSE
(FULL STRIP MINUS
½ TAB)

9" STARTER
(FULL SHINGLES MINUS 3"
WITH TABS INVERTED)

by the proper underlayment. The eaves should be properly prepared. Refer back to Fig. 5-26.

Starting at the rake   Use only the upper portion of the asphalt shingle. Cut off the tabs. Position it with the adhesive dots toward the eaves. The starter course should overhang the eaves and rake edges by ¼ inch. Nail it in a line 3 to 4 inches above the eaves. See Fig. 5-31.

Start the first course with a full strip. Overhang the drip edges at the eaves and rake by ¼ inch. Nail the strip in place. Drive the nails straight. The heads should be flush with the surface of the shingle.

Snap a chalk line along the top edge of the shingle. The line should be parallel with the eaves. Snap several others parallel with the first. Make them 10 inches apart. Use the lines to check alignment at every other course. Snap lines parallel with the rake at the shingle cutouts. Use the lines to check cutout alignment.

Start the second course with a full strip less 6 inches. This means half a tab is missing. Overhang the cut edge at the rake. Nail the shingle in place.

Start the third course with a full strip less a full tab.

Start the fourth course with half a strip.

Continue to reduce the length of the first shingle in each course by an additional 6 inches. The sixth course starts with a 6-inch strip.

Return to the eaves. Apply full shingles across the roof, finishing each course. Dormer, chimney, or vent pipe instructions are another matter. They will be found later in this chapter.

For best color distribution, lay at least four strips in each row. Do this before repeating the pattern up the roof.

Start the *seventh course* with a full shingle. Repeat the process of shortening. Each successive course of

shingles is shortened by an additional 6 inches. This continues to the twelfth course.

Return to the seventh course. Apply full shingles across the roof.

Starting at the Center (Hip Roof)

Snap a vertical chalk line at the center of the roof. See Fig. 5-32.

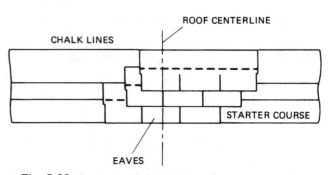

ROOF CENTERLINE

CHALK LINES

STARTER COURSE

EAVES

Fig. 5-32 *Starting asphalt shingles at the center.* (Bird and Son)

Put down starter strips along the eaves. Do this in each direction from the chalk line. Go slightly over the centerlines of the hips. Overhang the eaves by ¼ inch.

Align the butt edge of a full shingle with the bottom edge of the starter strip. Also, align it with its center tab centered on the chalk line.

Snap a chalk line along the top of the shingle parallel with the eaves. Snap several others parallel with the first. They should be 10 inches apart. Use the lines to check the alignment of alternate courses. Finish the first course with full shingles. Extend the shingles part way over the hips.

Finish the remaining courses with full shingles.

Valleys

There are three types of valleys. One is the open type. Here, the saturated felt can be seen after the shingles are applied. Another type is the woven valley. This one has the shingles woven. There is no obvious valley line. The other is the closed-cut valley. This one has a straight line where the roofs intersect.

Open valleys Use mineral-surfaced-material roll roofing for this valley. Match or contrast the color with that of the roof covering. The open valley method is shown in Fig. 5-33. The felt underlay is centered in the valley before shingles are applied. See Fig. 5-34.

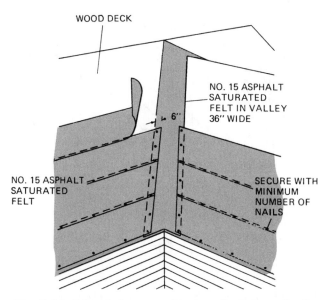

Fig. 5-34 *Felt underlay centered in the valley before valley linings are applied.* (Bird and Son)

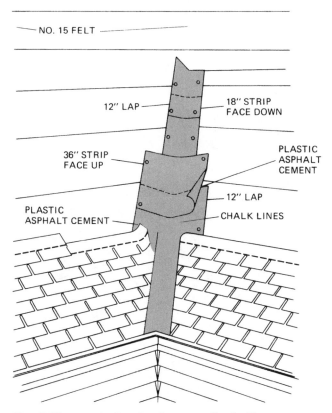

Fig. 5-33 *Use of roll roofing for open valley flashing.* (Bird and Son)

Center an 18-inch-wide layer of mineral-surfaced roll roofing in the valley. The surfaced side goes down. Cut the lower edge to conform to and be flush with the eave flashing strip. The ends of the upper segments overlap the lower segments in a splice. The ends are secured with plastic asphalt cement. See Fig. 5-33. Use only enough nails, 1 inch in from each edge, to hold the strip smoothly. Press the roofing firmly in place into the valley as you nail. Place another strip 36 inches wide on top of the first strip. This is placed surfaced side up. Center it in the valley. Nail it in the same manner as the underlying 18-inch strip.

Do this before the roofing is applied. Snap two chalk lines the full length along the valley, one line on each side of the valley. They should be 6 inches apart at the ridge, or 3 inches when measured from the center of the valley. The marks diverge at the rate of ⅛ inch per foot as they approach the eaves. A valley of 8 feet in length will be 7 inches wide at the eaves. One 16 feet long will be 8 inches wide at the eaves. The chalk line serves as a guide in trimming the last unit to fit the valley. This ensures a clean, sharp edge. The upper corner of each end shingle is clipped. See Fig. 5-34. This keeps water from getting in between the courses. The roofing material is cemented to the valley lining. Use plastic asphalt cement.

Woven and closed-cut valleys Some roofers prefer woven or closed-cut valleys. These are limited to strip-type shingles. Individual shingles cannot be used. Nails may be required at or near the center of the valley lining. Avoid placing a nail in an overlapped shingle too close to the center of the valley. It may sometimes be necessary to cut a strip short. That is done if it would otherwise end near the center. Continue from this cut end over the valley with a full strip. These methods increase the coverage of the shingles throughout the length of the valleys. This adds to the weather resistance of the roofs at these points.

Woven valleys There are two methods of weaving the shingles. See Fig. 5-35. They can be applied on both roof areas at the same time. This means you weave each course, in turn, over the valley. Or, you may cover each roof area first. Do this to a point about

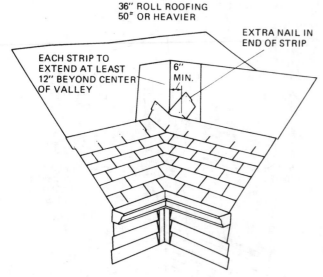

Fig. 5-35 *Weaving each course in turn to make a woven valley.*
(Bird and Son)

Fig. 5-36 *Worker installing shingles.*

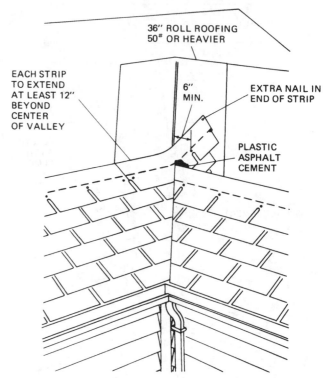

Fig. 5-37 *Closed-cut valley.* (Bird and Son)

3 feet from the center of the valley. Then weave the valley shingles in later.

In the first method, lay the first course. Place it along the eaves of one roof area up to and over the valley. Extend it along the adjoining roof area. Do this for a distance of at least 12 inches. Then lay the first course along the eaves of the intersecting roof area. Extend it over the valley. It goes on top of the previously applied shingle. The next courses go on alternately. Lay along one roof area and then along the other. Weave the valley shingles over each other. See Fig. 5-35. Make sure that the shingles are pressed tightly into the valley. Nail them in the normal manner. No nails are located closer than 6 inches to the valley centerline. Two nails are located at the end of each terminal strip. See Fig. 5-35.

Closed-cut valleys For a closed-cut valley, lay the first course of shingles along the eaves of one roof area up to and over the valley. Extend it along the adjoining roof section. The distance is at least 12 inches. Follow the same procedure when applying the next courses of shingles. See Fig. 5-36. Make sure that the shingles are pressed tightly into the valley. Nail in the normal manner, except that no nail is to be located closer than 6 inches to the valley centerline. Two nails are located at the end of each terminal strip. See Fig. 5-37.

Apply the first course of shingles. Do this along the eaves of the intersecting roofs. Extend it over previously applied shingles. Trim a minimum of 2 inches up from the centerline of the valley. Clip the upper corner of each end shingle. This prevents water from getting between courses. Embed the end in a 3-inch-wide strip of plastic asphalt cement. Other courses are applied and completed. See Fig. 5-37.

An open valley for a dormer roof A special treatment is needed where an open valley occurs at a joint between the dormer roof and the main roof through which it projects. See Fig. 5-38.

First apply the underlay. Main roof shingles are applied to a point just above the lower end of the valley. The course last applied is fitted. It is fitted close to and flashed against the wall of the dormer. The wall is under the projecting edge of the dormer eave. The first strip of valley lining is then applied. Do this the same way as for the open valley. The bottom end is cut so that it extends ¼ inch below the edge of the dormer deck. The lower edge of the section lies on the main deck. It projects at least 2 inches below the joining

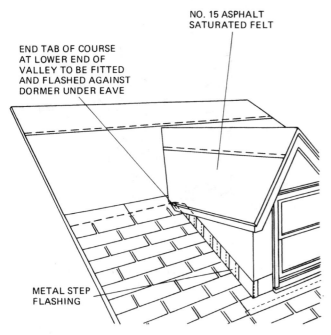

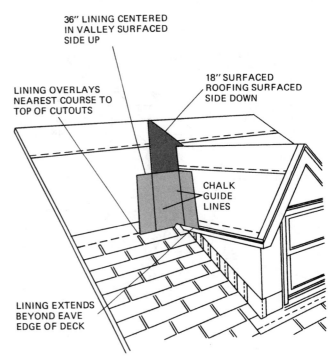

Fig. 5-38 *An open valley for a dormer roof. Shingles have been laid on main roof up to lower end of the valley.* (Bird and Son)

Fig. 5-39 *Application of valley lining for a dormer roof.* (Bird and Son)

roofs. Cut the second or upper strip on the dormer side. It should match the lower end of the underlying strip. Cut the side that lies on the main deck. It should overlap the nearest course of shingles. This overlap is the same as the normal lap of one shingle over another. It depends on the type of shingle being applied. In this case it extends to the top of the cutouts. This is a 12-inch-wide three-tab square-butt strip shingle.

The lower end of the lining is then shaped. See Fig. 5-39. It forms a small canopy over the joint between the two decks.

Apply shingles over the valley lining. The end shingle in each course is cut. It should conform to the guidelines. Bed the ends in a 3-inch-wide strip of plastic asphalt cement. Valley construction is completed in the usual manner. See Fig. 5-40.

Flashing Against a Vertical Wall

Step flashing is used when the rake of a roof abuts a vertical wall. It is best to protect the joint by using metal flashing shingles. They are applied over the end of each course of shingles.

The flashing shingles are rectangular in shape. They are from 5 to 6 inches long. They are 2 inches wider than the exposed face of the roofing shingles. When used with strip shingles laid 5 inches to the weather, they are 6 to 7 inches long. They are bent so as to extend 2 inches out over the roof deck. The remainder goes up the wall surface. Each flashing shingle is placed just uproof from the exposed edge of the

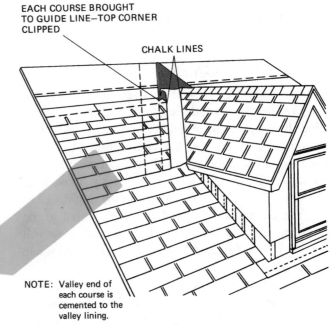

Fig. 5-40 *Dormer valley completed.* (Bird and Son)

single which overlaps it. It is secured to the wall sheathing with one nail in the top corner. See Fig. 5-41. The metal is 7 inches wide. The roof shingles are laid 5 inches to the weather. Each element of flashing will lap the next by 2 inches. See Fig. 5-41.

The finished siding is brought down over the flashing to serve as a cap flashing. However, it is held far enough away from the shingles. This allows the

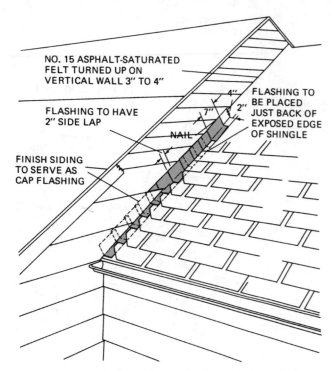

NO. 15 ASPHALT-SATURATED FELT TURNED UP ON VERTICAL WALL 3″ TO 4″

FLASHING TO HAVE 2″ SIDE LAP

FINISH SIDING TO SERVE AS CAP FLASHING

NAIL

4″
7″
2″

FLASHING TO BE PLACED JUST BACK OF EXPOSED EDGE OF SHINGLE

Fig. 5-41 *Use of metal flashing shingles to protect the joint between a sloping roof and a vertical wall.* (Bird and Son)

ends of the boards to be painted. Paint excludes dampness and prevents rot.

Chimneys

Chimneys are usually built on a separate foundation. This avoids stresses and distortions due to uneven settling. It is subject to differential settling. Flashing at the point where the chimney comes through the roof calls for something that will allow movement without damage to the water seal. It is necessary to use base flashings. They should be secured to the roof deck.

The counter or cap flashings are secured to the masonry. Figures 5-42 through 5-46 show how roll roofing is used for base flashing. Metal is used for cap flashing.

Apply shingles over the roofing felt up to the front face of the chimney. Do this before any flashings are placed. Make a saddle or cricket. See Fig. 5-42. This sits between the back face of the chimney and the roof deck. The cricket keeps snow and ice from piling up. It also deflects downflowing water around the chimney.

Apply a coat of asphalt primer to the brick work. This seals the surface. This is where plastic cement will later be applied. Cut the base flashing for the front. Cut according to the pattern shown in Fig. 5-43 (pattern A). This one is applied first. The lower section is laid over the shingles in a bed of plastic asphalt cement. Secure the upper vertical section against the ma-

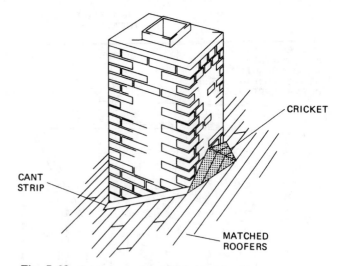

CRICKET

CANT STRIP

MATCHED ROOFERS

Fig. 5-42 *Cricket or saddle built behind the chimney.* (Bird and Son)

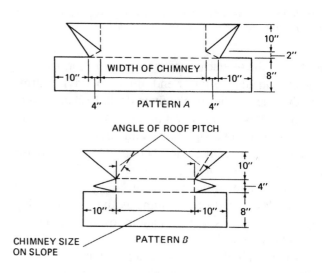

10″
2″
8″

WIDTH OF CHIMNEY

10″   10″

4″   PATTERN A   4″

ANGLE OF ROOF PITCH

10″
4″
8″

10″   10″

CHIMNEY SIZE ON SLOPE

PATTERN B

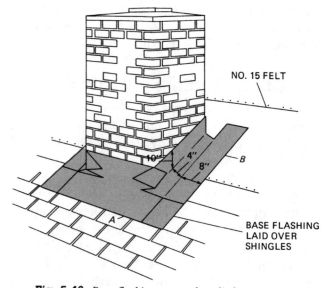

NO. 15 FELT

10″   4″   B
8″

A

BASE FLASHING LAID OVER SHINGLES

Fig. 5-43 *Base flashings cut and applied.* (Bird and Son)

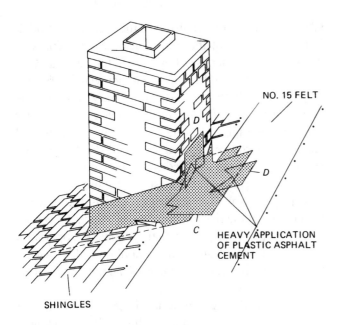

Fig. 5-44 *Flashing over the cricket in the rear of the chimney.*
(Bird and Son)

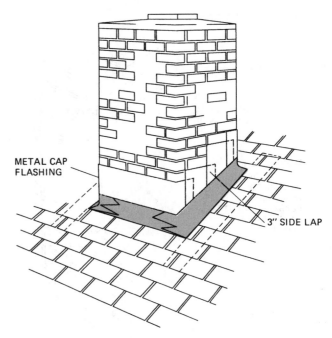

Fig. 5-46 *Metal cap flashing applied to cover the base flashing.*
(Bird and Son)

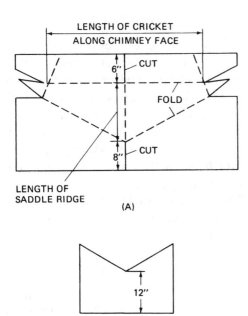

Fig. 5-45 *Flashing patterns.* (Bird and Son)

sonry with the same cement. Secure the upper vertical section against the masonry with the cement. Nails can also be used here, driven into the mortar joints. Bend the triangular ends of the upper section around the corners of the chimney. Cement in place.

Cut the side base flashings next. Use pattern *B* of Fig. 5-43. Bend them to shape and apply them as shown. Embed them in plastic asphalt cement. Turn the triangular ends of the upper section around the chimney corners. Cement them in place over the front base flashing.

Figure 5-44 shows the cutting and fitting of base flashings over the cricket. The cricket consists of two triangular pieces of board. The board is cut to form a ridge. The ridge extends from the centerline of the chimney back to the roof deck. The boards are nailed to the wood deck. They are also nailed to one another along the ridge. This is done before felt underlayment is applied. Cut the base flashing. See Fig. 5-45A. Bend it to cover the entire cricket. Extend it laterally to cover part of the side base flashing. Cut a second rectangular piece of roofing. See Fig. 5-45B. Make a cutout on one side to conform to the rear angle of the cricket. Set it tightly in plastic asphalt cement. Center it over that part of the cricket flashing extending up to the deck. This piece provides added protection where the ridge of the cricket meets the deck. Cut a second similar rectangular piece of flashing. Cut a V from one side. It should conform to the pitch of the cricket. Place it over the cricket ridge and against the chimney. Embed it in plastic asphalt cement.

Use plastic asphalt cement generously. Use it to cement all standing portions of the base flashing to the brickwork.

Cap flashings are shown in Fig. 5-46. They are made of sheet copper, 16-ounce or heavier. You can also make the caps of 24-gage galvanized steel. If steel is used, it should be painted on both sides.

Brickwork is secured to the cap flashing in Figs. 5-46 and 5-47. These drawings show a good method. Rake the mortar joint to a depth of 1½ inches. Insert the

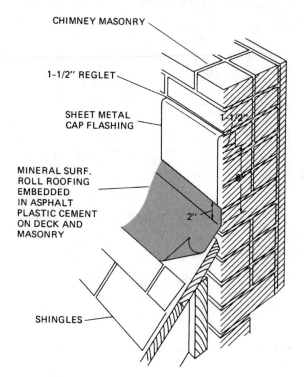

Fig. 5-47 *Method of securing cap flashing to the chimney.* (Bird and Son)

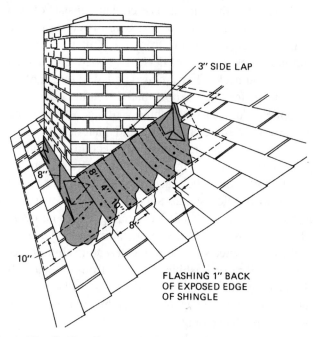

Fig. 5-48 *Alternative base flashing method.* (Bird and Son)

bent-back edge of the flashing into the cleared space between the bricks. It is under slight spring tension. It cannot easily be dislodged. Refill the joint with portland cement mortar. Or you can use plastic asphalt cement. This flashing is bent down to cover the base flashing. The cap lies snugly against the masonry.

The front unit of the cap flashing is one continuous piece. On the sides and rear, the sections are of similar

size. They are cut to conform to the locations of brick joints. The pitch of the roof is also needed. The side units lap each other. See Fig. 5-46. This lap is at least 3 inches. Figure 5-48 shows another way to flash a sloping roof abutting a vertical masonry wall. This is known as the *step flashing* method. Place a rectangular piece of material measuring 8 × 22 inches over the end tab of each course of shingles. Hold the lower edge slightly back of the exposed edge of the covering shingle. Bend it up against the masonry. Secure it with suitable plastic asphalt cement. Drive nails through the lower edge of the flashing into the roof deck. Cover the nails with plastic asphalt cement. Repeat the operation for each course. Flashing units should be wide enough to lap each other at least 3 inches. The upper one overlays the lower one each time.

Asphalt roofing can be used for step flashing. It simply replaces the base flashing already shown. Metal cap flashings must be applied in the usual manner. The metal cap completes a satisfactory job.

Soil Stacks

Most building roofs have pipes or ventilators through them. Most are circular in section. They call for special flashing methods. Asphalt products may be successfully used for this purpose. Figures 5-49 through 5-54 show a step-by-step method of flashing for soil pipes. A soil pipe is used as a vent for plumbing. It is made of cast iron or copper. The pipe gets its name from being buried in the soil as a sanitary sewer pipe.

An alternative procedure for soil stacks can be used. Obtain noncorrodible metal pipes. They should have adjustable flanges. These flanges can be applied as a flashing to fit any roof pitch.

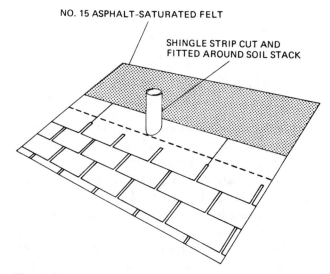

Fig. 5-49 *Roofing is first applied up to the soil pipe and fitted around it.* (Bird and Son)

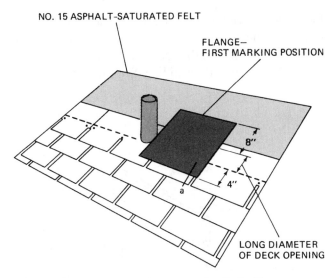

Fig. 5-50 *First step in marking an opening for flashing.* (Bird and Son)

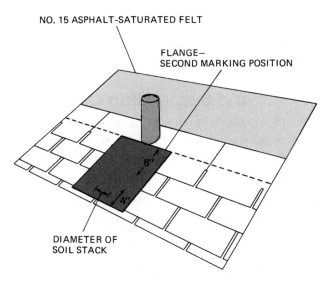

Fig. 5-51 *Second step in marking an opening for flashing.* (Bird and Son)

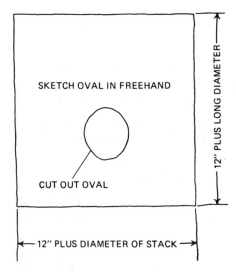

Fig. 5-52 *Cut oval in the flange.* (Bird and Son)

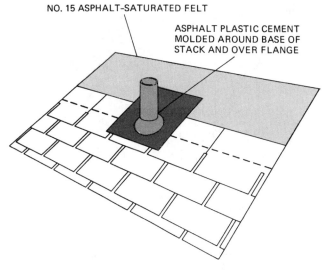

Fig. 5-53 *Cement the collar molded around the pipe.* (Bird and Son)

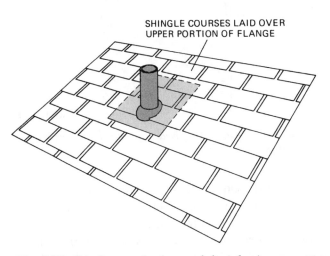

Fig. 5-54 *Shingling completed past and above the pipe.* (Bird and Son)

STRIP SHINGLES

Prepare the deck properly before starting to apply strip shingles. First place an underlayment down.

Deck Preparation

Metal drip edge   Use a metal drip edge, made of non-corrodible, nonstaining metal. Place it along the eaves and rakes. See Figs. 5-55 and 5-56. The drip edge is designed to allow water runoff to drip free into a gutter. It should extend back from the edge of the deck not more than 3 inches. Secure it with nails spaced 8 to 10 inches apart. Place the nails along the edge. Drip edges of other materials may be used. They should be of approved types.

Underlayment   Place a layer of no. 15 asphalt-saturated felt down. Cover the entire deck as shown. See Fig. 5-20.

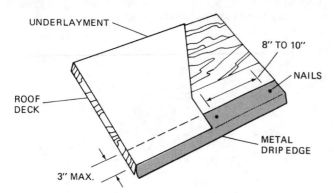

Fig. 5-55 *Application of the metal drip edge at eaves directly onto the deck.* (Bird and Son)

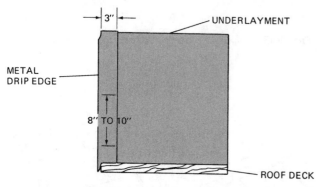

Fig. 5-56 *Application of the metal drip edge at the rakes over the underlayment.* (Bird and Son)

Chalk lines On small roofs, strip shingles may be laid from either rake. On roofs 30 feet or longer, it is better to start them at the center, then work both ways from a vertical line. This ensures better vertical alignment. It also provides for meeting and matching above a dormer or chimney. Chalk lines are used to control shingle alignment.

Eaves flashing Eaves flashing is required in cold climates. (January daily average temperatures of 25 degrees F or less call for eaves flashing.) In cold climates there is a possibility of ice forming along the eaves. If this happens, it causes trouble. Flashing should be used if there is doubt. Ice jams and water backup should be avoided. They can cause leakage into the ceiling below.

There are two flashing methods to prevent leakage. The methods depend on the slope of the roof. Possible severe icing conditions is another factor in choice of method.

Normal slope is 4 inches per foot or over. Install a course of 90-pound mineral-surfaced roll roofing. Or, apply a course of smooth roll roofing. It should not be less than 50 pounds. Install it to overhang the underlay and metal drip edge from ¼ to ⅜ inch. It should extend up to the roof. Cover a point at least 12 inches inside

the building's interior wall line. For a 36-inch eave overhang, the horizontal lap joint must be cemented. It should be located on the roof deck extending beyond the building's exterior wall line. See Fig. 5-57.

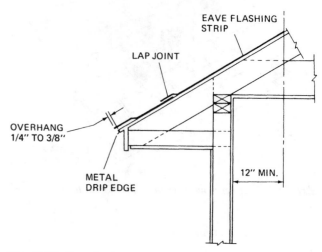

Fig. 5-57 *Eave flashing strip for normal-slope roof. This means 4 inches per foot or over.* (Bird and Son)

First and Succeeding Courses

Start the first course with a full shingle. Succeeding courses are started with full or cut strips. It depends upon the style of shingles being applied. There are three major variations for square-butt strip shingles:

1. Cutouts break at joints on the thirds. See Fig. 5-58.
2. Cutouts break at joints on the halves. See Fig. 5-59.
3. Random spacing. See Fig. 5-60.

Random spacing can be done by removing different amounts from the rake tab of succeeding courses. The amounts are removed according to the following scheme:

1. The width of any rake tab should be at least 3 inches.
2. Cutout centerlines should be located at least 3 inches laterally from other cutout centerlines. This means both the course above and the course below.
3. The rake tab widths should not repeat closely enough to cause the eye to follow a cutout alignment.

Ribbon Courses

Use a ribbon course to strengthen the horizontal roof lines. It adds a massive appearance that some people prefer. See Fig. 5-61.

One method involves special starting procedures. This is repeated every fifth course. Some people prefer this method.

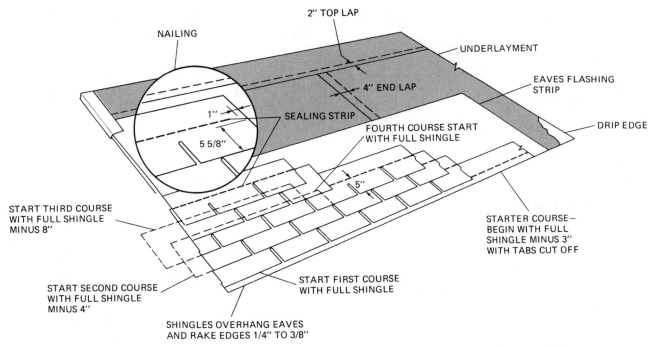

Fig. 5-58 *Applying three-tab square butt strips so that cutouts break the joints at thirds.* (Bird and Son)

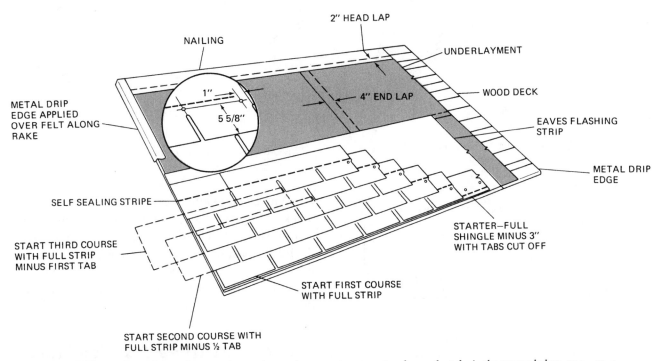

Fig. 5-59 *Applying three-tab square butt strips so that cutouts are centered over the tabs in the course below.* (Bird and Son)

1. Cut 4 inches off the top of a 12-inch-wide strip shingle. This will give you an unbroken strip 4 inches by 36 inches. You also get a strip 8 inches by 36 inches. Both strips contain the cutouts.

2. Lay the 4-×-36-inch strip along the eave.

3. Cover this with the 8-×-36-inch strip. The bottom of the cutouts is laid down to the eave.

4. Lay the first course of full (12-×-36-inch) shingles. It goes over layers *B* and *C*. The bottom of the cutouts is laid down to the eave. See Fig. 5-62.

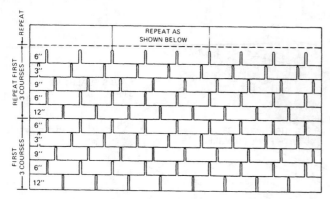

Fig. 5-60 *Random spacing of three-tab square butt strips.* (Bird and Son)

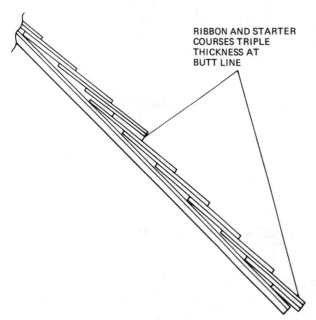

RIBBON AND STARTER
COURSES TRIPLE
THICKNESS AT
BUTT LINE

Fig. 5-61 *Ribbon courses. Side view.* (Bird and Son)

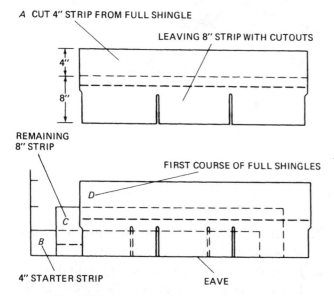

A CUT 4" STRIP FROM FULL SHINGLE

LEAVING 8" STRIP WITH CUTOUTS

4"

8"

REMAINING
8" STRIP

FIRST COURSE OF FULL SHINGLES

D

C

B

4" STARTER STRIP

EAVE

Fig. 5-62 *Laying the ribbon courses.* (Bird and Son)

Cutouts should be offset. This is done according to thirds, halves, or random spacing.

Wind Protection

High winds call for specially designed shingles. Cement the free tabs for protection against high winds. See Figs. 5-63 and 5-64.

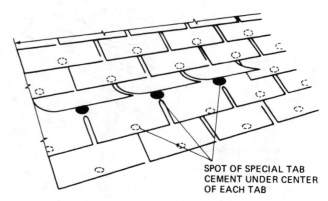

SPOT OF SPECIAL TAB
CEMENT UNDER CENTER
OF EACH TAB

Fig. 5-63 *Location of tab cement under square butt tabs.* (Bird and Son)

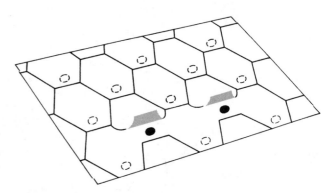

Fig. 5-64 *Location of tab cement under hex tabs.* (Bird and Son)

With a putty knife or caulking gun, apply a spot of quick-setting cement on the underlaying shingle. The cement should be about the size of a half-dollar. Press the free tab against the spot of cement. Do not squeeze the cement beyond the edge of the tab. Don't skip or miss any shingle tabs. Don't bend tabs back farther than needed.

Two- and Three-Tab Hex Strips

Nail two- and three-tab hex strips with four nails per strip. Locate the nails in a horizontal line 5¼ inches above the exposed butt edge.

Figure 5-65 shows how the two-tab strip is applied. Use one nail 1 inch back from each end of the strip. One nail is applied ¾ inch back from each angle of the cutouts.

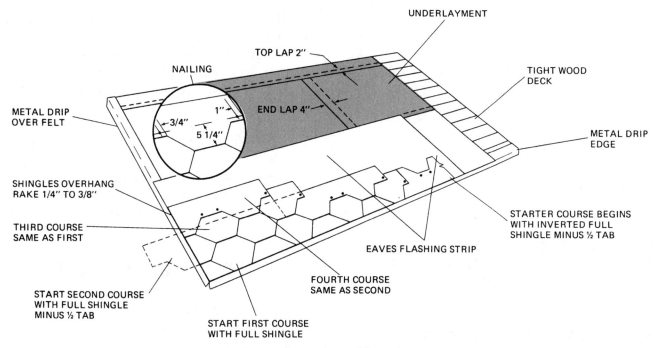

Fig. 5-65 *Application of two-tab hex strips.* (Bird and Son)

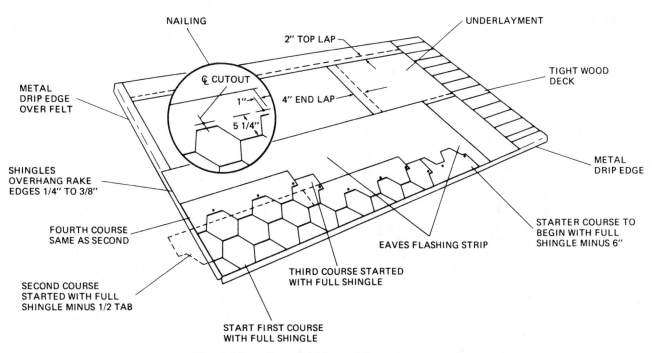

Fig. 5-66 *Application of three-tab hex strips.* (Bird and Son)

Three-tab shingles require one nail 1 inch back from each end. One nail is centered above each cutout. See Fig. 5-66.

Hips and Ridges

Use hip and ridge shingles to finish hips and ridges. They are furnished by shingle manufacturers. You can cut them from 12-×-36-inch square-butt strips. They should be at least 9 × 12 inches. One method of applying them is shown in Fig. 5-67.

Bend each shingle lengthwise down the center. This gives equal exposure on each side of the hip or ridge. Begin at the bottom of a hip. Or you can begin at one end of a ridge. Apply shingles over the hip or ridge. Expose them by 5 inches. Note the direction of the prevailing winds. This is important when you are placing ridge shingles. Secure each shingle with one

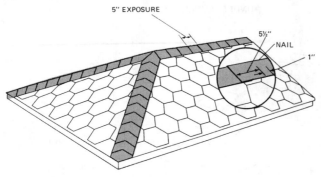

Fig. 5-67 *Hip-and-ridge shingles applied with hex strips.* (Bird and Son)

nail on each side. The nail should be 5½ inches back from the exposed end. It should be 1 inch up from the end. Never use metal ridge roll with asphalt roofing products. Corrosion may discolor the roof.

STEEP-SLOPE AND MANSARD ROOFS

New rooflines have caused changes recently. The mansard roofline requires some variations in shingle application. See Fig. 5-68.

Fig. 5-68 *Mansard roof.*

Excessive slopes give reduced results with factory self-sealing adhesives. This becomes obvious in colder or shaded areas.

Maximum slope for normal shingle application is 60° or 21 inches per foot.

Shingles on a steeper slope should be secured to the roof deck with roofing nails. See the maker's suggestions. Nails are placed 5⅝ inches above the butt. They should not be in or above the self-sealing strip.

Quick-setting asphalt adhesives should be applied as spots. The spots should be about the size of a quarter. They should be applied under each shingle tab. Do this immediately upon installation. Ventilation is needed to keep moisture-laden air from being trapped behind sheathing.

INTERLOCKING SHINGLES

Lockdown shingles are designed for windy areas. They have high resistance to high winds. They have an integral locking device. They can be classified into five groups. See Fig. 5-69.

These shingles generally do not require the use of adhesives. They may require a restricted use of cement. This is needed along the rakes and eaves. That is where the locking device may have to be removed.

Lock-type shingles can be used for both new and old buildings. Roof pitch is not too critical with these shingles. The designs can be classified into types 1, 2, 3, 4, and 5. Type 5 is a strip shingle with two tabs per strip.

Figure 5-70 shows how locking shingles work. Nail placement is suggested by the manufacturer. Nail placement is important for good results.

Figure 5-71 shows how the drip edge should be placed. The drip edge is designed to allow water runoff to drip free into gutters. The drip edge should not extend more than 3 inches back from the edge of the deck. Secure it with appropriate nails. Space the nails 8 to 10 inches apart along the inner edge.

Use the manufacturer's suggestions for starter course placement. Chalk lines will be very useful. Because it is very short, this type of shingle needs chalk lines for alignment.

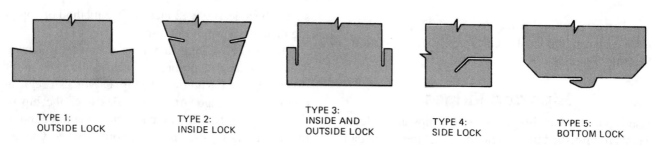

TYPE 1: OUTSIDE LOCK

TYPE 2: INSIDE LOCK

TYPE 3: INSIDE AND OUTSIDE LOCK

TYPE 4: SIDE LOCK

TYPE 5: BOTTOM LOCK

Fig. 5-69 *Locking devices used in interlocking shingles.* (Bird and Son)

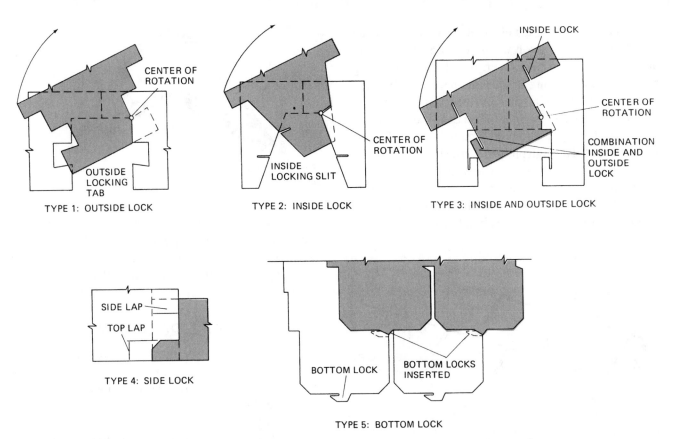

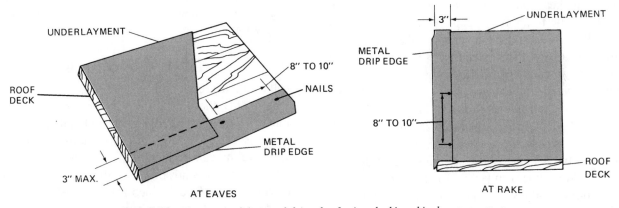

Fig. 5-70 *Methods of locking shingles types 1 through 5. In each case only the locking device is shown.* (Bird and Son)

Fig. 5-71 *Placement of the metal drip edge for interlocking shingles.* (Bird and Son)

Hips and Ridges

Hips and ridges require shingles made for that purpose. However, you can cut them from standard shingles. They should be 9 × 12 inches. Either 90-pound mineral-surfaced roofing or shingles can be used. See Fig. 5-72.

In cold weather, warm the shingles before you use them. This keeps them from cracking when they are bent to lock. Do not use metal hip or ridge materials. Metal may become corroded and discolor the roof.

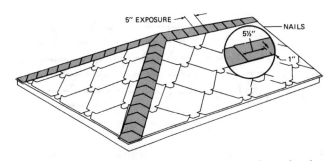

Fig. 5-72 *Hip-and-ridge application of interlocking shingles.* (Bird and Son)

Time is an important element in any job. You can speed up the roofing job when you don't have to cut and fit the hips and ridges with shingles you are using to cover the roof. You can speed things up and become more efficient by using pre-cut fiberglass high-profile hip and ridge caps. High profile means the cap sticks up higher than usual and gives a more "finished" quality to the roof. The high profile adds texture and shadow line to improve the appearance and enhance the beauty of any roof. They come in boxes with 48 pieces that cover 30 lineal feet of finished hip or ridge when applied with 8 inches to the weather. See Fig. 5-73.

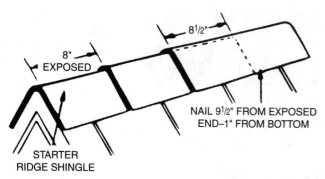

Fig. 5-73 *Back nail each shingle with two nails 9.5 inches from the exposed end.* (Ridglass)

The ridge and hip caps are back nailed with two nails 9.5 inches from the exposed end. See Fig. 5-74. Then continue to apply the caps with a 5-inch exposure. All nails should be covered by the succeeding shingle by at least 1 inch. See Fig. 5-75. In high-wind areas, seal down each installed ridge unit with elastomer adhesive or face nail each shingle with two nails, one on each side as shown in Fig. 5-75. In general, use 11- or 12-gage galvanized roofing nails with 7/16-inch heads that are long enough to penetrate the roof deck by ¾ inch. When applying the Ridglass® caps in cold weather (under 50° F), unpack the carton on the roof and allow the shingles to warm up before application.

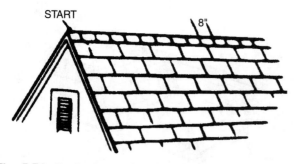

Fig. 5-74 *Continue to apply Ridglass with 9-inch exposure. All nails should be covered by the succeeding shingle by at least 1 inch.* (Ridglass)

Fig. 5-75 *Seal down each installed ridge unit with elastomeric adhesive and face nail each shingle with two nails.* (Ridglass)

These caps are made of fiberglass, not paper. They are SBS modified asphalt with no granule loss. There is no cutting, folding, or fabrication on the roof and matching colors are available for all manufacturers' shingles.

ROLL ROOFING

Roofing in rolls in some instances is very economical. Farm buildings and sheds are usually covered with inexpensive roll roofing. It is easier to apply and cheaper than strip shingles.

Do not apply roll roofing when the temperature is below 45 degrees F. If it is necessary to handle the material at lower temperatures, warm it before unrolling it. Warming avoids cracking the coating.

The sheet should be cut into 12-foot and 18-foot lengths. Spread them in a pile on a smooth surface until they flatten.

Windy Locations

Roll roofings are recommended for use in windy locations. Apply them according to the maker's suggestions. Use the pattern edge and blind nail. This means the nails can't be seen after the roofing is applied. Blind nailing can be used with 18-inch-wide mineral-surfaced or 65-pound smooth roofing. This can be used in windy areas. The 19-inch selvage double-coverage roll roofing is also suited for windy places.

Use concealed nailing rather than exposed nailing to apply the roll roofing. This ensures maximum life in service.

Use only lap cement or quick-setting cement. It should be cement the maker of the roofing suggests. Cements should be stored in a warm place until you are ready to use them. Place the unopened container in hot water to warm. Never heat asphalt cements directly over a fire. Use 11- or 12-gage hot-dipped galvanized nails. Nails should have large heads. This means at least ⅜-inch-diameter heads. The shanks should be ⅞ to 1 inch long. Use nails long enough to penetrate the wood below.

Exposed Nails—Parallel to the Rake

Exposed nailing, parallel to the eaves, is shown in Fig. 5-76. Figure 5-77 also shows the exposed nail method. It is parallel to the rake in this case. The overhang is ¼ to ⅜ inch over the rake. End laps are 6 inches wide and cemented down. Stagger the nails in rows, 1 inch apart. Space the nails on 4-inch centers in each row. Stagger all end laps. Do not have an end lap in one course over or adjacent to an end lap in the preceding course.

Hips and Ridges

For the method used to place a cap over the hips and ridge, see Fig. 5-78. Butt and nail sheets of roofing as they come up on either side of a hip or ridge. Cut strips of roll roofing 12 inches wide. Bend them lengthwise through their centers. Snap a chalk line guide parallel to the hip or ridge. It should be 5½ inches down on each side of the deck.

Cement a 2-inch-wide band on each side of the hip or ridge. The lower edge should be even with the chalk line. Lay the bent strip over the hip or ridge. Embed it in asphalt lap cement.

Secure the strip with two rows of nails. One row is placed on each side of the hip or ridge. The rows should be ¾ inch above the edges of the strip. Nails are spaced on 2-inch centers. Be sure the nails penetrate the cemented portion. This seals the nail hole with some of the asphalt.

WOOD SHINGLES

Wood shingles are the oldest method of shingling. In early U.S. history, pine and other trees were used for

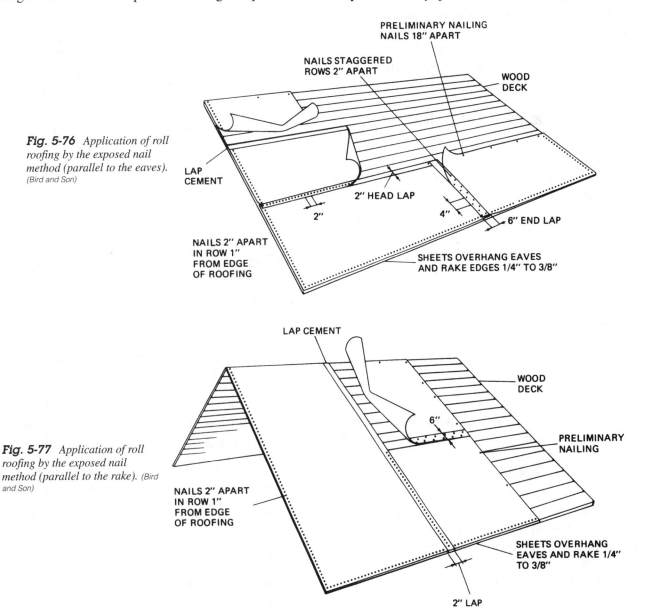

Fig. 5-76 *Application of roll roofing by the exposed nail method (parallel to the eaves).* (Bird and Son)

Fig. 5-77 *Application of roll roofing by the exposed nail method (parallel to the rake).* (Bird and Son)

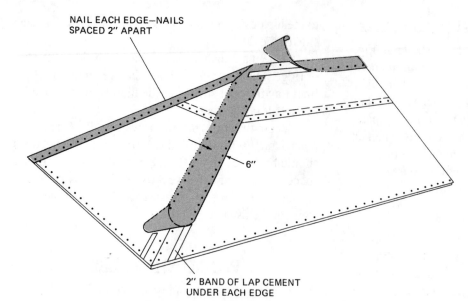

NAIL EACH EDGE—NAILS SPACED 2" APART

6"

2" BAND OF LAP CEMENT UNDER EACH EDGE

Fig. 5-78 *Hip-and-ridge application of roll roofing.* (Bird and Son)

shingles. Then the western United States discovered the cedar shingle. It is resistant to water and rot. If properly cared for, it will last at least 50 years. Application of this type of roofing material calls for some different methods.

Sizing Up the Job

You need to know a few things before ordering these shingles. First, find the pitch of your roof. See Fig. 5-79. Simply measure how many inches it rises for every foot it runs.

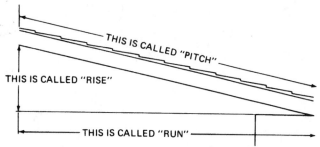

THIS IS CALLED "PITCH"

THIS IS CALLED "RISE"

THIS IS CALLED "RUN"

Fig. 5-79 *Figuring the pitch of a roof.* (Red Cedar Shingle & Handsplit Shake Bureau)

Remember that a square contains four bundles. It will cover 100 square feet of roof area. See Fig. 5-80.

Roof Exposure

Exposure refers to the area of the shingle that contacts the weather. See Fig. 5-81. Exposure depends upon roof pitch. A good shingle job is never less than three layers thick. See Table 5-6 for important information about shingles.

There are three lengths of shingles: 16 inches, 18 inches, and 24 inches.

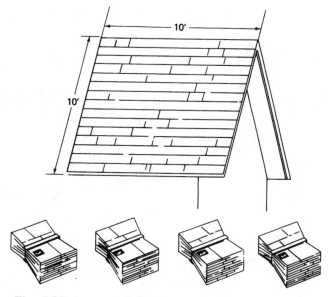

10'

10'

Fig. 5-80 *A square of shingle contains four bundles.* (Red Cedar Shingle & Handsplit Shake Bureau)

If the roof pitch is 4 inches in 12 inches or steeper (*three-ply roof*):

• For 16-inch shingles, allow a 5-inch exposure.
• For 18-inch shingles, allow a 5½-inch exposure.
• For 24-inch shingles, allow a 7½-inch exposure.

If the roof pitch is less than 4 inches in 12 inches but not below 3 inches in 12 inches (*four-ply roof*):

• For 16-inch shingles, allow a 3¾-inch exposure.
• For 18-inch shingles, allow a 4¼-inch exposure.
• For 24-inch shingles, allow a 5¾-inch exposure.

If the roof pitch is less than 3 inches in 12 inches, cedar shingles are not recommended. These exposures

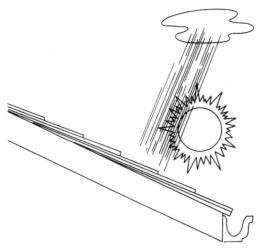

Fig. 5-81 *Shingle exposure to the weather. (Red Cedar Shingle & Hand-split Shake Bureau)*

are for no. 1 grade shingles. In applying no. 3 shingles, make sure you check with the manufacturer.

Estimating Shingles Needed

Determine the ground area of your house. Include the eaves and cornice overhang. Do this in square feet. If the roof pitch found previously:

- Rises 3 in 12, add 3 percent to the square foot total.
- Rises 4 in 12, add 5½ percent to the square foot total.
- Rises 5 in 12, add 8½ percent to the square foot total.
- Rises 6 in 12, add 12 percent to the square foot total.
- Rises 8 in 12, add 20 percent to the square foot total.
- Rises 12 in 12, add 42 percent to the square foot total.

Table 5-6 *Summary of Sizes, Packing, and Coverage of Wood Shingles*

Shake Type, Length, and Thickness, Inches	No. of Courses per Bundle	No. of Bundles per Square	Approximate Coverage (in Square Feet) of One Square, When Shakes are Applied with ½-Inch Spacing, at Following Weather Exposures (in Inches):								
			5½	6½	7	7½	8½	10	11½	14	16
18 × ½ medium resawn	9/9[a]	5[b]	55[c]	65	70	75[d]	85[e]	100[f]			
18 × ¾ heavy resawn	9/9[a]	5[b]	55[c]	65	70	75[d]	85[e]	100[f]			
24 × ⅜ handsplit	9/9[a]	5		65	70	75[g]	85	100[h]	115[i]		
24 × ½ medium resawn	9/9[a]	5		65	70	75[c]	85	100[j]	115[i]		
24 × ¾ heavy resawn	9/9[a]	5		65	70	75[c]	85	100[j]	115[i]		
24 × ½ to ⅝ tapersplit	9/9[a]	5		65	70	75[c]	85	100[j]	115[i]		
18 × ⅜ true-edge straight-split	14[k] straight	4								100	112[l]
18 × ⅜ straight-split	19[k] straight	5	65[c]	75	80	90[j]	100[i]				
24 × ⅜ straight-split	16[k] straight	5		65	70	75[c]	85	100[j]	115[i]		
15 starter-finish course	9/9[a]	5	Use supplementary with shakes applied with not over 10-inch weather exposure.								

[a]Packed in 18-inch-wide frames.
[b]Five bundles will cover 100 square feet of roof area when used as starter-finish course at 10-inch weather exposure; six bundles will cover 100 square feet wall area when used at 8½-inch weather exposure; seven bundles will cover 100 square feet roof area when used at 7½-inch weather exposure. [m]
[c]Maximum recommended weather exposure for three-ply roof construction.
[d]Maximum recommended weather exposure for two-ply roof construction; seven bundles will cover 100 square feet of roof area when applied at 7½-inch weather exposure. [m]
[e]Maximum recommended weather exposure for sidewall construction; six bundles will cover 100 square feet when applied at 8½-inch weather exposure. [m]
[f]Maximum recommended weather exposure for starter-finish course application; five bundles will cover 100 square feet when applied at 10-inch weather exposure. [m]
[g]Maximum recommended weather exposure for application on roof pitches between 4 in 12 in 8 in 12.
[h]Maximum recommended weather exposure for application on roof pitches of 8 in 12 and steeper.
[i]Maximum recommended weather exposure for single-coursed wall construction.
[j]Maximum recommended weather exposure for two-ply roof construction.
[k]Packed in 20-inch-wide frames.
[l]Maximum recommended weather exposure for double-coursed wall construction.
[m]All coverage based on ½-inch spacing between shakes.

- Rises 15 in 12, add 60 percent to the square foot total.
- Rises 18 in 12, add 80 percent to the square foot total.

Divide the number you have found by 100. The answer is the number of shingle "squares" you should order to cover your roof if the pitch is 4 inches in 12 inches or steeper. If the roof is of lesser pitch, allow one-third more shingles to compensate for reduced exposure.

Also, add 1 square for every 100 linear feet of hips and valleys.

Tools of the Trade

A shingler's hatchet speeds the work. See Fig. 5-82. Sneakers or similar traction shoes make the job safer. A straight board keeps your rows straight and true.

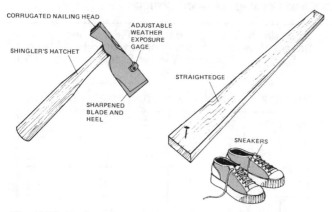

Fig. 5-82 Tools of the trade. (Red Cedar Shingle & Handsplit Shake Bureau)

APPLYING THE SHINGLE ROOF

Begin with a double thickness of shingles at the bottom edge of the roof. See Fig. 5-83. Let the shingles protrude over the edge to assure proper spillage into the eaves-trough or gutter. See *A* in Fig. 5-83.

Figure 5-84 shows how the nails are placed so that the next row above will cover the nails by not more than 1 inch. Use the board as shown in Fig. 5-85. Use

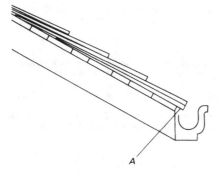

Fig. 5-83 Applying a double thickness of shingles at the bottom edge of the roof. (Red Cedar Shingle & Handsplit Shake Bureau)

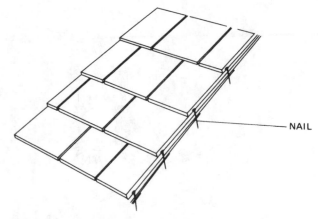

Fig. 5-84 Covering nails in the previous course. (Red Cedar Shingle & Handsplit Shake Bureau)

it as a straightedge to line up rows of shingles. Tack the board temporarily in place as a guide. It makes the work faster and the results look professional.

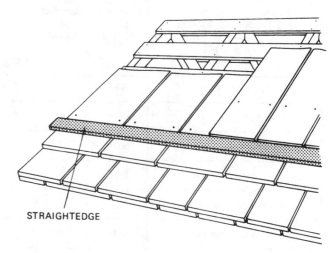

Fig. 5-85 Using a straightedge to keep the ends lined up. (Red Cedar Shingle & Handsplit Shake Bureau)

In Fig. 5-86 you can see the location of the nails. They should be placed no farther than ¾ inch from the edge of the shingle.

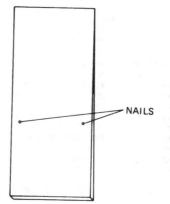

Fig. 5-86 Placement of nails in a single shingle. (Red Cedar Shingle & Handsplit Shake Bureau)

Figure 5-87 shows how the shingles are spaced ¼ inch apart to allow for expansion. Other simple rules are also shown in Fig. 5-88.

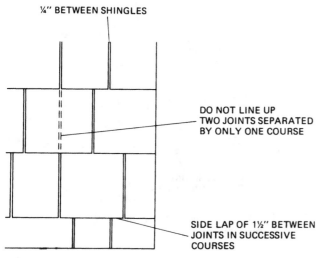

Fig. 5-87 *Spacing of shingle between courses.* (Red Cedar Shingle & Handsplit Shake Bureau)

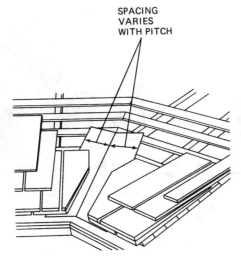

Fig. 5-88 *Spacing for valleys.* (Red Cedar Shingle & Handsplit Shake)

Valleys and Flashings

Extend the valley sheets beneath shingles. They should extend 10 inches on either side of the valley center. This is the case if the roof pitch is less than 12 inches in 12 inches. For steeper roofs, the valley sheets should extend at least 7 inches. See Fig. 5-88.

Most roof leaks occur at points where water is channeled for running off the roof. Or they occur where the roof abuts a vertical wall or chimney. At these points, use pointed metal valleys and flashings to assist the shingles in keeping the roof sound and dry. Suppliers will provide further information on which of

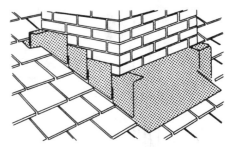

Fig. 5-89 *Flashing around a chimney.* (Red Cedar Shingle & Handsplit Shake Bureau)

the various metals to use. Figure 5-89 shows the flashing installed around a chimney.

Shingling at Roof Junctures

Apply the final course of shingles at the top of the wall. Install metal flashing (26-gage galvanized iron, 8 inches wide). Cover the top 4 inches of the roof slope. Bend the flashing carefully; avoid fracturing or breaking it. Make sure the flashing covers the nails that hold the final course. Apply a double starter course at the eave. Allow for a 1½-inch overhang of the wall surface. Complete the roof in the normal manner. See Fig. 5-90 for the convex juncture.

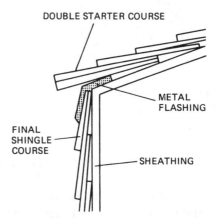

Fig. 5-90 *Convex roof juncture shingling.* (Red Cedar Shingle & Handsplit Shake Bureau)

For the concave juncture, apply the final course of shingles as shown in Fig. 5-91. Install the metal flashing to cover the last 4 inches of roof slope and bottom 4 inches of wall surface. Make sure the flashing covers the nails that hold the final course. Apply a double starter course at the bottom of the wall surface. Complete the shingling in the normal manner.

Before applying the final course of shingles, install 12-inch-wide flashing. This is to cover the top 8 inches of roof. Bend the remaining 4 inches to cover the top portion of the wall. See Fig. 5-92 for the treatment of apex junctures. Complete the roof shingling to

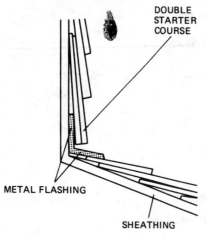

Fig. 5-91 Concave roof junction shingling. (Red Cedar Shingle & Handsplit Shake Bureau)

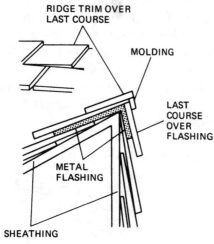

Fig. 5-92 Apex roof juncture shingling. (Red Cedar Shingle & Handsplit Shake Bureau)

cover the flashing. Allow the shingle tips to extend beyond the juncture. Complete the wall shingling. Trim the last courses to fit snugly under the protecting roof shingles. Apply a molding strip to cover the topmost portion of the wall. Trim the roof shingles even with the outer surface of the molding. Apply a conventional shingle "ridge" across the top edge of the roof. This is done in a single strip without matching pairs.

Applying Shingles to Hips and Ridges

The alternative overlap-type hip and ridge can be built by selecting uniform-width shingles. Lace them as shown in Fig. 5-93.

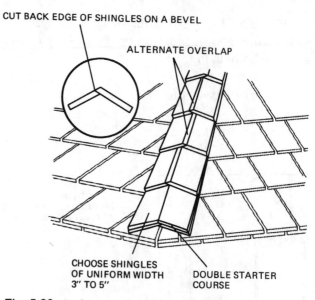

Fig. 5-93 Applying overlap-type roofing ridges. (Red Cedar Shingle & Handsplit Shake Bureau)

Fig. 5-94 Applying a factory-assembled hip or ridge unit. (Red Cedar Shingle & Handsplit Shake Bureau)

Time can be saved if factory-assembled hip-and-ridge units are used. These are shown in Fig. 5-94.

Nails for Wooden Shingles

Rust-resistant nails are very important. Zinc-coated or aluminum nails can be used. Don't skimp on nail quality. (See Fig. 5-95.)

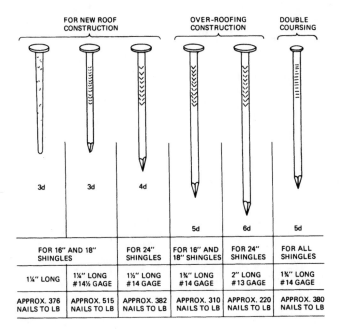

	FOR NEW ROOF CONSTRUCTION			OVER-ROOFING CONSTRUCTION		DOUBLE COURSING
	3d	3d	4d	5d	6d	5d
	FOR 16" AND 18" SHINGLES		FOR 24" SHINGLES	FOR 16" AND 18" SHINGLES	FOR 24" SHINGLES	FOR ALL SHINGLES
	1¼" LONG	1¼" LONG #14½ GAGE	1½" LONG #14 GAGE	1¾" LONG #14 GAGE	2" LONG #13 GAGE	1¾" LONG #14 GAGE
	APPROX. 376 NAILS TO LB	APPROX. 515 NAILS TO LB	APPROX. 382 NAILS TO LB	APPROX. 310 NAILS TO LB	APPROX. 220 NAILS TO LB	APPROX. 380 NAILS TO LB

Fig. 5-95 *Nails used for wood shingles*

6
CHAPTER

Installing Windows & Doors

WINDOWS AND DOORS PLAY AN important part in any type of house or building. They allow the air to circulate. They also allow passage in and out of the structure. Doors open to allow traffic in a planned manner. Windows are closed or open in design. There may be open and closed combinations, too. The design of a window or door is dictated by the building's use.

Buildings like those in Fig. 6-1 use the window as part of the design. The shape of a window may add to or detract from the design. The designer must be able to determine which is the right window for a building. The designer must also be able to choose a door that is architecturally compatible. Doors and windows come in many designs. However, they are limited in their function. This means some standards are set for the design of both. Most residential doors, for instance, are 6 feet 8 inches in height. If you are taller than that, you have to duck to pass through. Since most people are shorter, it is a safe height to use for doors.

Windows should be placed so that they will allow some view from either a standing or a sitting position. In most instances the window top and the door top are even. This way they look better from the outside.

In this unit you will learn how the carpenter installs windows and doors. You will learn how windows and doors are used to enhance a design. You will learn the various sizes and shapes of windows. You will learn the different sizes and hinging arrangements of doors. Locks will be presented so that you can learn how to install them.

Details for the placement of windows and doors are given. Things you will learn to do are:

- Prepare the window for installation
- Shim the window if necessary
- Secure the window in its opening
- Level and check for proper operation of the window
- Prepare a door for installation
- Shim the door if necessary
- Secure the door in its opening
- Level and check for proper operation

BASIC SEQUENCE

The carpenter should install a window in this order:

1. Check for proper window opening.
2. Uncrate the prehung window.
3. Remove the braces, if called for by the manufacturer.
4. Place builder's paper or felt between window and sheathing.
5. Place the window in the opening and check for level and plumb.
6. Attach the window at the corner or place one nail in the casing or flange (depending on the window design).
7. Check for proper operation of the window.
8. Secure the window in its opening.

The carpenter should install a door in this order:

1. Check the opening for correct measurements.
2. Uncrate the prehung door. If the door is not prehung, place the molding and trim in place first. Attach the door hinges by cutting the gains and screwing in the hinges.
3. Check for plumb and level.
4. Temporarily secure the door with shims and nails.
5. Check for proper operation.
6. Secure the door permanently.
7. Install the lock and its associated hardware.

Fig. 6-1 *Windows can add to or detract from a piece of architecture.* (Western Wood Product)

TYPES OF WINDOWS

There are many types of windows. Each hinges or swings in a different direction. They may be classified as:

1. Horizontal sliding windows
2. Awning picture windows
3. Double-hung windows
4. Casement windows

Most windows are made in factories today, ready to be placed into a rough opening when they arrive at the site. There are standards for windows. For instance, Commercial Standard 190 is shown in Fig. 6-2. The window requires a number of features, which are pointed out in the drawing. It must be weatherstripped to prevent air infiltration in excess of 0.75 cubic feet per minute per perimeter foot. This is under a 25-mph wind.

No more than two species of wood can be used in a unit. The wood used is ponderosa pine or a similar type of pine. Spruce, cedar, redwood, and cypress may also be used in the window frame, sill, and sash.

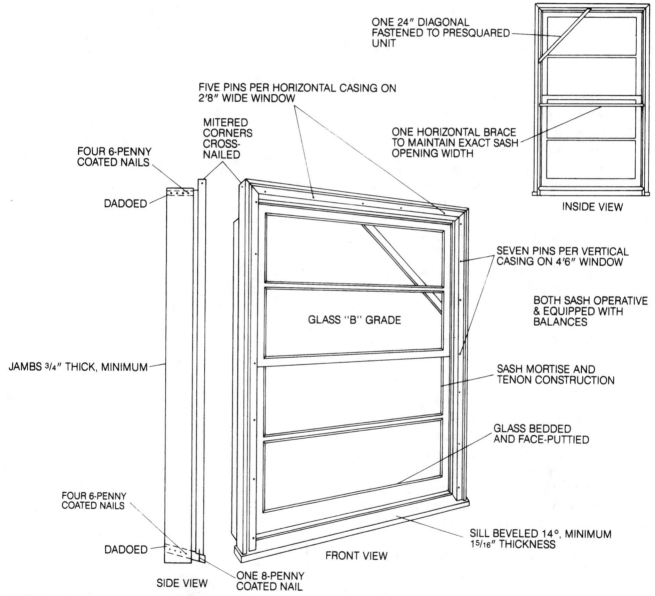

1. Weather-stripped to prevent air infiltration in excess of 0.75 cubic feet per minute per perimeter foot, under 25 MPH wind pressure.
2. No more than two species of wood per unit.
3. Chemically treated in accordance with NWMA minimum standards.
4. Finger-jointing permitted. See Commercial Standard 190, Par. 3.1.7.
5. Sash manufactured under Commercial Standard 163.
6. Ease of operation. See Commercial Standard 190. Par. 3.1.5.

Fig. 6-2 *Standard features of a window required to meet Commercial Standard 190.* (C. Arnold & Sons)

The wood has to be chemically treated in accordance with NWMA (National Window Manufacturers Association) minimum standards. Finger jointing is permitted. The sash has to be manufactured under Commercial Standard 163. Ease of operation is also spelled out in the written standard. Note how braces are specified to hold the window square and equally distant at all points. These braces are removed once the window has been set in place.

Horizontal sliding windows This type of window is fitted with a vinyl coating. The wood is not exposed at any point, which means less maintenance in the way of painting or glazing. The window is trimmed in vinyl so that it can be nailed into place on the framing around the rough opening. Figure 6-3 shows the window and details of its operation.

Figure 6-4 shows how the rough opening is made for a window. In this case the framing is on a 24-inch O.C. spacing. The large timber over the opening has to be large enough to support the roof without a stud where the window is placed. This prevents the window from buckling, which would stop the window from sliding. The cripple studs under the window opening are continuations of the studs that would be there normally. They are placed there to properly support the opening and to remove any weight from the window frame.

Sliding windows are available in a number of sizes. Figure 6-5 shows the possibilities. To find the overall unit dimension for a window which has a nonsupporting mullion, add the sum of the unit dimensions and subtract 2 inches. The mullion is the vertical bar between windows in a frame which holds two or more windows. The overall rough-opening dimension is equal to ¾ inch less than the overall unit dimension.

Double-hung window This window gets its name from the two windows that slide past one another. In this case they slide vertically. See Fig. 6-6. This is the most common type used today. The window shown is coated with plastic (vinyl) and can be easily attached to a stud through holes already drilled into the plastic around the frame. This plastic is called the flashing or the flange.

The double-hung window can be installed rather easily. See Fig. 6-7. The distance between the side jambs is checked to make sure they are even. Once the window is in the opening, place a shim where necessary. See Fig. 6-8. Note the placement of the nails here. Notice that this window does not have a vinyl flange around it. That is why the nails are placed as shown in Fig. 6-8. Figure 6-9 shows how shims are placed under the raised jamb legs and at the center of the long sills of a double window. Figure 6-10 shows a sash out of alignment. See the arrow. The sashes will not be parallel if the unit is out of square.

(A)

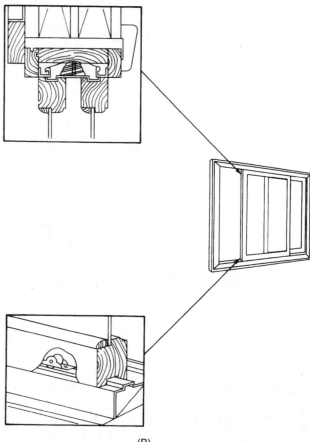
(B)

Fig. 6-3 *(A) Horizontal sliding window; (B) Details.* (Andersen)

WINDOW ON MODULE
6' WALL SECTION
STUDS 24 L.F.
JACKS 14 L.F.
CRIPPLES 3 L.F.
41 L.F.
24"
24"
24"
WINDOW OFF MODULE
6' WALL SECTION
STUDS 32 L.F.
JACKS 14 L.F.
CRIPPLES 6 L.F.
52 L.F.
(23% MORE FRAMING REQUIRED)
24"
24"
24"
L.F. = LINEAR FEET

Fig. 6-4 *Window openings in a house frame. Window on module and off module with 24" centers.* (American Plywood Association)

Once the window is in place and properly seated, you can pack insulation between the jambs and the trimmer studs. See Fig. 6-11. Figure 6-12 shows how 1¾-inch galvanized nails are placed through the vinyl

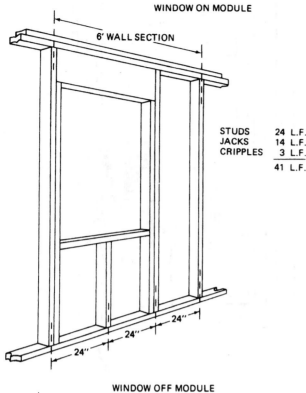

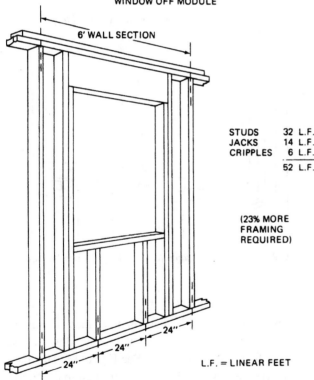

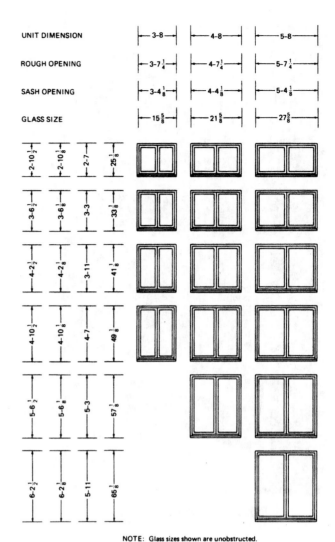

UNIT DIMENSION 3-8 4-8 5-8
ROUGH OPENING 3-7 1/4 4-7 1/4 5-7 1/4
SASH OPENING 3-4 1/8 4-4 1/8 5-4 1/8
GLASS SIZE 15 5/8 21 5/8 27 5/8
2-10 1/2 2-10 1/8 2-7 25 1/8
3-6 1/2 3-6 1/8 3-3 33 1/8
4-2 1/2 4-2 1/8 3-11 41 1/8
4-10 1/2 4-10 1/8 4-7 49 1/8
5-6 1/2 5-6 1/8 5-3 57 1/8
6-2 1/2 6-2 1/8 5-11 65 1/8

NOTE: Glass sizes shown are unobstructed.

Fig. 6-5 *Sliding window sizes.* (Andersen)

anchoring flange. This flange is then covered by the outside wall covering. The nails are not exposed to the weather.

Double-hung windows can be bought in a number of sizes. Figure 6-13 shows some of the sizes. In Fig. 6-14 the windows are all in place. The upperstory windows will be butted by siding. The downstairs windows are sitting back inside the brick. The upper windows will have part of the trim sticking out past the exterior siding.

Once the windows are in place and the house completed, the last step is to put the window dividers in place. They snap into the small holes in the sides of the window. The design may vary. See Fig. 6-15. The plastic grill patterns can be changed to meet the needs of the architectural style of the house. They can be removed by snapping them out. This way it is easier to clean the window pane. Figure 6-16 shows a house with the diamond light pattern installed in windows that swing out.

Fig. 6-6 *Double-hung window.* (Andersen)

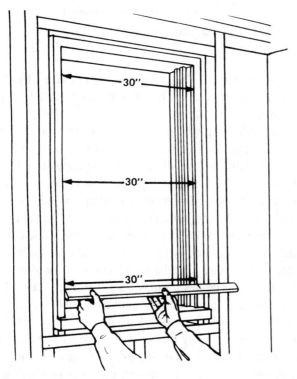

Fig. 6-7 *Measure the distance in at least three places to make sure the window is square.* (Andersen)

Fig. 6-8 *Shim the window where necessary. Side jambs are nailed through the shims.* (Andersen)

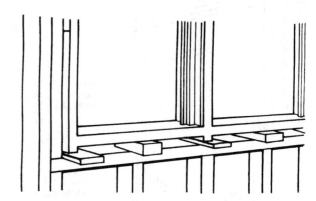

Fig. 6-9 *Shims raise the jamb legs and keep the window square.* (Andersen)

Casement window The casement window is hinged so that it swings outward. The whole window opens, allowing for more ventilation. See Fig. 6-17. This particular type has a vinyl flange for nailing it to the frame opening. It can be used as a single or in groups. This type of window can more easily be made weatherproof if it opens outward instead of inward.

Plastic muntins can be added to give a varied effect. They can be put into the windows as shown in Fig. 6-16. The diamond light muntins divide the glass space into small diamonds which resemble individual panes of glass.

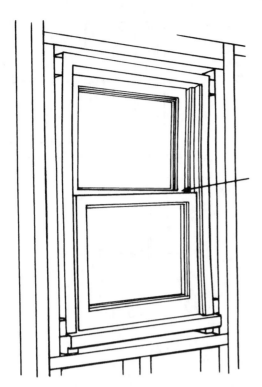

Fig. 6-10 *If the unit is not square, the sash rails will not be parallel. This can be spotted by eye.* (Andersen)

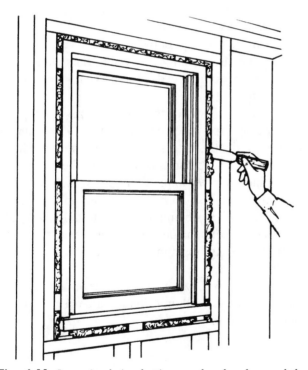

Fig. 6-11 *Loose insulation batting can be placed around the window to prevent drafts.* (Andersen)

The crank is installed so that the window opens outward with a twist of the handle. Figure 6-18 shows some of the ways this type of window may be operated.

Screens are mounted on the inside. Storm windows are mounted on the outside as in Fig. 6-19. In most

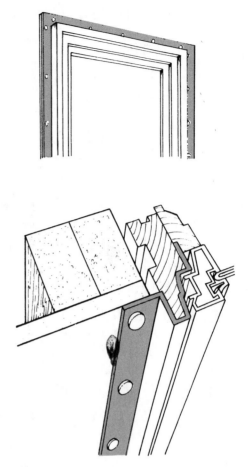

Fig. 6-12 *Installing the plastic flange around a prehung window with 1½-inch galvanized nails.* (Andersen)

cases, however, there is a thermopane used for insulation purposes. The thermopane is a double sheet of glass welded together with an air space between. This is then set into the sash and mounted as one piece of glass. See Fig. 6-20.

Multiple units are available. They may be movable or stationary. They may have one stationary part in the middle and two movable parts on the ends. Various combinations are available, as shown in Fig. 6-21.

Awning picture window This type of window has a large glass area. It also has a bottom panel which swings outward. A crank operates the bottom section. As it swings out, it has a tendency to form an awning effect—thus the name for this type. See Fig. 6-22. A number of sizes and combinations are available in this type of window. See Fig. 6-23. The fixed sash with an awning sash is also available in multiple units. Glass sizes are given in Fig. 6-23. If you need to find the overall basic unit dimension, add the basic unit to 2 ⅞ inch. The rough opening is the sum of the basic units plus ½ inch.

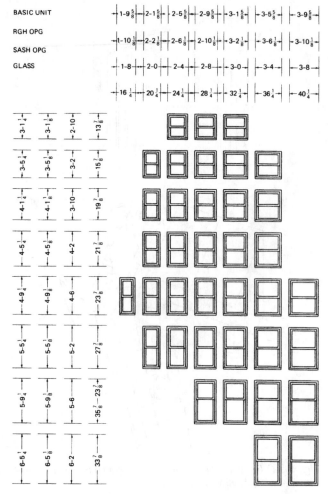

BASIC UNIT	1-9⅝	2-1⅝	2-5⅝	2-9⅝	3-1⅝	3-5⅝	3-9⅝
RGH OPG	1-10⅛	2-2⅛	2-6⅛	2-10⅛	3-2⅛	3-6⅛	3-10⅛
SASH OPG							
GLASS	1-8	2-0	2-4	2-8	3-0	3-4	3-8
	16¼	20¼	24¼	28¼	32¼	36¼	40¼

Fig. 6-13 *Various sizes of double-hung windows.* (Andersen)

Fig. 6-14 *Double-hung windows in masonry (bottom floor) and set for conventional sliding (top floor).*

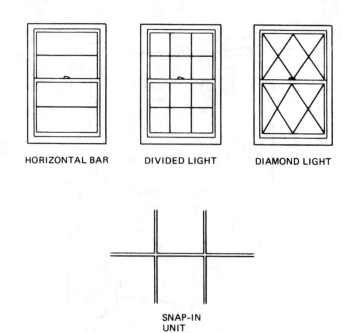

HORIZONTAL BAR   DIVIDED LIGHT   DIAMOND LIGHT

SNAP-IN UNIT

Fig. 6-15 *Plastic dividers make possible different windowpane treatments.* (Andersen)

Figure 6-24 shows the inswinging hopper type and the outswinging awning type of casement window. The awning sash type, shown in Fig. 6-22, has a bottom sash that swings outward. You have to specify which type of opening you want when you order. Specifying bottom-hinged or top-hinged is the quickest way to order.

Figure 6-25 shows the various sizes of hopper- and awning-type casements available. They can be stacked vertically. If this is done, the overall unit dimension for stacked units is the sum of the basic units plus ¾ inch for two units high, plus ¼ inch for three units high, and less 1¼ inches for four units high. To find the overall basic unit width of multiple units, add the basic unit dimensions plus 2⅞ inches to the total. To find the rough opening width, add the basic unit width plus ½ inch.

Figure 6-26 shows how a number of units may be stacked vertically. All units in this case open for maximum ventilation.

PREPARING THE ROUGH OPENING FOR A WINDOW

It is important that you consult the window manufacturer's specifications before you make the rough opening.

Fig. 6-16 *Diamond light pattern installed with plastic dividers.*

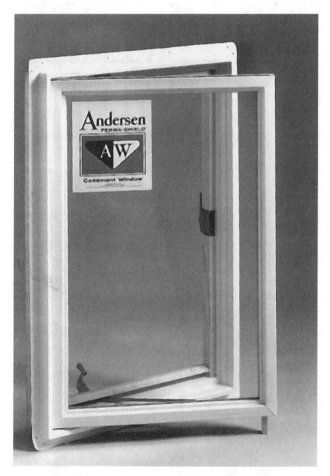

Fig. 6-17 *Casement window. (Andersen)*

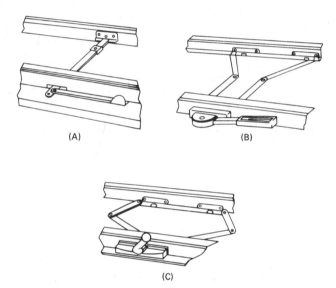

Fig. 6-18 *Methods of operating casement windows: (A) Standard push bar; (B) Lever lock; (C) Rotary gear. (Andersen)*

Installation techniques, materials, and building codes vary according to area. Contact the window dealer for specific recommendations.

The same rough opening preparation procedures are used for wood and Perma-Shield windows. Figures 6-27 and 6-28 show the primed wood window and the Perma-Shield windows made by Andersen. These will be the windows discussed here. The instructions for installation will show how a manufactured window is installed.

Figure 6-29 shows how the wood casement window operates. Note the operator and its location. Figure 6-30 gives the details of the Perma-Shield Narroline window.

Figure 6-31 shows some of the possible window arrangements available from a window manufacturer. There is a window for almost any use. Select the window

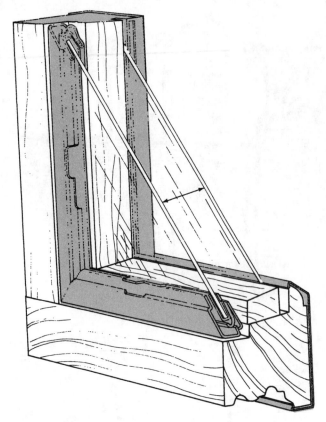

Fig. 6-19 *Double glass insulation. A $\frac{13}{16}$-inch air space is placed between panes.* (Pella)

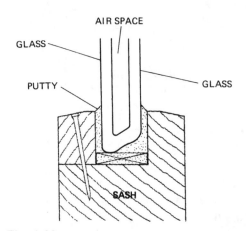

Fig. 6-20 *Welded glass or thermopane windows.*

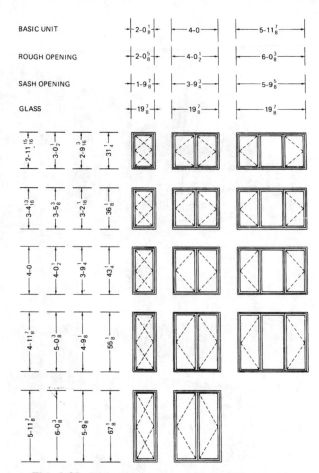

Fig. 6-21 *Various sizes of casement windows.* (Andersen)

needed and follow the instructions or similar steps for your window.

Brick veneer with a frame backup wall is similar in construction to the frame wall in the following illustrations.

When the opening must be enlarged, make certain the proper size header is used. Contact the dealer for the proper size header. To install a smaller size window, frame the opening as in new installation.

Fig. 6-22 *Picture window with awning bottom.* (Andersen)

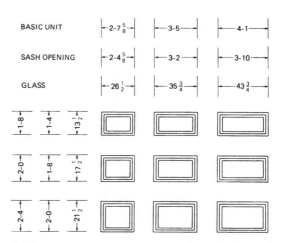

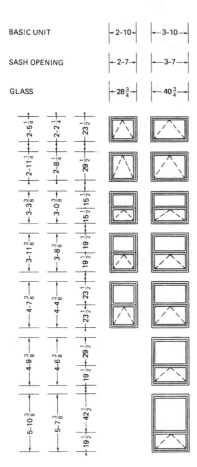

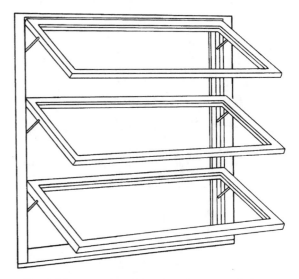

Fig. 6-25 *Various sizes of out swinging and in swinging windows.* (Andersen)

Fig. 6-23 *Various sizes of the fixed sash with awning sash windows.* (Andersen)

IN SWINGING HOPPER

OUT SWINGING AWNING

Fig. 6-24 *In swinging hopper and the out swinging awning types of window. These are casement windows.*

Fig. 6-26 *Vertical stacking of out swinging windows.*

Steps in Preparing the Rough Opening

In some remodeling jobs, this must be done:

1. Lay out the window-opening width between regular studs to equal the window rough opening width plus

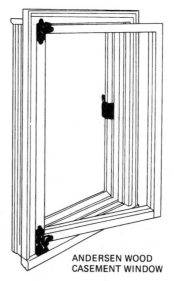

ANDERSEN WOOD CASEMENT WINDOW

Fig. 6-27 *Primed-wood window.* (Andersen)

PERMA-SHIELD®
NARROLINE® WINDOW

Fig. 6-28 *Perma-Shield window.* (Andersen)

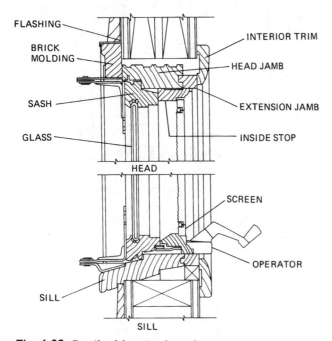

FLASHING

BRICK
MOLDING

INTERIOR TRIM

HEAD JAMB

SASH

EXTENSION JAMB

GLASS

INSIDE STOP

HEAD

SCREEN

OPERATOR

SILL

SILL

Fig. 6-29 *Details of the primed-wood casement window.* (Andersen)

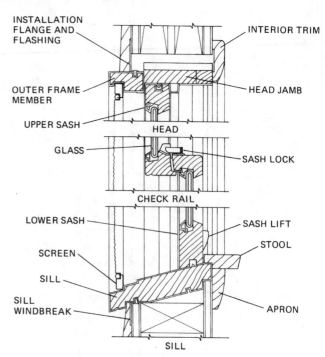

INSTALLATION
FLANGE AND
FLASHING

INTERIOR TRIM

OUTER FRAME
MEMBER

HEAD JAMB

UPPER SASH

HEAD

GLASS

SASH LOCK

CHECK RAIL

LOWER SASH

SASH LIFT

SCREEN

STOOL

SILL

SILL
WINDBREAK

APRON

SILL

Fig. 6-30 *Details of the Perma-Shield window.* (Andersen)

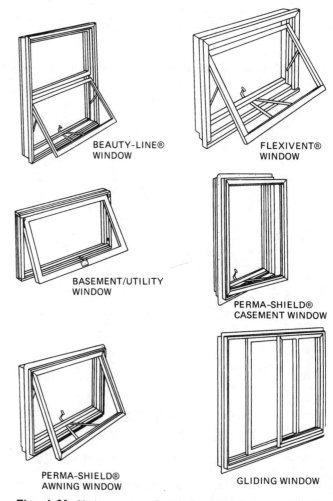

BEAUTY-LINE®
WINDOW

FLEXIVENT®
WINDOW

BASEMENT/UTILITY
WINDOW

PERMA-SHIELD®
CASEMENT WINDOW

PERMA-SHIELD®
AWNING WINDOW

GLIDING WINDOW

Fig. 6-31 *Various types of manufactured windows ready for quick installation.* (Andersen)

the thickness of two regular studs. See Fig. 6-32. Normally, in new construction the rough opening is already there, so all you have to do is install the window in it.

2. Cut two pieces of window header material to equal the rough opening of the window plus the thickness of two jack or trimmer studs. Nail the two header members together using an adequate spacer so that the header thickness equals the width of the jack or trimmer stud. See Fig. 6-33.

3. Cut the jack or trimmer studs to fit under the header for support. Nail the jack or trimmer studs to the regular studs. See Fig. 6-34.

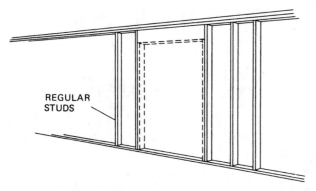

Fig. 6-32 *How to locate a window rough opening. (Andersen)*

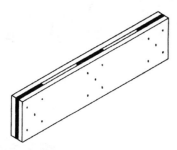

Fig. 6-33 *Making the header outside of the window opening.* (Andersen)

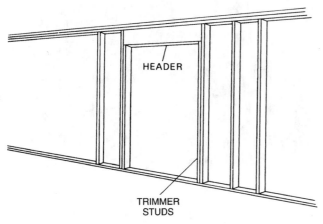

Fig. 6-34 *Placement of the jack studs. (Andersen)*

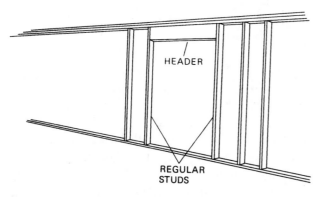

Fig. 6-35 *Placing the header where it belongs. (Andersen)*

4. Position the header at the desired height between the regular studs. Nail through the regular studs into the header to hold the header in place until the next step is completed. See Fig. 6-35.

5. Measure the rough opening height from the bottom of the header to the top of the rough sill. Cut 2-×-4-inch cripples and the rough sill to the proper length. See Fig. 6-36. The rough sill length is equal to the rough opening width of the window. Assemble the cripples by nailing the rough sill to the ends of the cripples.

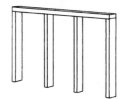

Fig. 6-36 *Assembling the cripples for easy placement.* (Andersen)

6. Fit the rough sill and cripples between the jack studs. See Fig. 6-37. Toenail the cripples to the bottom plate and the rough sill to the jack studs at the sides. See the round insert in Fig. 6-37.

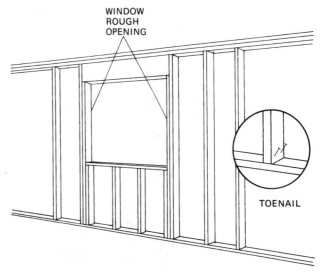

Fig. 6-37 *Placing the rough sill and cripples between the jack studs. Note the insert showing the toenailing. (Andersen)*

7. Apply the exterior sheathing (fiberboard, plywood, etc.) flush with the rough sill, header, and jack or trimmer stud framing members. See Fig. 6-38.

INSTALLING A WOOD WINDOW

The installation of a wood window is slightly different from that of a Perma-Shield window. However, there are many similarities. The following steps will show you how the windows are installed in the

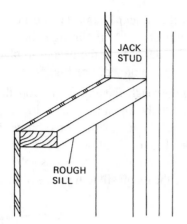

Fig. 6-38 *Applying the exterior sheathing. (Andersen)*

Fig. 6-40 *Use 3½-inch nails to partially secure one corner through the head casing. (Andersen)*

rough opening you just made from the preceding instructions.

1. Set the window in the opening from the outside with the exterior window casing overlapping the exterior sheathing. Locate the unit on the rough sill and center it between the side framing members (jack studs). Use 3½-inch casing nails and partially secure one corner through the head casing. See Fig. 6-39. Drive the nail at a slight upward angle, through the head casing into the header. See Fig. 6-40.

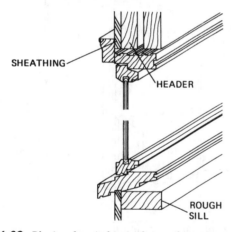

Fig. 6-39 *Placing the window in the rough opening. (Andersen)*

(A)

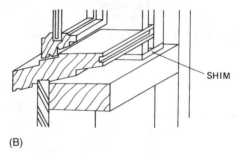

(B)

Fig. 6-41 *(A) Level the window across the casing and nail through the opposite corner (Andersen). (B) Location of the shim that holds the window level.*

2. Level the window across the casing and nail it through the opposite corner with a 3½-inch casing nail. It may be necessary to shim the window under the side jambs at the sill to level it. This is done from the interior. See Fig. 6-41A and B.

3. Plumb (check the vertical of) the side jamb on the exterior window casing and drive a nail into the lower corner. See Fig. 6-42. Complete the installation by nailing through the exterior casing with 3½-inch nails. Space the nails about 10 inches apart.

4. Before you finally nail in the window, make sure you check the sash to see that it operates easily.

5. Apply a flashing with the rigid portion over the head casing. See Fig. 6-43. Secure this flashing

Fig. 6-42 *Check for plumb and nail the lower corner.*

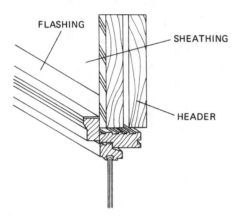

Fig. 6-43 *Apply the flashing with the rigid part over the head casing.* (Andersen)

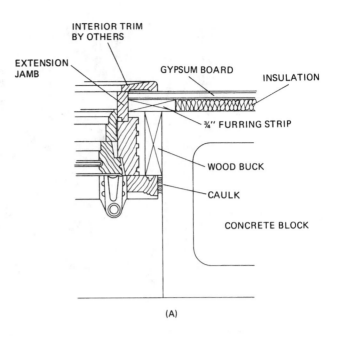

(A)

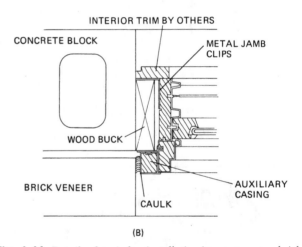

(B)

Fig. 6-44 *Details of a window installation in a masonry or brick veneer wall. (A) Wood window; (B) Perma-Shield window.* (Andersen)

with 1-inch nails through the flexible vinyl into the sheathing. Do not nail into the head casing.

6. Caulk around the perimeter of the exterior casing after the exterior siding or brick is applied.

Masonry or brick veneer wall This type of window can be installed in masonry wall construction. Fasten the wood buck to the masonry wall and nail the window to the wood buck using the procedures just shown for frame wall construction.

In Fig. 6-44A you see the wood window installed in a masonry wall. Figure 6-44B shows a Perma-Shield window installed with metal jam clips in a masonry wall with brick veneer. The metal jam clips and the auxiliary casing are available when specified.

Keep in mind that when brick veneer is used as an exterior finish, adequate clearance must be left for caulking between the window sill and the masonry.

This will prevent damage and bowing of the sill. The bowing is caused by the settling of the structural member. Shrinkage will also cause damage. Shrinkage takes place as the rough-in lumber dries after enclosure and the heat is turned on in the house.

Installing Windows by Nailing the Flange to the Sheathing

A simple procedure is used to place windows into the rough openings left in the framing of the house for such purposes. See Fig. 6-45. Most windows come with a flange that can be nailed to the sheathing or window framing. See Fig. 6-46. This eliminates cold air seepage in the winter and some noise generated by

Fig. 6-45 *A double-hung window mounted by nailing the flange to the sheathing and structural frame of the house.*

Nail

Nail

Nail

Fig. 6-46 *Using a light-weight plastic "glass block" window that fits as a unit and can be mounted by nailing the flange to the frame.*

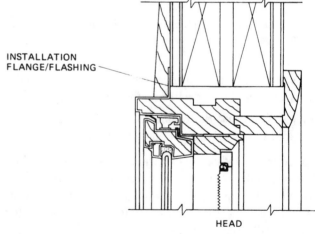

INSTALLATION
FLANGE/FLASHING

HEAD

Fig. 6-47 *Details of how the flange or flashing is covered by the exterior wall covering.*

brisk winds. Examine Fig. 6-47 for details of how the flange or flashing is covered by the exterior wall covering.

In most instances, it is possible for the window unit to be handed out the opening to an outside carpenter who can nail it in place with little effort while the inside carpenter holds the unit in place or levels it with shims.

Figure 6-48 shows a house with windows installed using the methods just described.

Fig. 6-48 *Windows help make the house attractive.*

SKYLIGHTS

Having windows in the walls is not enough today. Houses also have windows in the ceiling. This new demand has been popular where more light is needed and an air of openness is desired. The skylight seems to be the answer to the demands of today's lifestyles. Bathrooms and kitchens are the rooms most often fitted with skylights.

Skylights can be installed when the house is built or they can be added later. In our examples here, we have selected the second approach since it does encompass both methods and can be easily adapted for original construction. Describing how to do it in original construction, however, would not necessarily serve those who want to install skylights after the house is built.

There are four basic types of skylights shown here, ranging from flush-mount to venting types. Figure 6-49 shows the flush-mount type. The low-profile flush-mount skylight provides the most economical solution to skylight installation. The flush-mount includes two heavy gage domes that are formed to provide a built-in deck mounting flange. This flange reduces installation error and allows fast and easy roof attachment when used with the mounting clips pre-packaged in the carton. The frameless/seamless feature eliminates potential leakage and provides airtight reliability. This one is designed specifically for residential use. It is designed to be installed on a pitched roof of 20° or more. It is available in four roof opening sizes.

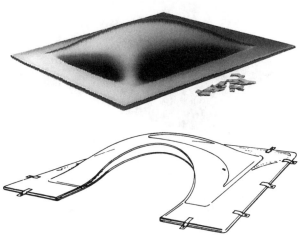

Fig. 6-49 *Flush-mount skylight.* (Novi)

Figure 6-50 is a curb-mount model and is ideal for locations where water, leaves, or snow collect on a roof, making an elevated skylight desirable. It is designed to utilize a wood curb put in place by the installer. This type can be installed on either a flat or pitched roof. It also is available in four roof opening sizes.

Fig. 6-50 *Curb-mount skylight.* (Novi)

Figure 6-51 shows a self-curbing type of skylight, which eliminates on-roof curb construction. This feature allows simple, time-saving installation while eliminating the potential for leakage around a wood curb. The premanufactured curbing is fully insulated and includes an extra-wide deck mounting flange with predrilled holes. This unit is easily installed by placing it directly over the roof opening and fastening it through the flange. It also has built-in condensation

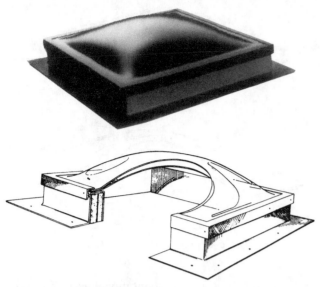

Fig. 6-51 *Self-mount skylight. (Novi)*

channels that get rid of unwanted moisture for maintenance-free operation. It, too, can be installed on a flat or pitched roof and comes in four roof opening sizes.

If you want something in the venting type, Fig. 6-52 has it. A built-in crank system allows operation with a hand crank or optional extension pole. The venting system is chain driven for better control and provides airtight closure when required. It can be installed on a flat or pitched roof and comes in two roof opening sizes. This one not only provides light but also gives you a way to allow for hot air to escape in the summer.

Installing the Skylight

Determine the roof opening location for the skylight on an inside ceiling or attic surface. See Fig. 6-53. Roof openings should be positioned between rafters whenever possible to keep rafter framing to a minimum. Take a look at Fig. 6-54 for framing diagrams for both 16-inch O.C. and 24-inch O.C. rafter spacing.

In a room with a cathedral ceiling, use a drill and wire probe to determine rafter spacing. When working in an attic, position the roof opening so that it is relative to the proposed opening. See Fig. 6-53. Make sure the installation area does not have plumbing or electrical wires inside the opening area.

Preparing the Roof Opening

From inside the house, square the finished roof opening dimensions for the skylight between the roof rafters. If a roof rafter does not cross the proposed opening, mark the four corner points of the roof opening. See Fig. 6-55.

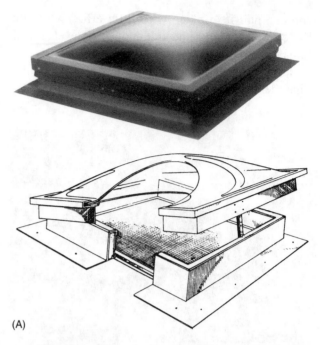

(A)

(B)

Fig. 6-52 *(A) Venting self-mount skylight. (Novi) (B) Using a crank pole to open the skylight. (Velux-America)*

When a rafter must be cut away from the opening, measure 1½ inch beyond the finished roof opening dimensions on the upper and lower sides of the proposed opening and mark the four corner points. See Fig. 6-55B. This extra measurement allows for lumber that will frame the openings.

Fig. 6-53 *Measuring the opening for the skylight.*

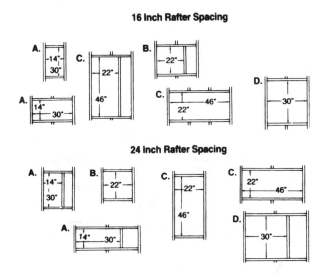

16 Inch Rafter Spacing

24 Inch Rafter Spacing

MODEL	KEY	OVERALL SIZE	FINISHED ROOF OPENING DIMENSION
16 x 32	A	24" x 40"	14" x 30"
24 x 24	B	32" x 32"	22" x 22"
24 x 48	C	32" x 56"	22" x 46"
32 x 32	D	40" x 40"	30" x 30"

Fig. 6-54 *Framing diagrams.* (Novi)

Cutting the Roof Opening

Drive a nail up through the roof at each designated corner point. Go to the roof and remove the shingles covering the proposed opening as indicated by the nails. Remove the shingles 12 to 14 inches beyond the proposed opening at the top and both sides, leaving the bottom row of shingles in place. If the roof felt underlays the shingles, it is not necessary to remove it from the installation area.

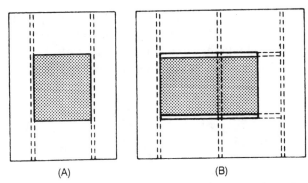

Fig. 6-55 *(A) Roof rafter does not cross the proposed opening. (B) Rafter must be cut to make room for the opening.* (Novi)

Fig. 6-56 *Cutting the hole in the roof.* (Novi)

Draw connecting lines between the nails and cut a hole in the roof using a saber saw or circular saw set to a depth of about one inch. See Fig. 6-56.

The rafter sections that remain in the opening must be cut away perpendicular to the roof deck surface with a handsaw. Temporary rafter supports should be installed to maintain structural alignment of the rafters.

Framing the Roof Opening

Frame the roof opening at the top and bottom by cutting two sections of lumber to fit between the existing rafters under the roof deck. Refer to the framing diagrams in Fig. 6-54.

Lumber used in framing should be the same size as the existing roof rafter lumber.

When working with a hole such as shown in Fig. 6-55, position each header under the roof deck to align with the top and bottom edges of the roof opening and secure it with nails.

If a rafter has been removed from the opening, shown in Fig. 6-55B, secure each header between the rafters and into the cut rafter ends with nails. See Fig. 6-57. Nail a 1½-inch-wide sheathing patch over the top of each header to make the opening level with the roof.

Fig. 6-57 *Framing the roof opening. (Novi)*

Install a side header where the roof opening does not align with an existing roof rafter. The finished rafter frame should align with the finished roof opening dimension.

Mounting the Skylight

Working on the roof, apply a layer of roof mastic (¼ inch deep) around the outer edge of the roof opening, 3 to 4 inches wide. Keep the mastic 1 inch away from the edge of the roof opening to prevent oozing. See Fig. 6-58.

Fig. 6-58 *Spreading mastic around the opening. (Novi)*

For installations made on a pitched roof you have to rotate the skylight until the runoff diverter strip is positioned at the top side of the roof openings, as shown in Fig. 6-59.

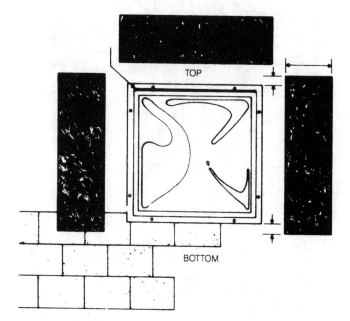

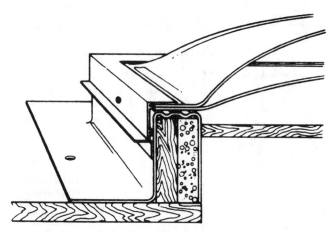

Fig. 6-59 *Making sure the diverter strip is properly located. (Novi)*

Set the skylight into the mastic to align with the roof opening. Be sure that the lower skylight flange overlaps the bottom row of shingles by at least 1 inch. Secure the skylight to the roof with screws using the predrilled flange holes.

Sealing the Installation

Select a flashing material that will seal three sides of the installation area. You may want to use metal, aluminum, or asphalt.

If you use asphalt, select a minimum 30-pound (#) rolled asphalt. Begin by cutting three sheets 5 inches longer than each exterior deck flange located at the top and both sides of the skylight. The width of each sheet should be cut to measure 8 to 10 inches. See Fig. 6-59.

Apply a layer of roof mastic to cover the exterior deck flange and roof deck at the top and both sides of the skylight. Spread the mastic to completely cover the deck flange and 8 to 10 inches of the roof deck around three sides of the installation. See Fig. 6-60.

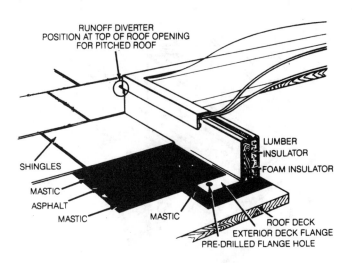

Fig. 6-60 *Applying mastic to cover the exterior deck flange. (Novi)*

Fig. 6-61 *Applying asphalt or tar paper or roofing paper. (Novi)*

Working with either side of the skylight, center the asphalt over the deck flange and press it into the mastic. Asphalt should extend 2½ inches over the flange at the top and bottom for adequate coverage. Continue to apply asphalt to the opposite side, and then to the top. Asphalt at the top must overlap asphalt on both sides. See Fig. 6-61.

Replacing the Shingles

After the asphalt has been set into place, apply a layer of mastic to completely cover the asphalt around the

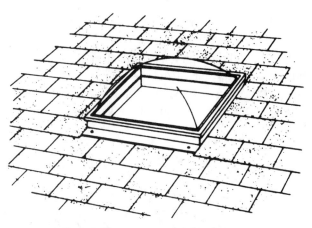

Fig. 6-62 *Replacing the shingles. (Novi)*

skylight. Starting at the bottom, replace each row of shingles. Trim the shingles around the skylight where necessary. See Fig. 6-62.

Preparing the Ceiling Opening

Drop a plumb line from each inside corner of the framed roof opening to the ceiling and mark four corner points. See Fig. 6-63. Remove the insulation between the joists 4 to 6 inches beyond the proposed ceiling opening.

If you want a ceiling opening that is larger than the roof opening or if an angled shaft is desired, some more measuring will have to be done at this time.

To angle the base of the skylight shaft beyond a parallel ceiling opening, pull the plumb line taut to the floor at the desired angle and mark each point. Using a tape measure and carpenter's square, determine the exact size and location of the proposed ceiling opening.

Fig. 6-63 *Dropping a plumb line to find the ceiling opening. (Novi)*

Tap a nail through the ceiling at each corner point. Find the locator nails from the room below and draw connecting lines between each point. Cut through the ceiling along the lines and remove the section. See Fig. 6-64.

Fig. 6-64 *Removing the ceiling section. (Novi)*

Framing the Ceiling Opening

The ceiling opening may be framed using procedures that apply to the framing of any roof opening. Using the same dimensional lumber as existing ceiling joists, cut the headers to fit between the joists and secure them in place with nails. The finished inside frame should align with the ceiling opening.

Constructing the Light Shaft

The light shaft is constructed with bevel-cut 2-×-4-inch lumber hung vertically from the corners of the roof frame to the corners of the ceiling frame. Right angles are formed using two 2 × 4s at each corner of the proposed shaft to provide a nailing surface for the shaft liner. Additional 2 × 4 nailers are spaced to reinforce the shaft frame. See Fig. 6-65.

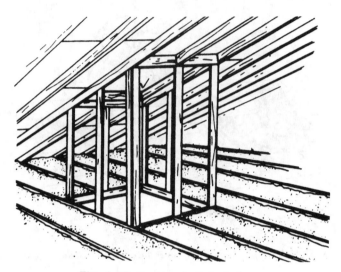

Fig. 6-65 *Framing the light shaft. (Novi)*

Cut the drywall or plywood to the size of the inside shaft walls and nail into place. The shaft may be insulated from the attic for maximum efficiency. See Fig. 6-66. The interior light shaft surfaces may be finished to match the room decor. Place trim around the opening in the ceiling and finish to match the room decor.

Fig. 6-66 *Finishing or closing up the light shaft. (Novi)*

OPERATION AND MAINTENANCE OF SKYLIGHTS
Condensation

Drops of condensation may appear on the inner dome surfaces with sudden temperature changes or during periods of high humidity. These droplets are condensed moisture and do not indicate a water leak from outside moisture. Condensation will evaporate as conditions of temperature and humidity normalize.

Figure 6-67 shows how light shaft installations can be used to present the light from the skylight to various parts of the room below. In Figs. 6-68 and 6-69 you will find how the original installations are made in houses under construction. The details and basic sizes are given, along with the roof pitch/slope chart. These will help you plan the installation from the start.

Care and Maintenance

If the dome is made of plastic, the outer dome surface may be polished with paste wax for added protection from outdoor conditions. If it is made of glass, you may want to wash it before installation and then touch up the finger marks after it is in place. Roofing mastic can be removed with rubbing alcohol or lighter fluid. Avoid petroleum-based or abrasive cleaners, especially on the clear plastic domes. Roof inspection should be conducted every two years to determine potential loosening of screws, cracked mastic, and other weather-related problems that may result from the normal exposure to outdoor conditions.

Tube-type Skylights

The newer tube-type skylights can be installed during construction or after. They are designed to provide maximum light throughput from a relatively small unit. They are right for areas where a larger, standard skylight may not be practical. See Fig. 6-70. The tube shaft can be designed to reflect 95 percent of available sunlight. The low profile ceiling diffuser spreads natural light evenly through interior space. Early morning and late afternoon light can be captured and used by the dome to provide good illumination even during winter months in northern locations. See Fig. 6-71.

The tube-type skylight comes in kit form with everything needed, including illustrated instructions for the do-it-yourselfer. It installs in a few hours with basic hand tools. There is no framing, dry-walling, mudding, or painting required. It is available in both 10-inch and 14-inch diameters and therefore fits easily between standard 16-inch and 24-inch on-center rafters. See Fig. 6-72.

Most people are concerned with skylights because they have heard of them leaking, especially during the winter with snow pileup and melting. The illustrated skylight has a one-piece roof flashing that eliminates leaks. Flashing is specific to the roof type and ensures a perfect fit. The 14-inch spreads light up to 300 square feet. There is also an electric light kit available that makes the skylight into a standard light fixture at night and during dark periods of the day. It is designed to work from a wall switch and is a UL-approved installation. See Fig. 6-73.

Installation To start, locate the diffuser position on the ceiling. Check the attic for any obstructions or wiring. Locate the position on the roof for flashing and

SUGGESTED LIGHT SHAFT INSTALLATIONS

Where a roof window is installed above a flat ceiling, a light shaft will be needed. Typical installations are shown below. Flaring the shaft will give broader light distribution. Shaft construction by others.

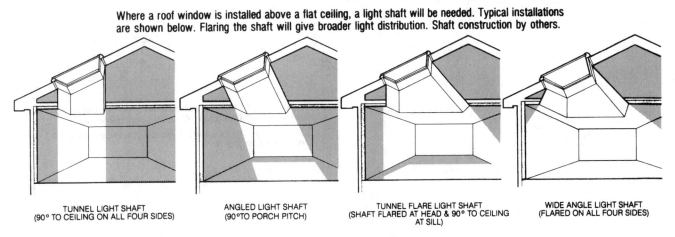

TUNNEL LIGHT SHAFT (90° TO CEILING ON ALL FOUR SIDES)

ANGLED LIGHT SHAFT (90°TO PORCH PITCH)

TUNNEL FLARE LIGHT SHAFT (SHAFT FLARED AT HEAD & 90° TO CEILING AT SILL)

WIDE ANGLE LIGHT SHAFT (FLARED ON ALL FOUR SIDES)

Fig. 6-67 *Suggested light shaft installations.* (Andersen)

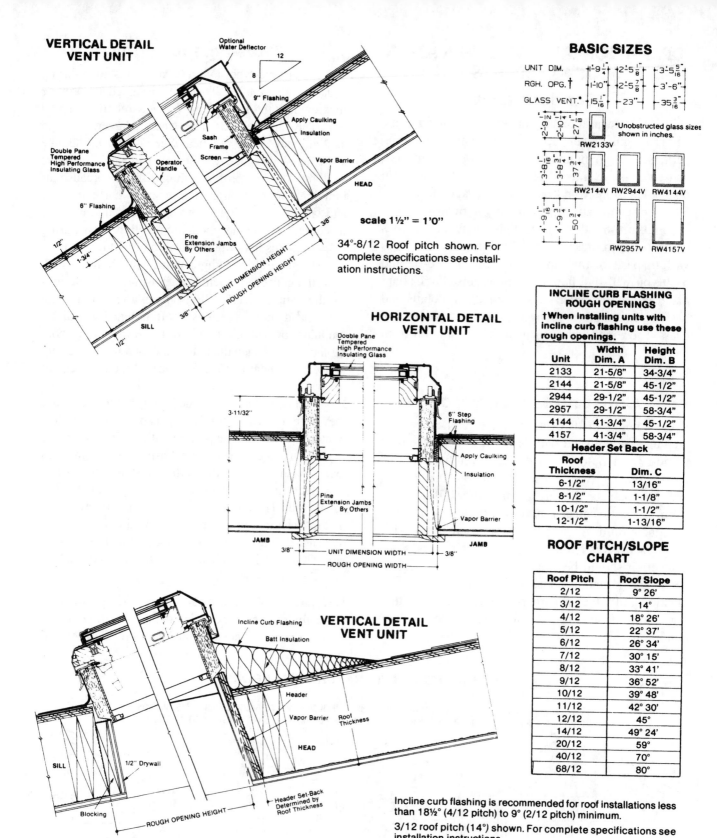

VERTICAL DETAIL VENT UNIT

Optional Water Deflector

9" Flashing

Apply Caulking

Insulation

Vapor Barrier

HEAD

Sash Frame

Screen

Operator Handle

Double Pane Tempered High Performance Insulating Glass

6" Flashing

Pine Extension Jambs By Others

SILL

scale 1½" = 1'0"

34°-8/12 Roof pitch shown. For complete specifications see installation instructions.

HORIZONTAL DETAIL VENT UNIT

Double Pane Tempered High Performance Insulating Glass

3-11/32"

6" Step Flashing

Apply Caulking

Insulation

Vapor Barrier

Pine Extension Jambs By Others

JAMB

JAMB

UNIT DIMENSION WIDTH

ROUGH OPENING WIDTH

VERTICAL DETAIL VENT UNIT

Incline Curb Flashing

Batt Insulation

Header

Vapor Barrier

Roof Thickness

HEAD

SILL

1/2" Drywall

Blocking

Header Set-Back Determined by Roof Thickness

ROUGH OPENING HEIGHT

BASIC SIZES

UNIT DIM.	1'-9¼"	2'-5⅝"	3'-5 5/16"
RGH. OPG. †	1'-10"	2'-5⅞"	3'-6"
GLASS VENT.*	15 1/16"	23"	35 3/16"

RW2133V

RW2144V   RW2944V   RW4144V

RW2957V   RW4157V

*Unobstructed glass sizes shown in inches.

INCLINE CURB FLASHING ROUGH OPENINGS

†When installing units with incline curb flashing use these rough openings.

Unit	Width Dim. A	Height Dim. B
2133	21-5/8"	34-3/4"
2144	21-5/8"	45-1/2"
2944	29-1/2"	45-1/2"
2957	29-1/2"	58-3/4"
4144	41-3/4"	45-1/2"
4157	41-3/4"	58-3/4"

Header Set Back

Roof Thickness	Dim. C
6-1/2"	13/16"
8-1/2"	1-1/8"
10-1/2"	1-1/2"
12-1/2"	1-13/16"

ROOF PITCH/SLOPE CHART

Roof Pitch	Roof Slope
2/12	9° 26'
3/12	14°
4/12	18° 26'
5/12	22° 37'
6/12	26° 34'
7/12	30° 15'
8/12	33° 41'
9/12	36° 52'
10/12	39° 48'
11/12	42° 30'
12/12	45°
14/12	49° 24'
20/12	59°
40/12	70°
68/12	80°

Incline curb flashing is recommended for roof installations less than 18½° (4/12 pitch) to 9° (2/12 pitch) minimum.

3/12 roof pitch (14°) shown. For complete specifications see installation instructions.

Fig. 6-68 *Roof window vent unit, in place.* (Andersen)

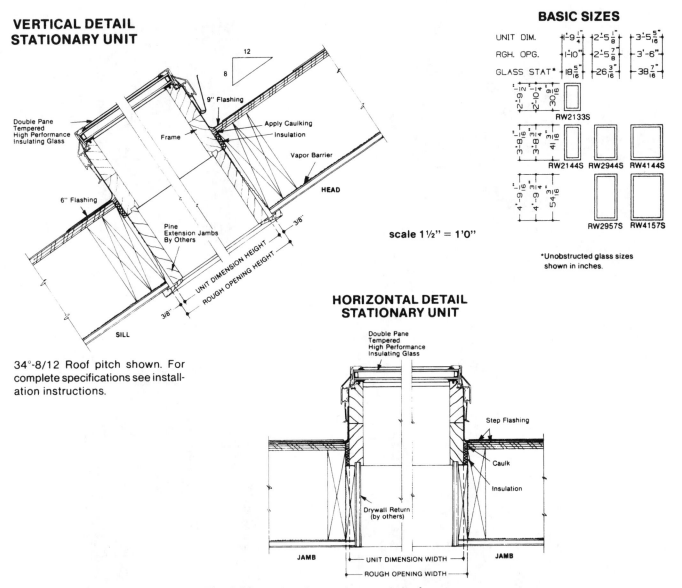

VERTICAL DETAIL STATIONARY UNIT

12
8

Double Pane Tempered High Performance Insulating Glass

9" Flashing

Frame

Apply Caulking

Insulation

Vapor Barrier

HEAD

6" Flashing

Pine Extension Jambs By Others

3/8"

UNIT DIMENSION HEIGHT

ROUGH OPENING HEIGHT

3/8"

SILL

34°-8/12 Roof pitch shown. For complete specifications see installation instructions.

BASIC SIZES

UNIT DIM.	1'-9¼"	2'-5⅛"	3'-5⁵⁄₁₆"
RGH. OPG.	1'-10"	2'-5⅞"	3'-6"
GLASS STAT*	18⁵⁄₁₆	26³⁄₁₆	38⁷⁄₁₆

2'-9½" 2'-10¼" 30¹⁄₁₆

RW2133S

3'-8¹⁄₁₆" 3'-8¾" 41¹⁄₁₆

RW2144S RW2944S RW4144S

4'-9¹⁄₁₆" 4'-9¾" 54³⁄₁₆

RW2957S RW4157S

scale 1½" = 1'0"

*Unobstructed glass sizes shown in inches.

HORIZONTAL DETAIL STATIONARY UNIT

Double Pane Tempered High Performance Insulating Glass

Step Flashing

Caulk

Insulation

Drywall Return (by others)

JAMB

UNIT DIMENSION WIDTH

ROUGH OPENING WIDTH

JAMB

Fig. 6-69 *Roof window, stationary unit, in place.* (Andersen)

dome. If the skylight is being installed in new construction, you can make sure plumbing and electrical take the skylight into consideration during the construction phase. Measure and cut an opening in the roof. Loosen shingles and install the flashing. (In new construction, it may be best to install the flashing before shingles are in place.) Insert the adjustable tube. Attach the dome. See Fig. 6-74. Measure and cut an opening in the ceiling. Install the ceiling trim ring. Attach the diffuser. In the attic, assemble, adjust, and install the tubular components. In colder climates it is necessary to insulate the tube shaft.

TERMS USED IN WINDOW INSTALLATION

Now is a good time to review the terms associated with the installation of a window. This will make it possible for you to understand the terminology when you work with a crew installing windows.

Plumb The act of checking the vertical line of a window when installing it in a rough opening.

Level The act of checking the horizontal line of a window when installing it in a rough opening.

Regular stud A vertical frame member that runs from the bottom plate on the floor to the top plate at the ceiling. In normal construction, this is a 2 × 4 approximately 8 feet long.

Jack or trimmer stud A vertical frame member that forms the window rough opening at the sides and supports the header. It runs from the bottom plate at the floor to the underside of the header.

Header A horizontal framing member located over the window rough opening supported by the jack studs.

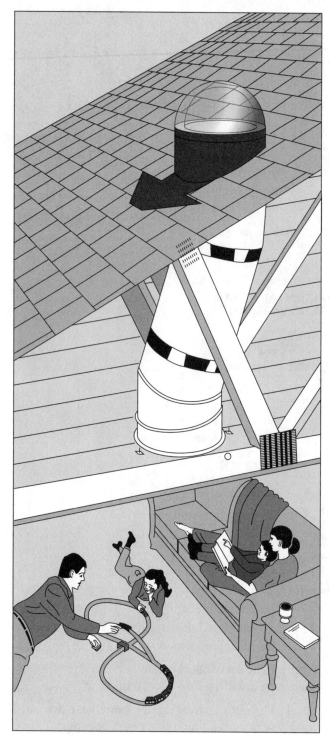

Fig. 6-70 *Skylight installation.* (ODL Inc.)

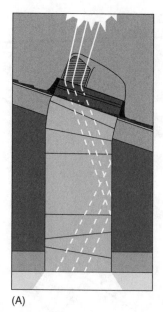

(A)

Fig. 6-71A *The dome above the roof line.* (ODL Inc.)

(B)

Fig. 6-71B *Dome reflects the sunlight coming from any angle throughout the day in any season.* (ODL Inc.)

Depending upon the span, headers usually are double 2 × 6s, 2 × 8s, or 2 × 10s in frame wall construction, or steel I beams in heavier masonry construction.

Rough sill A horizontal framing member, usually a single 2 x 4, located across the bottom of the window rough opening. The window unit rests on the rough sill.

Cripples Short vertical framing members spaced approximately 16 inches O.C., located below the rough sill across the width of the rough opening. Also used between the header and the top plate, depending upon the size of the headers required.

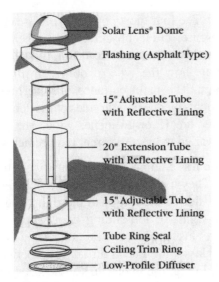

Fig. 6-72 *Exploded view of the skylight.* (ODL Inc.)

- Solar Lens® Dome
- Flashing (Asphalt Type)
- 15" Adjustable Tube with Reflective Lining
- 20" Extension Tube with Reflective Lining
- 15" Adjustable Tube with Reflective Lining
- Tube Ring Seal
- Ceiling Trim Ring
- Low-Profile Diffuser

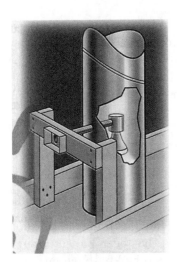

Fig. 6-73 *Conversion of the skylight to a light fixture.* (ODL Inc.)

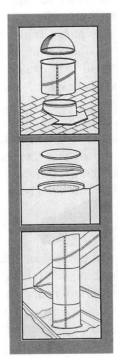

Fig. 6-74 *Installation of the skylight.* (ODL Inc.)

Shim   An angled wood member—wedge shaped—used as a filler at the jamb and sill. (Wood shingle makes a good shim.)

Wood buck   A structural wood member secured to a masonry opening to provide an installation frame for the window unit.

PREHUNG DOORS
Types of Doors

Exterior doors are made in many sizes and shapes. See Fig. 6-75. They may be solid with a glass window. They may have an **X** shape at the bottom, in which case they are referred to as a *cross-buck*. These doors are made in a factory and crated and shipped to the site. There they are unpacked and placed in the proper opening. There is little to do with them other than level them and nail them in place. The hardware is already mounted on the door.

Figure 6-76 shows a door that was prehung and shipped to the site. Note how it sticks out from the sheathing so that the siding can be applied and butted to the side jamb. Doors are chosen for their contribution to the architecture of the building. They must harmonize with the design of the house. Figure 6-77 shows a door that adds to the design of the house. The door facing or trim adds to the column effect of the porch.

Fig. 6-75 *Various door designs.* (National Woodwork Manufacturers)

Fig. 6-76 *A prehung exterior door with three panels of glass. Note how the trim sticks out sufficiently for the siding to butt against it.*

Fig. 6-77 *The proper door can do much to improve the looks of the house.*

Flush doors Flush doors are made of plywood or some facing over a solid core. The core may be made of a variety of materials. In some instances where the door is used inside, the inside of the door is nothing more than a mesh or strips. See Fig. 6-78. Wood is usually preferred to metal for exterior doors of homes. Wood is nature's own insulator. While metal readily conducts heat and cold, wood does not. Wood is 400 times more effective an insulator than steel and 1800 times more effective than aluminum.

Stock wood flush doors come in a wide variety of sizes, designs, and shapes. Standard wood door frames will accommodate wood combination doors, storm doors, and screen doors without additional framing expense.

Panel doors This type of door has solid vertical members, rails, and panels. Many types are available. See Figs. 6-79 and 6-80. The amount of wood and glass varies. Many people want a glass section in the front door. The four most popular types are clear, diamond obscure, circle obscure, and amber Flemish. Figure 6-80 shows some of the decorative variations in doors. Note how the type of door can improve the architecture. The main entrance may be highlighted with sidelights on one or both sides of the door. See Fig. 6-81. These are 12 or 14 inches wide. The panels of glass are varied to meet different requirements.

Sliding doors Sliding doors are just what the name suggests. They usually have tempered or safety glass. They can slide to the left or to the right. You have to specify a right- or left-sliding door when ordering. Most have insulating construction. They have two panes of glass with a dead-air space in between. See Fig. 6-82. In Fig. 6-83 you can see the sizes available. Also note the arrow which indicates the direction in which the door slides.

French doors This is usually two or more doors grouped to open outward onto a patio or veranda. They have glass panes from top to bottom. They may be made of metal or wood. Later in this chapter you will see two- and three-window groupings mounted step by step.

INSTALLING AN EXTERIOR DOOR

The door frame has to be installed in an opening in the house frame before the door can be hung. Figure 6-84 shows the parts of the door frame. Note the way it goes together. Figure 6-85 shows the frame in position with a wire on the right. It has been pulled through for the installation of the doorbell push button. Note the spacing of the hinges. The door in the background is a six-panel type that is already hung. The siding has not yet been butted against the door casing. It will be placed as close as possible and then caulked to prevent moisture from damaging the wood over a period of time.

Figure 6-86 shows the general information you need to be able to identify the parts of a door. It is important that the door be fitted so that there is a uniform ⅛-inch clearance all around to allow for free swing. Allow ½ to ¾ inch at the bottom for a better fit with carpet.

There are nine steps to installing an exterior door:

1. See Fig. 6-86B. Trim the height of the door. Many doors are made with extra long stiles. Before proceeding, cut the horns off the *top* of the door, even with the top of the top rail. When cutting, start with the saw at the outside edge to avoid splintering the edges of the door.

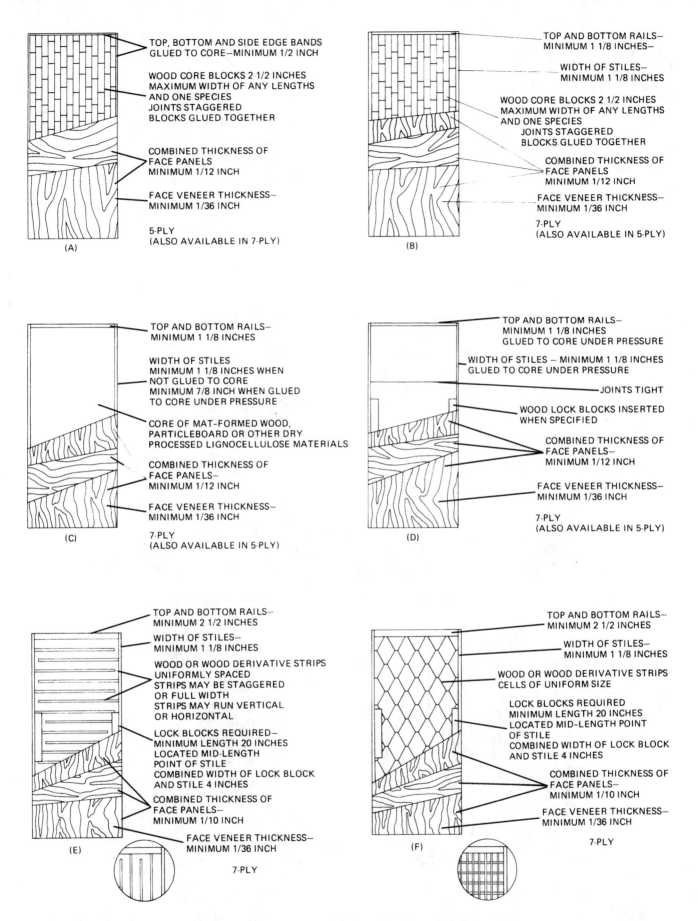

Fig. 6-78 *Various types of materials are used to fill the interior space in flush doors.* (National Woodwork Manufacturers)

Fig. 6-79 *Add-on panels and lights give designers options with flush doors.* (General Products)

2. See Fig. 6-86C. On the inside of the room, place the door into the opening upside down, and against the door jamb. Keep the hinge stile tight against the hinge jamb A1 of the door frame. This edge must be kept straight to ensure that hinges will be parallel when installed. Place two ¼-inch blocks under the door. These will raise the door to allow for ⅛-inch clearance at both top and bottom when cut. Mark the door at C, the top of the door frame opening, for cutting.

3. See Fig. 6-86D. After cutting the door to the proper height, place the door into the opening again—this

CHATEAU 8-PANEL

WILLIAMSBURG 6-PANEL

COLONIAL 9-LIGHT

CROSS BUCK 9-LIGHT

Fig. 6-80 *The four most popular door styles. They have clear, diamond obscure, circle obscure, and amber Flemish safety glass.* (General Products)

Fig. 6-81 *Sidelights are designed for fast installation as integrated units in wood or plastic models. This insulated safety glass comes in 12-inch or 14-inch widths.* (General Products)

time right side up, with ⅛-inch blocks under the door for clearance. With the door held tightly against the hinge jamb of the door frame, have someone mark a pencil line along the lock stile of the door from outside the door opening (line 1 to 1A), holding a ⅛-inch block between the pencil and the door frame. This will automatically allow for the necessary ⅛-inch clearance needed.

4. See Fig. 6-86E. Trim the width of the door. If the amount of wood to be removed from the door (line 1 to 1A) is more than ¼ inch, it will be necessary to trim both edges of the door. Trim one-half the width of the wood to be removed from each edge of the door. Use a smooth or jack plane.

Fig. 6-82 *Sliding door.*

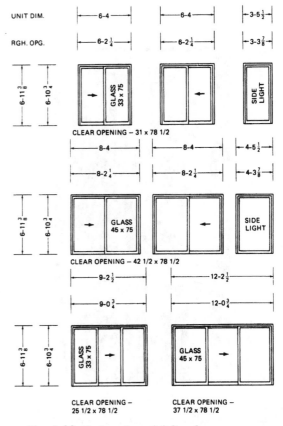

Fig. 6-83 *Various sizes of sliding doors.* (Andersen)

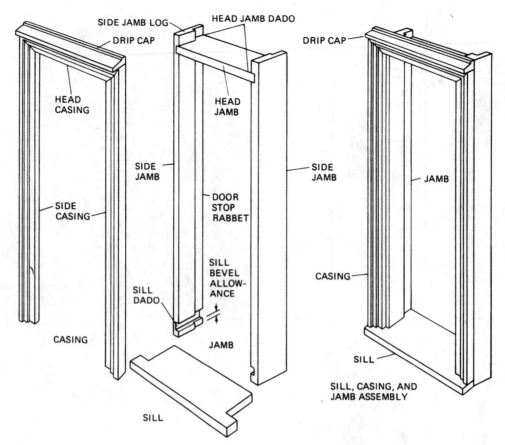

Fig. 6-84 *Assembling a door frame (left to right).*

Labels in figure 6-84:
SIDE JAMB LOG
HEAD JAMB DADO
DRIP CAP
DRIP CAP
HEAD CASING
HEAD JAMB
SIDE JAMB
SIDE CASING
SIDE JAMB
JAMB
DOOR STOP RABBET
CASING
SILL BEVEL ALLOWANCE
SILL DADO
SILL
CASING
JAMB
SILL
SILL, CASING, AND JAMB ASSEMBLY

Fig. 6-85 *Prehung door in place. Note the wires hanging out on the left side. They indicate where an outside light will go.*

5. See Fig. 6-86F. Bevel the lock stile of the door. The lock stile of the door should be planed to about a 3° angle so that it will clear the door frame when the door is swung shut.

6. See Fig. 6-86G. Install the hinges. Be sure that markings for hinges are uniform for all hinges used. The mortises (cutouts) for hinges should be of uniform depth (thickness of the hinge). Measure 7 inches down from the top of the door and 7⅛ inches down from the underside of the door frame top. Mark the locations of the top edge of the upper hinge. Placement of hinges will be unnecessary if the door is prehung.

7. See Fig. 6-86H. The bottom hinge is 9 inches up from the door bottom (9⅛ inches up from the threshold). The middle hinge is centered in the door height. Attach the hinge leafs to the door and door frame. Hang the door in the opening. If the mortises are cut properly but the hinges still bind, the frame jamb may be distorted or bowed. It may be necessary to place a thin shim under one edge of the frame hinge leaf to align it parallel and relieve the binding.

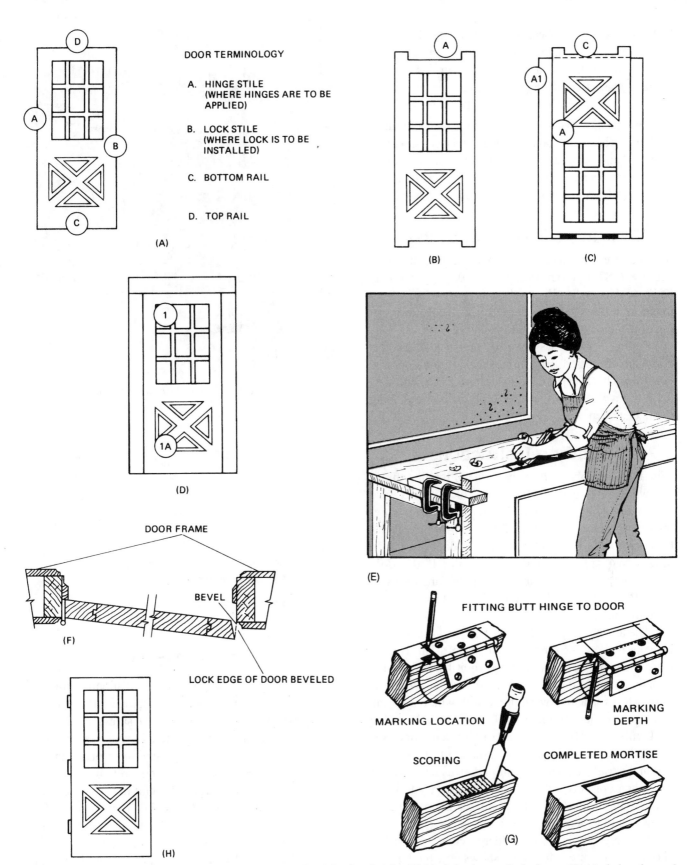

Fig. 6-86 *(A) Terms used with doors* (Grossman Lumber). *(B) Trim the door height. (C) Mark the door to fit the frame. (D) Mark the other end of the door. (E) Trim the width of the door. (F) Bevel the lock stile of the door. (G) Install the hinges. (H) Check the height of the hinges.*

8. Install the lockset. Because of the variety of styles of locksets, there is no one way to install them. This subject will be discussed in detail later in this chapter. Each lockset comes with a complete set of instructions. The best advice is to follow these instructions.

9. Finish the door. Care must be taken to paint or seal all door surfaces. Top, bottom, edges, and faces should be sealed and painted. Weather and moisture can hurt a door and decrease its performance. A properly treated door will give many years of satisfactory service.

Hanging a Two-Door System

In some cases the two-door system is the main entrance. See Fig. 6-87. In other cases the two door system may be French doors. See Fig. 6-88. This type of door comes in pieces and has to be assembled. The details for hanging of this type of door are shown in Fig. 6-89.

Figure 6-90 shows how the active and inactive doors are identified first. In most cases both doors do not open. This is especially the case when the entrance door is involved. In the case of French doors, both doors can open.

Handing Instructions

With the assembly kit furnished, one of the first things you will want to check is the "hand" of the door. Hand of doors is always determined by the *outside*. Inswinging doors are more common. See Fig. 6-91. The right-hand symbol is RH, and it means the door swings on hinges that are mounted on the right. If the door swings out, it is a right-hand reverse door and the symbol is RHR. It is still hinged on the right looking at it from the outside.

The left-hand symbol is LH. That means that on an inswinging door the hinges are on the left. This is most convenient for persons who are right-handed. If the door is outswinging, the left-hand reverse symbol is LHR. Doors usually swing into a wall where they can rest against the adjoining wall. They usually swing back only 90°. The traffic pattern also determines the way a door swings. Doors swing out in most commercial, industrial, and school buildings. This lets a person open the door outward so that it will be easy to leave the building if there is a fire. Safety is the prime consideration in this case.

Figure 6-92 shows how energy conservation has entered the picture. The figure shows how the top and bottom of the door are fitted to make sure air does not leak through the door. The door may be made of metal. Because metal conducts heat and cold, the door is insulated. See Fig. 6-93.

Fig. 6-87 *Double doors installed. The one on the left is the active door.*

Fig. 6-88 *Double door ready for assembly.* (General Products)

Metal Doors

Metal doors may also be used for residential houses. In some cases they are used to replace old or poorly fitting wooden doors. Figure 6-94 shows how doors with metal frames are designed for ease of installation. In Fig. 6-95 you can see how the metal frame is attached to concrete, wood, and concrete blocks.

One of the advantages of metal doors is their fire resistance. Frames for 2'8" and 3'0" doors carry a 1½-hour label. The frames for 3'6" and double doors are not labeled as to fire resistance.

In the case of metal frames, the frame has to be installed before the walls are constructed. The frame requires a rough opening 4½ inches wider and 2¼ inches higher than nominal. The stock frame is usually 5¾ inches wide.

Step 1

Attach Astragal to Inactive Door

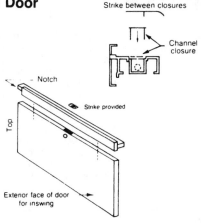

Remove plastic filler plates from deadbolt and lock locations. Remove appropriate metal knock-out at deadbolt location.

Place door on edge, place astragal with *notch to door top* as shown. Compress outer flange against face of door, install bolt retainer spring at top and bottom of astragal, and secure astragal with five (5) self-tapping screws using power driver or drill.

Place two nylon screw bosses under the strike route and secure strike with 2 No. 8 screws provided.

Strike has tab that can be adjusted to assure proper closing while hanging door.

Place the bolt assembly in top and bottom of astragal. Adjust the bolt retainer spring to proper position and secure bolt retainer with Allen wrench provided.

Snap in 2 ea. channel closures in bolt recess above and below strike location. (proper lengths provided).

Tape one pile pad to interior face of inactive door for installation on astragal after door is installed in opening.

Step 2

Attach Sweep

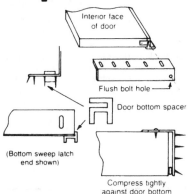

(Bottom sweep latch end shown)

Place inactive door on prehang table, interior face up. Pick up sweep with flush bolt hole, place door bottom spacer on latch end of sweep as shown.

Place sweep on bottom of door, flush latch end with latch edge of astragal.

Compress tightly against door bottom and drive self drilling screws into door skin at extreme bottom of slot. Tighten screws moderately to hold sweep in up position.

Check operation of bolt assembly.

Turn door over (Interior face down)

Place active door on table (interior face up) and install sweep in same manner as first door.

Turn door over and proceed with frame prehanging. (Step 3)

Step 3

Attach Hinge Jambs

Place hinge jambs beside edge of doors as shown and attach hinges to doors with No. 10 machine screws.

Apply caulking tape to jambs at threshold locations. Make sure it follows contour of vinyl threshold part.

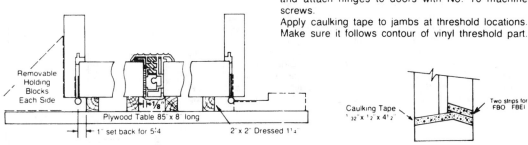
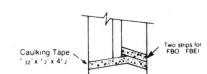

Use Same Table For 6'0

Fig. 6-89 *Details for hanging a two-door system.*

INSTALLING FOLDING DOORS

Folding doors are used to cover closets with any number of interesting patterns. They may be flush, and plain or mirrored. They may have two panels or four panels. The sizes and door widths vary to suit the particular application. See Table 6-1.

Figure 6-96 shows the openings and the details of fitting the metal bifold door.

Figure 6-97 shows the details of the four panels to be installed and different panel styles available. Note the names given to the parts so you can follow the installation instructions.

1. Carefully center the top track lengthwise in the finished opening. This should suit both flush and recessed mountings. Attach the track with No. 10 × 1¼-inch screws in the provided holes. The bifold

Step 4   Installing Header and Threshold

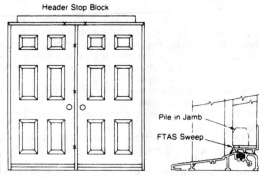

Stick ⅛" PAK-WIK spacers—2 ea. on header jambs, and 3 ea. on the astragal side at locations marked "X", to maintain ⅛" clearance between doors and jambs. Place header jamb against header stop block. Line up doors with header jamb and press doors firmly against it. Raise jambs up, make sure they are flush with header jamb at top corners. Drive 3 ea. 2¼" long staples in each corner of frame.

Assemble vinyl and aluminum threshold parts. Place threshold in frame and secure with #10 x 1½" screws through pre-drilled holes. Back edge to be flush with frame. (Make sure pile is firmly in contact with threshold and weatherstripping.) If not, remove and reposition.

Step 5   Attaching Brickmold

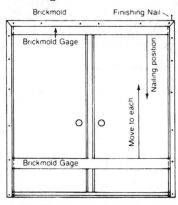

NOTE: If door is to be outswing, proceed with bracing shown in Step 6. Then turn unit upside down and install bolt strike as in Step 6. Proceed with brickmold in Step 5. NOTE: Outswing frame requires ⅝" longer header brickmold than inswing. Place brickmold gage at top and bottom of jambs. Position miter joints of jamb and header brickmold. Align fit of mitered corners and properly space reveal. Tack each corner, nail header brickmold, move gage down jambs and nail brickmold. Use 6 ea. No. 10 x 2½" finishing nails as shown, drive 2 ea. No. 10 x 2½" finishing nails in two corners as shown.

Step 6   Installing Flush Bolt Strike and Bracing Frame

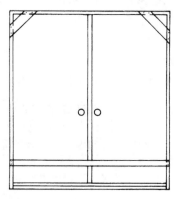

Turn unit upside down and replace on table. Place pencil mark at flush bolt locations on header and threshold. Open inactive door, place thin dab of putty at bolt pencil marks on header and threshold. Close inactive door. Move bolts to mark the putty. Open inactive door. Center punch top and bottom bolt locations with nail. Remove putty. Drill ⅝" hole in header and threshold. Place bolt strike on header and align with drilled hole. Install 2 ea. No. 6 x 1" screws provided. Close inactive door and secure bolts. Close active door. Tack corner braces as shown. Tack strip of wood across frame, approximately 12" above threshold as shown. Use 8d coated box nails. On outswing unit, cut bracing to fit between brickmold.

Fig. 6-89 Continued.

track is assembled for four-panel door installation. The track may be separated at the center without the use of tools when a two-panel door is installed. See Fig. 6-98. Knobs, screws, and rubber stops are packaged for two-panel installation.

2. Place the bottom track, either round edge or square edge, toward the room. Plumb the groove with the top track. See Fig. 6-99. Screw the track to the floor with ½-inch screws or fasten it to a clean floor with 3M double-coated tape no. 4432.

3. On all two-panel sections, lower the bottom pivot rod until it projects ½ inch below the edge of the door. Make it 1¼ inch if the carpet is under the door.

4. Attach the doorknobs.

5. Lift one door set. Insert the bottom pivot rod (threaded) into the bottom pivot bracket. Pull down the top spring-loaded pivot rod. Insert it into the pivot bracket in the top rack. Insert the top and bottom nylon glide rod tips into the track (Fig. 6-100).

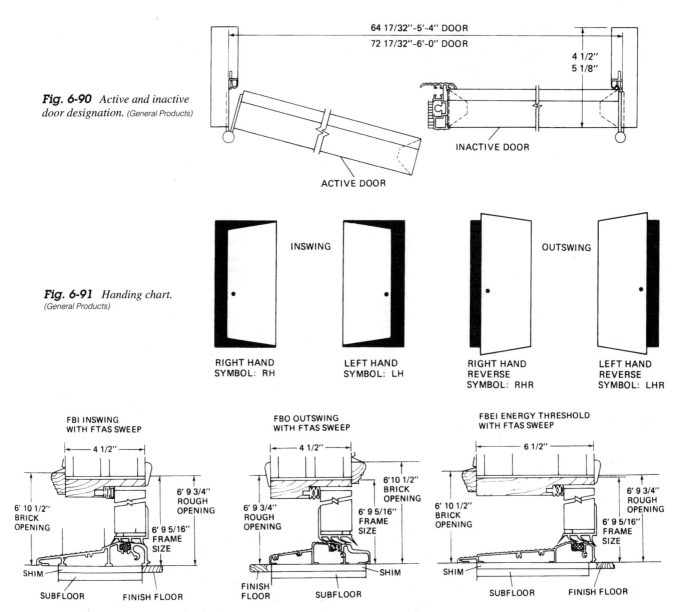

Fig. 6-90 *Active and inactive door designation.* (General Products)

64 17/32″–5′-4″ DOOR
72 17/32″–6′-0″ DOOR

4 1/2″
5 1/8″

INACTIVE DOOR

ACTIVE DOOR

Fig. 6-91 *Handing chart.* (General Products)

INSWING

OUTSWING

RIGHT HAND
SYMBOL: RH

LEFT HAND
SYMBOL: LH

RIGHT HAND
REVERSE
SYMBOL: RHR

LEFT HAND
REVERSE
SYMBOL: LHR

FBI INSWING
WITH FTAS SWEEP

4 1/2″

6′ 10 1/2″
BRICK
OPENING

6′ 9 3/4″
ROUGH
OPENING

6′ 9 5/16″
FRAME
SIZE

SHIM

SUBFLOOR   FINISH FLOOR

FBO OUTSWING
WITH FTAS SWEEP

4 1/2″

6′10 1/2″
BRICK
OPENING

6′ 9 3/4″
ROUGH
OPENING

6′ 9 5/16″
FRAME
SIZE

SHIM

FINISH
FLOOR   SUBFLOOR

FBEI ENERGY THRESHOLD
WITH FTAS SWEEP

6 1/2″

6′ 10 1/2″
BRICK
OPENING

6′ 9 3/4″
ROUGH
OPENING

6′ 9 5/16″
FRAME
SIZE

SHIM

SUBFLOOR   FINISH FLOOR

Fig. 6-92 *Finished dimensions on a double-hung door.* (General Products)

6. Install the second door set the same way.

7. Insert the rubber stop in the center of the top and bottom tracks. Make sure that the stop seats firmly in the track. For a two-panel installation, cut the rubber stop to the proper length.

8. Because of their design, the bifold doors are rigid enough to operate smoothly without a full bottom track. This permits better carpeting in the closet. Saw off a 4-inch section of the bottom track. Place this on the floor. Use a plumb bob to pivot the bottom points with the top pivot. See Fig. 6-101. Fasten the section to the floor with two ½-inch screws. Remove the bottom glide rods from the doors. Single-track installation is not recommended for 8'0"-high doors, 7'0"-wide four-panel doors, or 3'6"-wide two-panel doors.

Final adjustments To raise or lower the doors to the desired height, turn the threaded bottom pivot rod with a screwdriver. Make sure the doors are even and level across the top. Tighten the locknut.

Doors should close snugly against the rubber stop. For horizontal (lateral) alignment, loosen the screw holding the top or bottom pivot brackets in the track. Adjust the door in or out. Retighten the screw.

Keep all glides and track free from paint and debris. The aluminum track is already lubricated to ensure smooth operation. Occasionally repeat the lubrication with silicon spray, paraffin, or soap. This keeps door operation free and easy.

These instructions are for a particular make of door. However, most manufacturers' instructions are basically the same. There are some minor adjustments

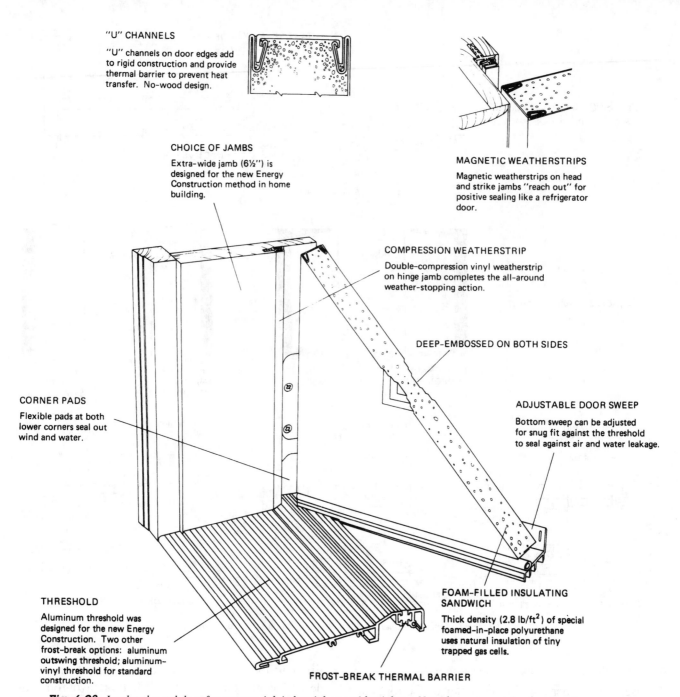

"U" CHANNELS

"U" channels on door edges add to rigid construction and provide thermal barrier to prevent heat transfer. No-wood design.

CHOICE OF JAMBS

Extra-wide jamb (6½") is designed for the new Energy Construction method in home building.

MAGNETIC WEATHERSTRIPS

Magnetic weatherstrips on head and strike jambs "reach out" for positive sealing like a refrigerator door.

COMPRESSION WEATHERSTRIP

Double-compression vinyl weatherstrip on hinge jamb completes the all-around weather-stopping action.

DEEP-EMBOSSED ON BOTH SIDES

CORNER PADS

Flexible pads at both lower corners seal out wind and water.

ADJUSTABLE DOOR SWEEP

Bottom sweep can be adjusted for snug fit against the threshold to seal against air and water leakage.

THRESHOLD

Aluminum threshold was designed for the new Energy Construction. Two other frost-break options: aluminum outswing threshold; aluminum-vinyl threshold for standard construction.

FOAM-FILLED INSULATING SANDWICH

Thick density (2.8 lb/ft^2) of special foamed-in-place polyurethane uses natural insulation of tiny trapped gas cells.

FROST-BREAK THERMAL BARRIER

Fig. 6-93 *Insulated metal door for commercial, industrial, or residential use. Note the energy-saving features.* (General Products)

you will have to make for each manufacturer. Make sure you follow the manufacturer's recommendations.

DOOR AND WINDOW TRIM
Interior Door Trim

Most inside or interior doors have two hinges. They usually come in a complete package. Once they are set in place, the casing has to be applied. See Fig. 6-102 for the location of the casing around an interior door. Note

how the jamb is installed. The stop is attached with nails and has a bevel cut at the bottom of the door. It prevents the door from swinging forward more than it should.

In Fig. 6-103 you will find two of the most commonly used moldings applied to the trim of a door. These two are colonial and ranch casing moldings. These are the names you use when ordering them. They are ordered from a lumberyard or mill.

Figure 6-104 shows how a molded casing is mitered at the corner. It is secured with a nail through the 45° cut. In the other part of this figure you see the

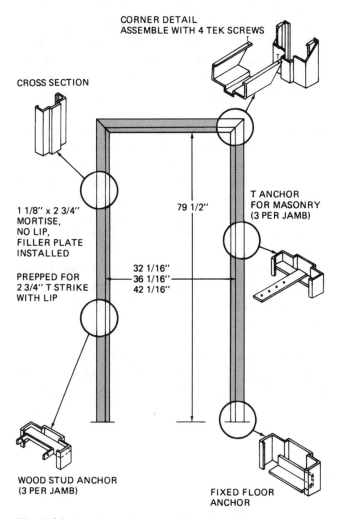

CORNER DETAIL
ASSEMBLE WITH 4 TEK SCREWS

CROSS SECTION

1 1/8" x 2 3/4"
MORTISE,
NO LIP,
FILLER PLATE
INSTALLED

PREPPED FOR
2 3/4" T STRIKE
WITH LIP

79 1/2"

32 1/16"
36 1/16"
42 1/16"

T ANCHOR
FOR MASONRY
(3 PER JAMB)

WOOD STUD ANCHOR
(3 PER JAMB)

FIXED FLOOR
ANCHOR

Fig. 6-94 *Putting together a metal frame for a door.* (General Products)

butt joint. This is where the casing meets at the side and top. Notice the way the nail is placed to hold the two pieces securely. Also notice the other nail locations. Why do you need to drill the nail hole for the toenailed side?

Table 6-1 *Finished Opening Sizes for Bifold Doors.*

| Door Width | Number of Panels | Door opening, Inches* | | Actual Door Width, Inches |
		6'8"	8'0"	
1'6"	2	18¹/₂ × 80³/₄	18¹/₂ × 95¹/₄	17⁷/₁₆
2'0"	2	24¹/₂ × 80³/₄	24¹/₂ × 95¹/₄	23⁷/₁₆
2'6"	2	30¹/₂ × 80³/₄	30¹/₂ × 95¹/₄	29⁷/₁₆
3'0"	2	36¹/₂ × 80³/₄	36¹/₂ × 95¹/₄	35⁷/₁₆
3'0"	4	36¹/₂ × 80³/₄	36 × 95¹/₄	35
3'6"	2	42¹/₂ × 80³/₄	42¹/₂ × 95¹/₄	41⁷/₁₆
4'0"	4	48 × 80³/₄	48 × 95¹/₄	47
5'0"	4	60 × 80³/₄	60 × 95¹/₄	59
6'0"	4	72 × 80³/₄	72 × 95¹/₄	71
7'0"	4	84 × 80³/₄	84 × 95¹/₄	83

*Finished opening width shown provides ¹/₂ inch clearance each side of door. Finished opening width may be reduced by ¹/₂ inch provided finished opening is square and plumb. This will require cutting track. Finished opening heights shown provide ³/₈ inch clearance between door and track—top and bottom. (This makes ⁷/₈ inch between door and floor.) Doors can be raised to have 1¹/₈ inch clearance door to floor without increasing opening height.

Installation of the strike plate in the side jamb is shown in Fig. 6-105. It has to be routed or drilled out. This allows the door locking mechanism to move into the hole. Figure 6-106 shows how the strike plate is mounted onto the door jamb.

Window Trim

Windows have to be trimmed. This completes the installation job. See Fig. 6-107. There are a couple of ways to trim a window. Take a look at Fig. 6-108 and note the difference. Shown is a trimmed window with casing at the bottom instead of a stool and apron. This is a quicker and simpler method of finishing a window. There is no need for a stool to overlap the apron or casing in some instances. This is the choice of the architect or the owner of the home. There are problems with the apron and stool method. The apron and stool will pull away from the inside casing. This leaves a gap of up to ¼ inch. It can become unsightly in time.

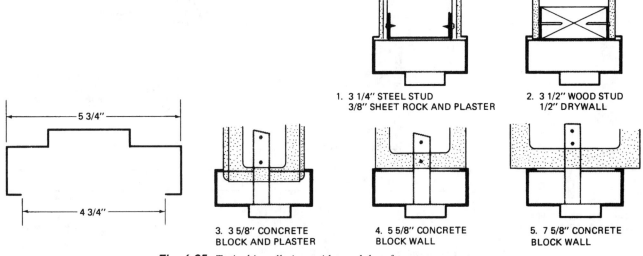

5 3/4"

4 3/4"

1. 3 1/4" STEEL STUD
3/8" SHEET ROCK AND PLASTER

2. 3 1/2" WOOD STUD
1/2" DRYWALL

3. 3 5/8" CONCRETE
BLOCK AND PLASTER

4. 5 5/8" CONCRETE
BLOCK WALL

5. 7 5/8" CONCRETE
BLOCK WALL

Fig. 6-95 *Typical installations with metal door frames.* (General Products)

Figure 6-109 shows some of the moldings that can be used in trimming windows, doors, or panels. These moldings are available in prepared lengths of 8 feet and 12 feet. Generally speaking, the simpler the molding design, the easier it is to clean. Many depressions or designs in a piece of wood can allow it to pick up dust. Some are very difficult to clean.

INSTALLING LOCKS

There are seven simple steps to installing a lock in a door. Figure 6-110 shows them in order.

In some cases you might want to reverse the lock. This may be the case when you change the lock from one door to another. The hand of the door might be different. In Fig. 6-111 you can see how simple it is to change the hand of the lock. In some cases you may have bought the lock without noticing how it should fit. This way you are able to make it fit in either direction.

There are a number of locks available. Figure 6-112 shows how 18 different locksets can be replaced by National Lock's locksets or lever sets. Figure 6-113 shows some of the designs available for strikes. The strike is always supplied with the lockset. Figure 6-114 shows the latch bolts. They may have round or square corners. They may have or may not have deadlock capability.

Entrance handle locks Most homes have an elaborate front door handle. In Fig. 6-115 you can see two of the types of escutcheons used to decorate the doorknob.

Door handles also become something of a decorative item. They come in a number of styles. Each lock manufacturer offers a complete line. See Fig. 6-116 for an illustration of two such handles. These handles are usually cast brass.

A number of lockset designs are available for entrance and interior doors. They may lock, then require a nail or pin to be opened. Or they may require a key. See Fig. 6-117.

Auxiliary locks Auxiliary locks are those placed on exterior doors to prevent burglaries. They are called deadlocks. They usually have a 1-inch bolt that projects

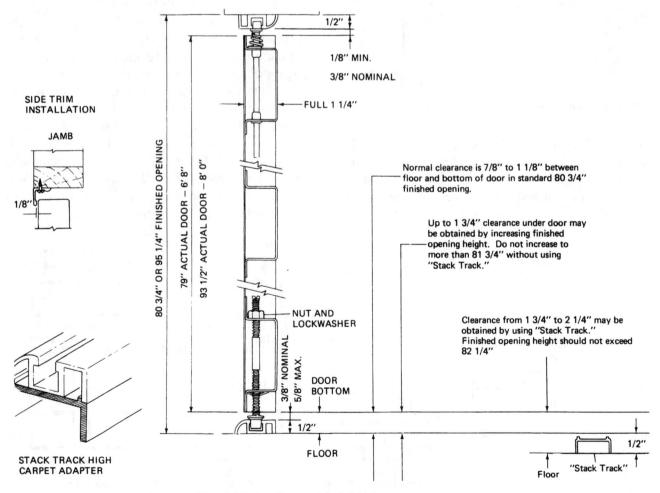

Fig. 6-96 *Installation of the bifold door.* (General Products)

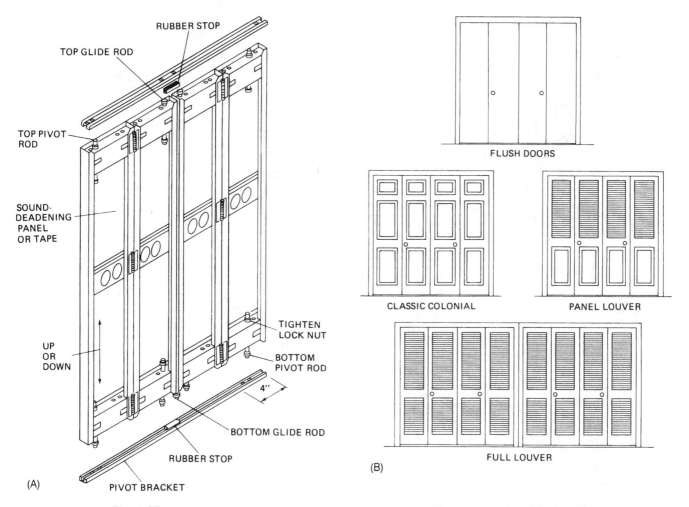

Fig. 6-97 (A) Details of the metal bifold door. *(General Products)* (B) Different designs for bifold doors.

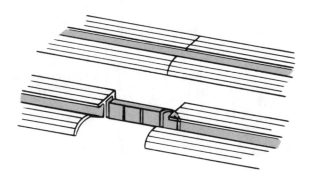

Fig. 6-98 *Vinyl connectors let you snap apart four-panel track instantly to install two-panel bifolds.* (General Products)

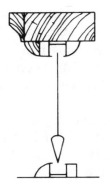

Fig. 6-99 *Using a plumb bob to make sure the tracks line up.* (General Products)

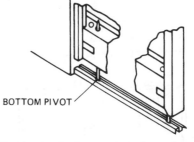

Fig. 6-100 *Pivoting the bottom pin in the track.* (General Products)

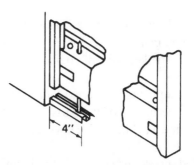

Fig. 6-101 *You can remove most of the bottom track if you don't want it on the floor. This lets carpeting run straight through to the closet wall.* (General Products)

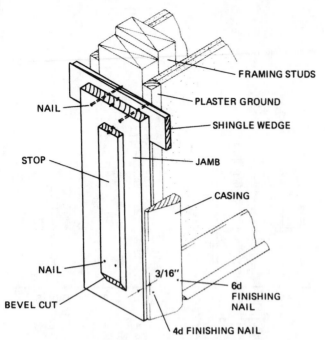

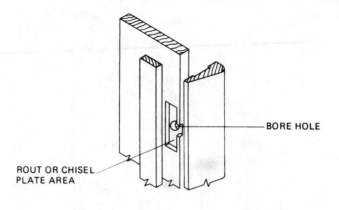

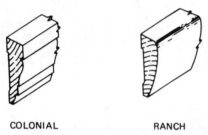

Fig. 6-102 *Trim details for a door frame.*

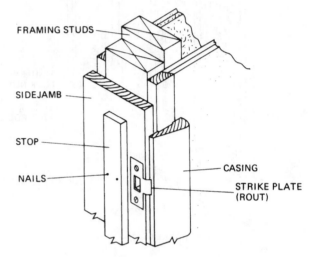

Fig. 6-103 *Two popular types of molding used for trim around a door.*

Fig. 6-105 *Installing a strike plate for the lockset.*

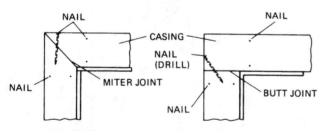

Fig. 6-104 *Two methods of joining trim over a door.*

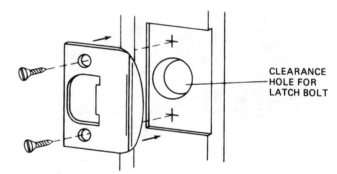

Fig. 6-106 *Installing the strike plate on the door jamb.*

past the door. It fits into the door jamb (Fig. 6-118). In Fig. 6-119 you can see three of the various deadlock designs used. The key side, of course, goes on the outside of the door.

A standard type of lock is exploded for you in Fig. 6-120. Note the names of the parts. These locks have keys that fit into a cylinder. The keys lock or unlock the latch bolt.

Exposed brass, bronze, or aluminum parts are buffed or brushed. They are protected with a coat of lacquer. Aluminum is brushed and anodized.

Construction keying There are a couple of methods used for keying locksets. One of them makes it possible for a construction supervisor to get into a number of buildings with one key. Once the building is occupied, the lock is converted. The lock's key and no other will then operate it. See Fig. 6-121.

Builders use a short four-pin tumbler key to operate the lock during the construction period. A nylon wafer is inserted in the keyway at the factory. This blocks operation of the fifth and sixth tumblers. Accidental

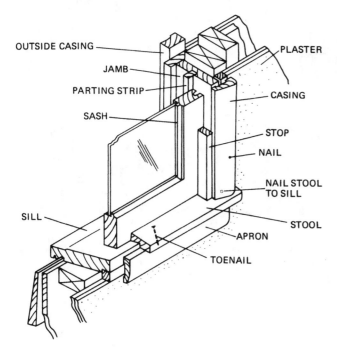

Fig. 6-107 *Window trim. Note the apron and stool.*

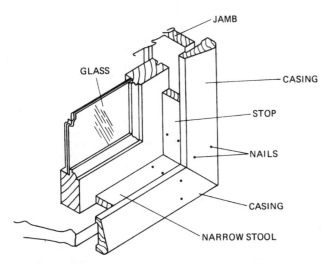

Fig. 6-108 *Window trim. Note the absence of the apron.*

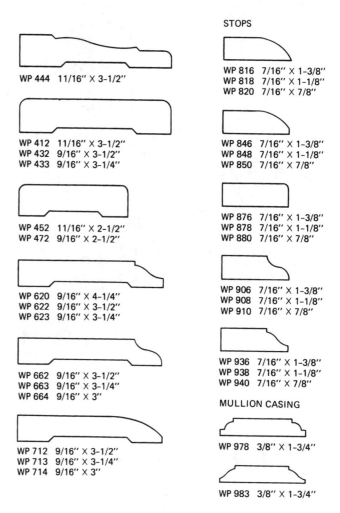

STOPS

WP 444	11/16″ × 3-1/2″

WP 412	11/16″ × 3-1/2″
WP 432	9/16″ × 3-1/2″
WP 433	9/16″ × 3-1/4″

WP 452	11/16″ × 2-1/2″
WP 472	9/16″ × 2-1/2″

WP 620	9/16″ × 4-1/4″
WP 622	9/16″ × 3-1/2″
WP 623	9/16″ × 3-1/4″

WP 662	9/16″ × 3-1/2″
WP 663	9/16″ × 3-1/4″
WP 664	9/16″ × 3″

WP 712	9/16″ × 3-1/2″
WP 713	9/16″ × 3-1/4″
WP 714	9/16″ × 3″

WP 816	7/16″ × 1-3/8″
WP 818	7/16″ × 1-1/8″
WP 820	7/16″ × 7/8″

WP 846	7/16″ × 1-3/8″
WP 848	7/16″ × 1-1/8″
WP 850	7/16″ × 7/8″

WP 876	7/16″ × 1-3/8″
WP 878	7/16″ × 1-1/8″
WP 880	7/16″ × 7/8″

WP 906	7/16″ × 1-3/8″
WP 908	7/16″ × 1-1/8″
WP 910	7/16″ × 7/8″

WP 936	7/16″ × 1-3/8″
WP 938	7/16″ × 1-1/8″
WP 940	7/16″ × 7/8″

MULLION CASING

WP 978	3/8″ × 1-3/4″

WP 983	3/8″ × 1-3/4″

Fig. 6-109 *Different designs of molding used as trim for windows, doors, and other parts of the house.*

deactivation of the builder's key is unlikely. This is because a conscious effort to apply a 10- to 15-pound force is required to dislodge the nylon wafer the first time the five-tumbler key is used. The owner's keys are packed in specially marked, sealed envelopes.

When construction is completed, the unit is ready for occupancy. The homeowner inserts the regular five-pin tumbler key to move the nylon wafer. This makes the fifth tumbler operative. It automatically deactivates the four-pin tumbler arrangement, making the builder's key useless. Now the locksets can be operated only by the owner's keys.

Other lockmakers have different methods for this key operation. See Fig. 6-122. Figure 6-123 shows another method of key operation of locks. In this case the whole cylinder of the new lock is removed. The construction worker inserts a different cylinder. This cylinder works with the master key. Once the job is finished, the original cylinder is reinserted. The construction worker's key no longer operates this lock. Only the homeowner can operate the lock.

STORM DOORS AND WINDOWS

In most windows thermopane is used. It consists of two pieces of glass welded together with an air space. (See Fig. 6-20.) Some windows have the space evacuated so that a vacuum exists inside. This cuts down on the transfer of heat from the inside of a heated building to the cold outside, or the reverse during the summer. See Fig. 6-124.

Metal transfers heat faster than wood. Wood is 400 times more effective than steel as an insulator. It is 1800 times more effective than aluminum.

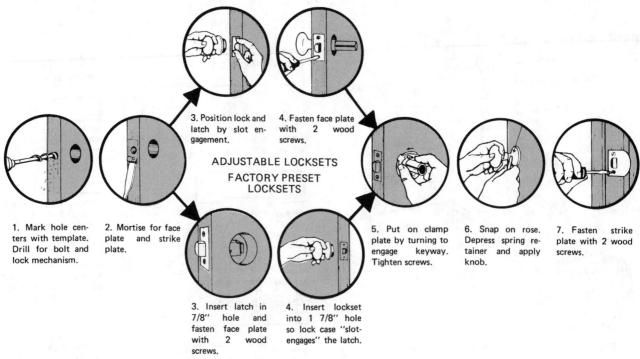

3. Position lock and latch by slot engagement.

4. Fasten face plate with 2 wood screws.

ADJUSTABLE LOCKSETS
FACTORY PRESET LOCKSETS

1. Mark hole centers with template. Drill for bolt and lock mechanism.

2. Mortise for face plate and strike plate.

3. Insert latch in 7/8" hole and fasten face plate with 2 wood screws.

4. Insert lockset into 1 7/8" hole so lock case "slot-engages" the latch.

5. Put on clamp plate by turning to engage keyway. Tighten screws.

6. Snap on rose. Depress spring retainer and apply knob.

7. Fasten strike plate with 2 wood screws.

Fig. 6-110 *Seven simple steps to install a lockset.* (National Lock)

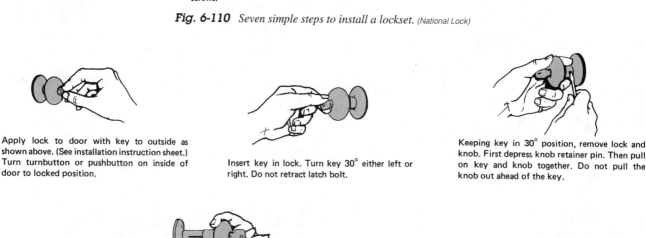

Apply lock to door with key to outside as shown above. (See installation instruction sheet.) Turn turnbutton or pushbutton on inside of door to locked position.

Insert key in lock. Turn key 30° either left or right. Do not retract latch bolt.

Keeping key in 30° position, remove lock and knob. First depress knob retainer pin. Then pull on key and knob together. Do not pull the knob out ahead of the key.

Be sure plug is in locked position. Then remove key approximately halfway out of plug. Turn entire plug, key, and cylinder in knob so bitting of key is up.

Apply knob on cam by engaging tab on knob with slot in tube. Keep key in vertical position. Push knob on tube while rotating key and plug gently. You will feel the proper engagement of locating tab on lock plug with slot in tube.

Depress knob retainer pin and push on knob as far as possible.

Push key all the way in keyway. Turn key to 30° position and push knob until retainer pin engages slot in knob shank.

Fig. 6-111 *Reversing the hand of a lock.* (National Lock)

Storm doors come in a wide variety of shapes and designs. They usually have a combination of screen and glass. The glass is removed in the summer and a screen wire panel is inserted in its place. This way the storm door serves year-round. See Fig. 6-125. They are delivered prehung and ready for installation. All that has to be done to install them is to level the door and add screws in the holes around the edges. A door closer is added to make sure the door closes after use. In some cases a spring adjustment device is added to the top so that the door closer and the door are protected from wind gusts. Most storm doors are made of metal. However, they are available in wood or plastic.

EASY LOCKSET INSTALLATION

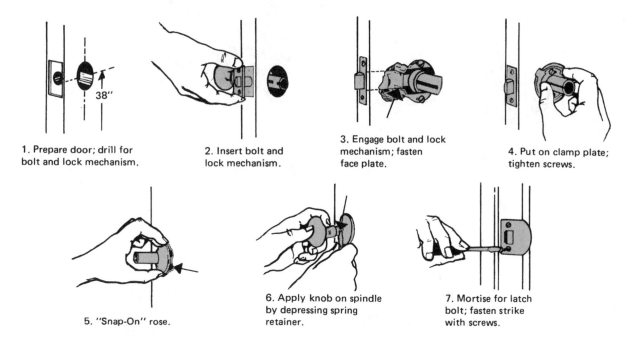

1. Prepare door; drill for bolt and lock mechanism.

2. Insert bolt and lock mechanism.

3. Engage bolt and lock mechanism; fasten face plate.

4. Put on clamp plate; tighten screws.

5. "Snap-On" rose.

6. Apply knob on spindle by depressing spring retainer.

7. Mortise for latch bolt; fasten strike with screws.

EASY LOCKSET REPLACEMENT

With just a screwdriver, the following 18 residential lockset brands can be replaced by National Lock locksets or lever sets.

ARROW COMET CORBIN DONNER ELGIN HARLOC KWIKSET LOCKWOOD MEDALIST NATIONAL RUSSWIN SARGENT SCHLAGE TROJAN TROY WEISER WESLOCK YALE

Fig. 6-112 Replacement of a lockset. (National Lock)

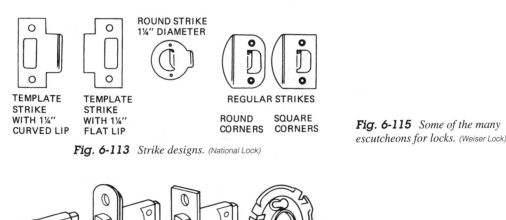

TEMPLATE STRIKE WITH 1¼" CURVED LIP

TEMPLATE STRIKE WITH 1¼" FLAT LIP

ROUND STRIKE 1¼" DIAMETER

REGULAR STRIKES

ROUND CORNERS

SQUARE CORNERS

Fig. 6-113 Strike designs. (National Lock)

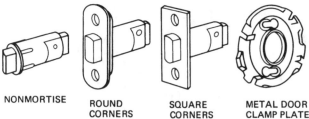

NONMORTISE

ROUND CORNERS

SQUARE CORNERS

METAL DOOR CLAMP PLATE

Fig. 6-114 Latch bolt designs. (National Lock)

Standard sizes are for openings from 35¾ to 36⅜ inches wide and from 79¾ to 81¼ inches high. There is an extender Z bar available for openings up to 37⅛ inches.

Fig. 6-115 Some of the many escutcheons for locks. (Weiser Lock)

According to the company's testing lab, this plastic (polypropylene) door has 45 percent more heat retention than an aluminum door. See Fig. 6-126.

INSTALLING A SLIDING DOOR

Sliding doors are a common addition to a house today. The doors slide open so that the patio can be reached

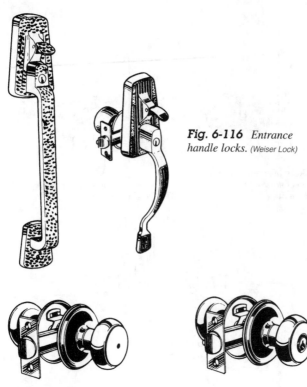

Fig. 6-118 *A deadlock projects out from the door and fits into the door jamb to make a secure door.* (Weiser Lock)

Fig. 6-116 *Entrance handle locks.* (Weiser Lock)

Fig. 6-117 *Locksets for interior and exterior doors.* (Weiser Lock)

Fig. 6-119 *Auxiliary locks. Chain and bolt, and two types of deadbolts.* (Weiser Lock)

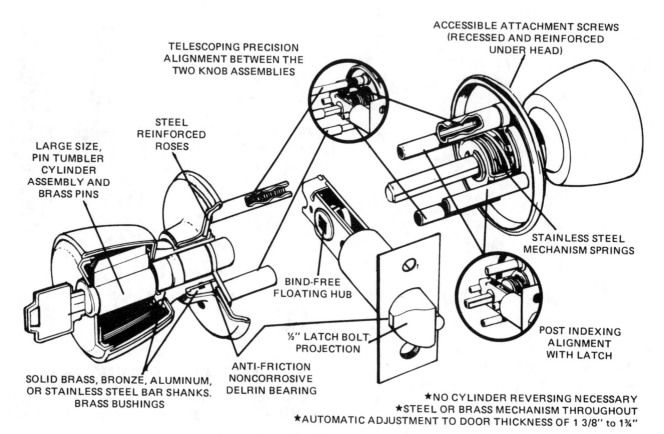

ACCESSIBLE ATTACHMENT SCREWS (RECESSED AND REINFORCED UNDER HEAD)

TELESCOPING PRECISION ALIGNMENT BETWEEN THE TWO KNOB ASSEMBLIES

STEEL REINFORCED ROSES

LARGE SIZE, PIN TUMBLER CYLINDER ASSEMBLY AND BRASS PINS

STAINLESS STEEL MECHANISM SPRINGS

BIND-FREE FLOATING HUB

½" LATCH BOLT PROJECTION

POST INDEXING ALIGNMENT WITH LATCH

SOLID BRASS, BRONZE, ALUMINUM, OR STAINLESS STEEL BAR SHANKS. BRASS BUSHINGS

ANTI-FRICTION NONCORROSIVE DELRIN BEARING

★NO CYLINDER REVERSING NECESSARY
★STEEL OR BRASS MECHANISM THROUGHOUT
★AUTOMATIC ADJUSTMENT TO DOOR THICKNESS OF 1 3/8" to 1¾"

Fig. 6-120 *Exploded view of a lockset.* (Weiser Lock)

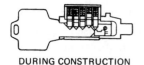

DURING CONSTRUCTION

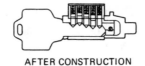

AFTER CONSTRUCTION

Fig. 6-121 *Construction keying of locksets.* (National Lock)

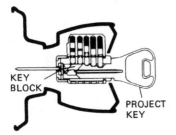

KEY BLOCK

PROJECT KEY

Lock cylinder is operated by the special "project key." The last two pins in the cylinder are held inoperative by the key block.

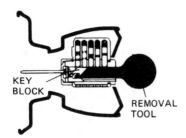

KEY BLOCK

REMOVAL TOOL

The special "project key" is canceled out by removal of the key block. A key block removal tool is furnished with the master keys for the locks. Simply push the removal tool into the keyway. Upon withdrawal, the key block will come out of the keyway. Thereafter, the "project key" no longer will operate the lock cylinder.

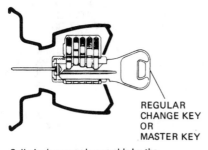

REGULAR CHANGE KEY OR MASTER KEY

Cylinder is now only operable by the regular change key or master key.

Fig. 6-122 *Another method of construction keying.* (Weiser Lock)

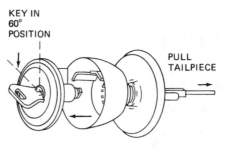

KEY IN 60° POSITION

PULL TAILPIECE

REMOVE REGULAR KEY CYLINDER

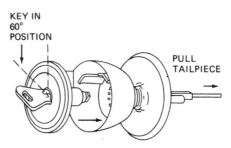

KEY IN 60° POSITION

PULL TAILPIECE

INSERT CONSTRUCTION CYLINDER

Fig. 6-123 *Some types of construction cylinders have to be removed.* (Weiser Lock)

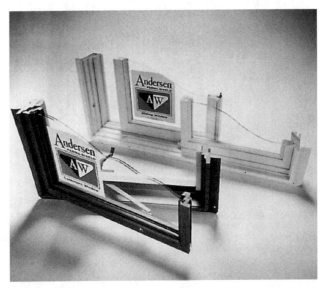

Fig. 6-124 *Examples of thermopane windows.* (Andersen)

easily. It takes some special precautions to make sure the sliding doors will operate correctly. Most of these doors are made by a manufacturer like Andersen. They require a minimum of effort on the part of the carpenter. However, some very special steps are required. This portion of the chapter will deal primarily with the primed-wood type of gliding door and the Andersen Perma-Shield® type of gliding door.

The primed-wood gliding door (see Fig. 6-127) does not have a flange around it for quick installation. It requires some special attention. You will get an idea of how it fits into the rough opening from Fig. 6-128.

The Perma-Shield® gliding door is slightly different from the primed-wood type. See Fig. 6-129. In Fig. 6-130 you will find the details of the Perma-Shield door so that you can see the differences between the two types.

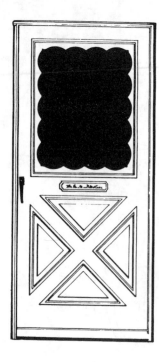

Fig. 6-125 Storm door. (EMCO)

Installation of both types of doors requires a rough opening in the frame structure of the house or building. The rough opening is constructed the same way for both types of doors.

Preparation of the Rough Opening

Installation techniques, materials, and building codes vary from area to area. Contact your local material supplier for specific recommendations for your area.

The same rough opening procedures can be used for both doors. There are, however, variations to note in gliding door installation procedures. These will be looked at more fully as we go along here.

If you need to enlarge the opening, make sure you use the proper-size header. Header size is usually given by the manufacturer of the door, or you can obtain it from your local supplier.

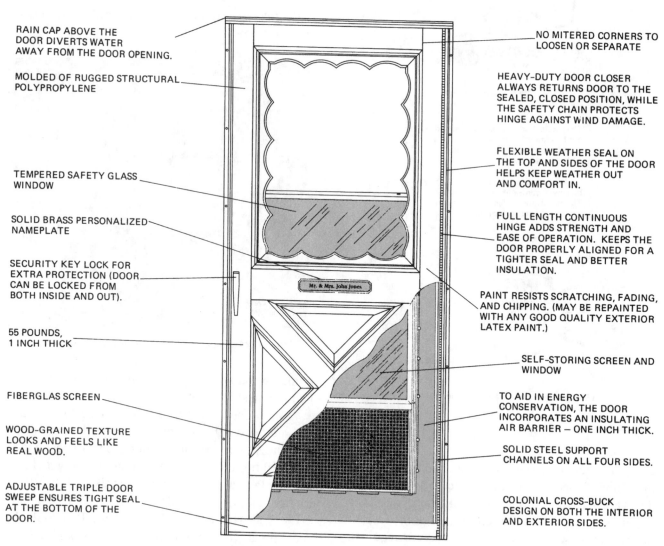

RAIN CAP ABOVE THE DOOR DIVERTS WATER AWAY FROM THE DOOR OPENING.

MOLDED OF RUGGED STRUCTURAL POLYPROPYLENE

TEMPERED SAFETY GLASS WINDOW

SOLID BRASS PERSONALIZED NAMEPLATE

SECURITY KEY LOCK FOR EXTRA PROTECTION (DOOR CAN BE LOCKED FROM BOTH INSIDE AND OUT).

55 POUNDS, 1 INCH THICK

FIBERGLAS SCREEN

WOOD-GRAINED TEXTURE LOOKS AND FEELS LIKE REAL WOOD.

ADJUSTABLE TRIPLE DOOR SWEEP ENSURES TIGHT SEAL AT THE BOTTOM OF THE DOOR.

NO MITERED CORNERS TO LOOSEN OR SEPARATE

HEAVY-DUTY DOOR CLOSER ALWAYS RETURNS DOOR TO THE SEALED, CLOSED POSITION, WHILE THE SAFETY CHAIN PROTECTS HINGE AGAINST WIND DAMAGE.

FLEXIBLE WEATHER SEAL ON THE TOP AND SIDES OF THE DOOR HELPS KEEP WEATHER OUT AND COMFORT IN.

FULL LENGTH CONTINUOUS HINGE ADDS STRENGTH AND EASE OF OPERATION. KEEPS THE DOOR PROPERLY ALIGNED FOR A TIGHTER SEAL AND BETTER INSULATION.

PAINT RESISTS SCRATCHING, FADING, AND CHIPPING. (MAY BE REPAINTED WITH ANY GOOD QUALITY EXTERIOR LATEX PAINT.)

SELF-STORING SCREEN AND WINDOW

TO AID IN ENERGY CONSERVATION, THE DOOR INCORPORATES AN INSULATING AIR BARRIER — ONE INCH THICK.

SOLID STEEL SUPPORT CHANNELS ON ALL FOUR SIDES.

COLONIAL CROSS-BUCK DESIGN ON BOTH THE INTERIOR AND EXTERIOR SIDES.

Mr. & Mrs. John Jones

Fig. 6-126 Energy saving plastic storm door. (EMCO)

Preparation of the rough opening should follow these steps:

1. Lay out the gliding-door opening width between the regular studs to equal the gliding-door rough opening width plus the thickness of two regular studs. See Fig. 6-131.

2. Cut two pieces of header material to equal the rough opening width of the gliding door plus the

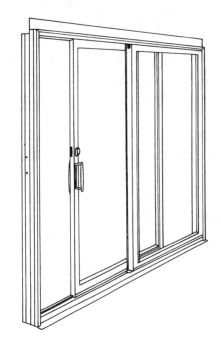

Fig. 6-129 *Perma-Shield® gliding door.* (Andersen)

Fig. 6-127 *Primed-wood gliding door.* (Andersen)

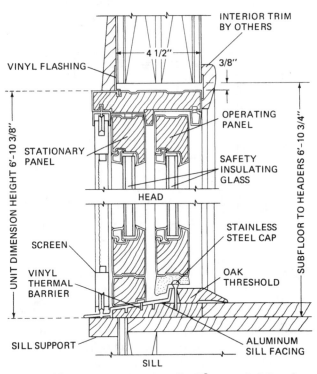

Fig. 6-130 *Details of the Perma-Shield® type of gliding door.* (Andersen)

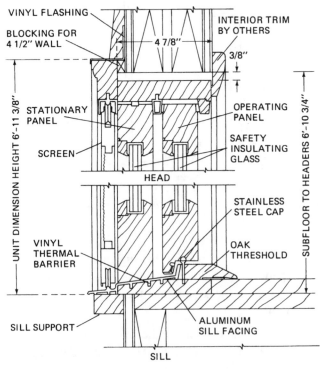

Fig. 6-128 *Details of the primed-wood gliding door.* (Andersen)

thickness of two trimmer studs. Nail two header members together using an adequate spacer so that the header thickness equals the width of the trimmer stud. See Fig. 6-132.

3. Position the header at the proper height between the regular studs. Nail through the regular studs into the header to hold the header in place until the next step is completed. See Fig. 6-133.

4. Cut the jack or trimmer studs to fit under the header. This will support the header. Nail the jack or trimmer studs to the regular studs. See Fig. 6-134.

5. Apply the exterior sheathing (fiberboard, plywood, etc.) flush with the header and jack stud members. See Fig. 6-135.

Installation of a Wood Gliding Door

Keep in mind that all these illustrations are as viewed from the outside. Be sure the subfloor is level and the rough opening is plumb and square before installing the gliding-door frame. If you follow these steps closely, the door should be properly installed.

1. Run caulking compound across the opening to provide a tight seal between the door sill and the floor. Remove the shipping skids from the sill of the frame if the gliding door has been shipped set up. Follow the instructions included in the package if the frame is not set up. See Fig. 6-136.

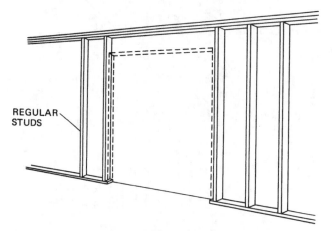

Fig. 6-131 *Layout of a rough opening for a gliding door.* (Andersen)

2. Position the frame in the opening from the outside. See Fig. 6-137. Apply pressure to the sill to properly distribute the caulking compound. The sill must be level. Check carefully and shim if necessary.

3. After leveling the sill, secure it to the floor by nailing along the inside edge of the sill with 8d coated nails spaced approximately 12 inches apart. See Fig. 6-138.

4. The jamb must be plumb and straight. Temporarily secure it in the opening with 10d casing nails through each side casing into the frame members. Using a straightedge, check the jambs for bow and shim. Shim solidly (five per jamb) between side jambs and jack studs.

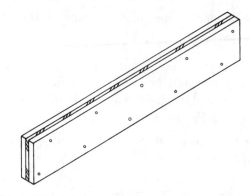

Fig. 6-132 *Header for a gliding-door opening.* (Andersen)

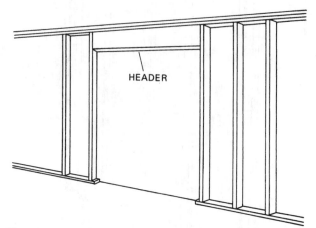

Fig. 6-133 *Placement of the header in the rough opening.* (Andersen)

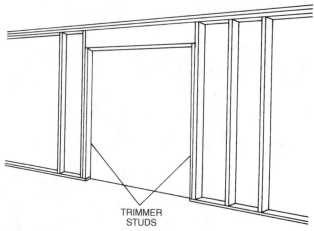

Fig. 6-134 *Placement of jack or trimmer studs in the rough opening.* (Andersen)

5. Complete the exterior nailing of the unit in the opening by nailing through the side and head casings into the frame members with 10d casing nails. See Fig. 6-139.

6. Position the flashing on the head casing and secure it by nailing through the vertical leg. The vertical center brace may now be removed from the frame.

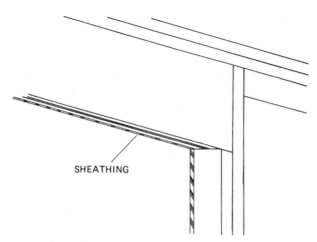

Fig. 6-135 *Application of exterior sheathing over the header.*
(Andersen)

SHEATHING

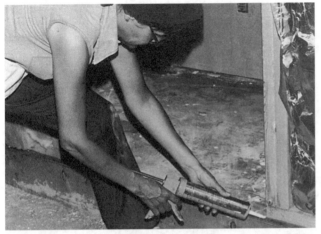

Fig. 6-136 *Running a bead of caulking to seal the sill and door for a sliding door.*

Fig. 6-137 *Leveling the sill.*

Be sure to remove and save the head and sill brackets. See Fig. 6-140.

7. Apply the treated wood sill support under the protruding metal sill facing. See Fig. 6-141. Install it tight to the underside of the metal sill with 10d casing nails.

Fig. 6-138 *Securing the sill to the floor by nailing.*

Fig. 6-139 *Exterior nailing of the unit.* (Andersen)

Fig. 6-140 *Securing the flashing on the head casing by nailing.*

Fig. 6-141 *Applying the sill support under the metal sill facing.*

8. Position the stationary door panel in the outer run. Be sure the bottom rail is straight with the sill. Force the door into the run of the side jamb with a 2-×-4 wedge. See Fig. 6-142. Check the position by aligning the screw holes of the door bracket with the holes in the sill and head jamb. Repeat the above procedure for stationary panels of the triple door (if one is used here). Before the left-hand stationary panel is installed in a triple door, be sure to remove the screen bumper on the sill. Keep in mind, however, that only a double door is shown here.

9. Note the mortise in the bottom rail for a bracket. Secure the bracket with No. 8 one-inch flathead screws through the predrilled holes. See Fig. 6-143

Fig. 6-142 *Positioning the stationary door panel in the outer run by using a 2-×-4 wedge.*

for details. Align the bracket with the predrilled holes in the head jamb and secure it with No. 8 one-inch flathead screws. See Fig. 6-144. Repeat the procedure for the stationary panels of a triple door. The head stop is now removed if the unit has been shipped set up.

10. Apply security screws. Apply the two 1½-inch No. 8 flathead painted head screws through the parting stop into the stationary door top rail. See Fig. 6-145. Repeat for the stationary panels of a triple door.

11. Place the operating door on the rib of the metal sill facing, and tip the door in at the top. See Fig. 6-146. Position the head stop and apply with 1⁹⁄₁₆-inch No. 7 screws. See Fig. 6-147.

12. Check the door operation. If the door sticks or binds or is not square with the frame, locate the two adjustment sockets on the outside of the bottom rail. See Fig. 6-148. Simply remove the caps, insert the screwdriver, and turn to raise or lower the door. Replace the caps firmly.

Fig. 6-143 *Securing the bottom bracket with a screw.* (Andersen)

Fig. 6-144 *Securing the top bracket with a screw.* (Andersen)

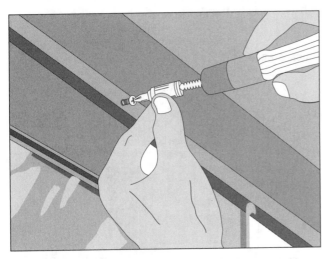

Fig. 6-145 *Applying the security screws.* (Andersen)

Fig. 6-146 *Placing the operating door on the rib of the metal sill facing.* (Andersen)

Fig. 6-147 *Positioning the head stop.* (Andersen)

13. If it is necessary to adjust the "throw" of the latch on two-panel doors, turn the adjusting screw to move the latch in or out. See Fig. 6-149. The lock may be adjusted on triple doors by loosening the screw to move the lock plate. See Fig. 6-150.

Masonry or Brick-Veneer Wall Installation of a Gliding Door

Gliding doors can be installed in masonry wall construction. Fasten a wood buck to the masonry wall and nail the sliding door to the wood buck using the procedures shown for frame wall construction.

Fig. 6-148 *Adjusting the door for square.* (Andersen)

Fig. 6-149 *The throw of the door is adjusted by this screw.* (Andersen)

Fig. 6-150 *Lock adjustment on a triple door.* (Andersen)

In Fig. 6-151 a wood gliding door is installed in a masonry wall. Figure 6-152 shows a Perma-Shield® door installed with metal wall plugs or extender plugs in the masonry wall with brick veneer. Metal wall plugs or extender plugs and auxiliary casing can be specified when the door is ordered.

Keep in mind that when brick veneer is used as an exterior finish, adequate clearance must be left for caulking between the frame and the masonry. This will prevent damage and bowing caused by shrinkage and settling of the structural lumber.

Installation of a Perma-Shield Gliding Door

The Perma-Shield type of door is installed in the same way as has just been described, with some exceptions. The exceptions follow:

1. Note that wide vinyl flanges which provide flashing are used at the head and side jambs. See Fig. 6-153A. Locate the side member flush with the bottom of the sill with an offset leg pointing toward the inside of the frame. Tap with the hammer using a wood block to firmly seat the flashing in the groove. See Fig. 6-153B. Apply the head member similarly. Overlap the side flange on the outside.

2. After securing the frame to the floor with nails through the sill, apply clamps to draw the flanges tightly against the sheathing. See Fig. 6-154.

3. Temporarily secure the door in the opening with 10d casing nails through each side casing into the frame members. Using a straightedge, check the jambs for bow and shim. The jamb must be plumb

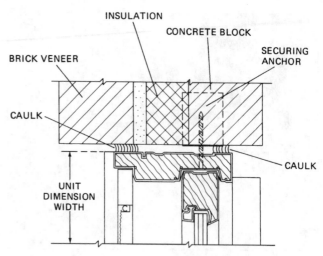

Fig. 6-152 Perma-Shield® *door installed with metal wall plugs or extender plugs in masonry wall with a brick veneer.* (Andersen)

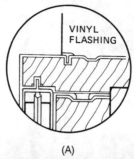

(A)

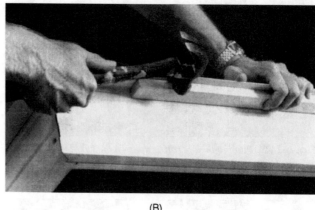

(B)

Fig. 6-153 (A) Location of the vinyl flashing after it has been applied (Andersen). (B) Using a wooden block to apply the vinyl flashing. (Andersen)

and straight. Shim solidly using five shims per jamb between the side jambs and the jack studs.

4. Side members on the Perma-Shield® gliding doors have predrilled holes to receive 2½-inch No. 10 screws. See Fig. 6-155. Shim at all screw holes between the door frame and the studs. Drill pilot holes into the studs. Secure the door frame to the studs with screws.

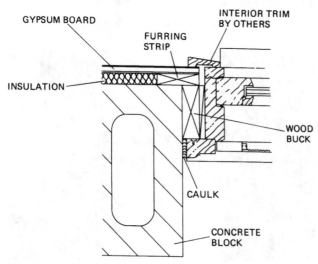

Fig. 6-151 Wood gliding door installed in masonry wall. (Andersen)

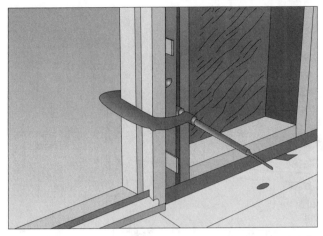

Fig. 6-154 *Clamps draw the flanges tightly against the sheathing.* (Andersen)

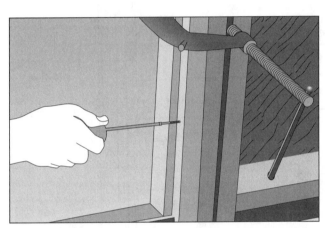

Fig. 6-155 *Securing the door frame to the studs with screws.* (Andersen)

5. The head jamb also has predrilled holes to receive the 2½-inch No. 10 screws. Shim at all screw holes between the door frame and header. Drill the pilot holes into the header. Insert the screws through the predrilled holes and draw up tightly. Do not bow the head jamb. See Fig. 6-156.

Figure 6-157 shows a Perma-Shield® door completely installed. Interior surfaces of the panel and frame should be primed before or immediately after installation for protection.

INSTALLING THE GARAGE DOOR

Garage doors are available in metal and wood. Wood doors are available in a high-grade hemlock/fir frame construction and recessed hardboard or raised redwood panels. Rough sawn, flush wood doors are also available. They come ready to be primed, painted, or stained to match the house finish. Steel doors come with a primer and need to be painted with a second coat to match the owner's preference in color or finish.

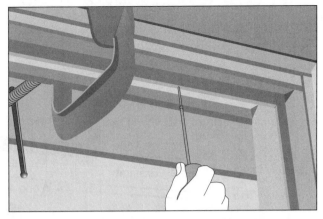

Fig. 6-156 *Placing screws in predrilled holes in the door header.* (Andersen)

Fig. 6-157 *Finished installation of a Perma-Shield® gliding door.*

Figures 6-158 and 6-159 show the types of springs used to aid in the raising of the garage door. The torsion springs are usually used for heavy doors for a two-car garage. A garage door with extension springs is usually involved with single-car garage doors.

A word of caution is usually sufficient: to avoid installation problems that could result in personal injury or property damage, use only the track specified and supplied with the door unit.

Some large doors weigh as much as 400 pounds when the spring tension is released. A single door weighs up to 200 pounds, and two people should work on it to prevent damage.

One of the primary concerns about installing a door is the headroom. Headroom is the space needed above the top of the door for the door, the overhead tracks, and the springs. Measure to check that there is

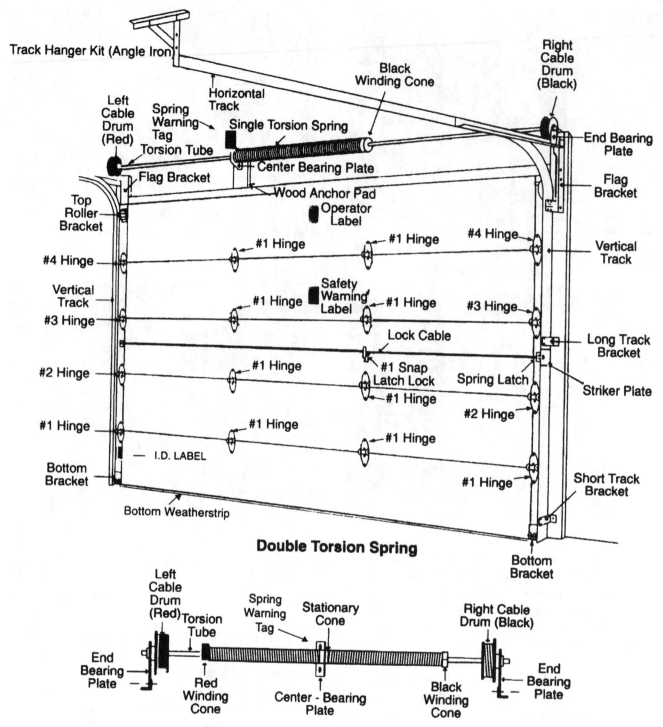

Fig. 6-158 *A garage door with torsion springs.* (Clopay)

no obstructions in your garage within that space. The normal space requirement is shown in Table 6-2. The backroom distance is measured from the back of the door into the garage, and should be at least 18 inches more than the height of the garage door. A minimum sideroom of 3.75 inches (5.5 inch EZ-Set Spring®) should be available on each side of the door on the interior wall surface to allow for attachment of the vertical track assembly. See Fig. 6-160.

Track radius is another important concern in the installation of the track. The radius of the track can be determined by measuring the dimension *R* in Fig. 6-161. If the dimension *R* measures 11 to 12 inches, then you have a 12-radius track. If *R* equals 14 to 15 inches, then you have a 15-inch radius track. See Fig. 6-161. About 3 inches of additional headroom height at the center plus additional backroom is needed to install an automatic garage door opener.

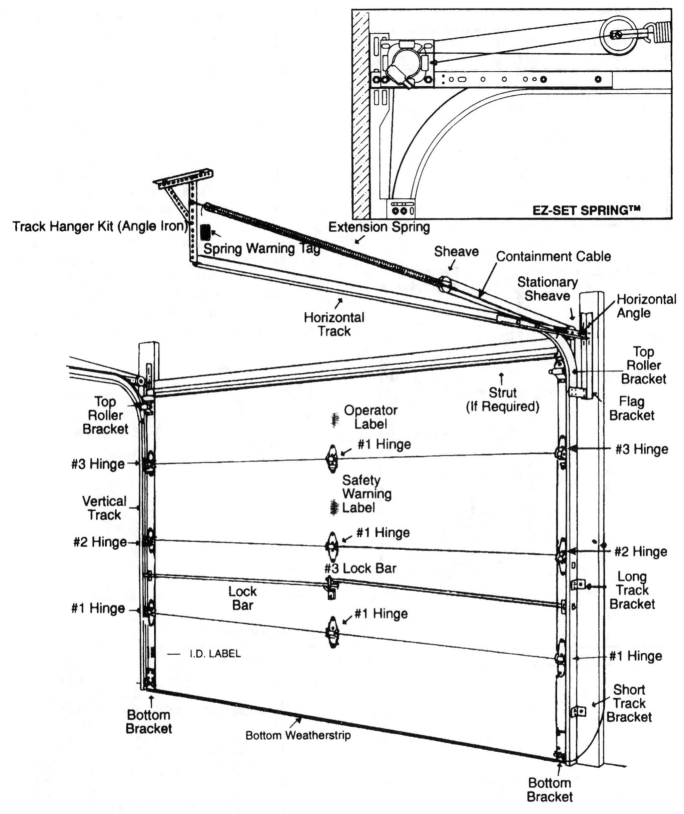

Track Hanger Kit (Angle Iron)

Spring Warning Tag

Extension Spring

Sheave

Containment Cable

Stationary Sheave

Horizontal Angle

Horizontal Track

EZ-SET SPRING™

Top Roller Bracket

Flag Bracket

Strut (If Required)

Top Roller Bracket

Operator Label

#1 Hinge

#3 Hinge

#3 Hinge

Safety Warning Label

Vertical Track

#2 Hinge

#1 Hinge

#2 Hinge

#3 Lock Bar

Long Track Bracket

Lock Bar

#1 Hinge

#1 Hinge

#1 Hinge

I.D. LABEL

Short Track Bracket

Bottom Bracket

Bottom Weatherstrip

Bottom Bracket

Fig. 6-159 *A garage door with extension springs.* (Clopay)

Check the door opener instructions to be sure. If the headroom is low there are a few options to compensate. For instance, a double track low headroom reduces the headroom requirement to 4.5 inches on

EZ-Set Spring® and extension springs (9.5 inches on front mount torsion springs). Instructions are provided with the track. There is a low headroom conversion kit that reduces the required headroom to 4.5

Table 6-2 *Headroom Requirement Chart*

SPRING TYPE	TRACK RADIUS	HEADROOM REQUIRED
EZ-Set Spring™/ Extension	12"	10"
EZ-Set Spring™/ Extension	15"	12"
Torsion	12"	12"
Torsion	15"	14"

with any other low headroom option. This is used in place of the existing top roller. Instructions are included with the kit.

The next step is to prepare the opening. Figure 6-163 shows the rough opening for the door and the necessary additions. An option is stop molding featuring a built-in weather seal as shown in Fig. 6-164.

Next, prepare for installing the door sections. Spread the hardware on the floor in groups so that you can easily find the parts. Assemble as the directions require. One thing to keep in mind is if the door is going to be equipped with an automatic garage door opener, make sure that the door is always unlocked when the operator is being used. This will avoid damage to the door. Instructions come with the door in the form of a pamphlet with detailed line drawings.

Assemble and install the track. Follow the directions for the specific door being used. Pay special attention and use adequate length screws to fasten the

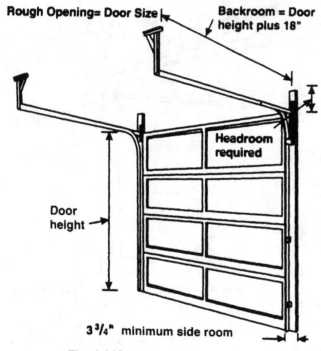

Fig. 6-160 *Required headroom.* (Clopay)

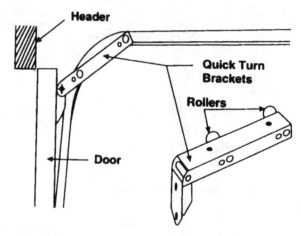

Fig. 6-162 *Quick-Turn bracket for low headroom.* (Clopay)

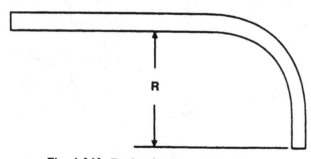

Fig. 6-161 *Track radius measurement.* (Clopay)

inches. This option is designed to modify the standard track. Instructions are provided in the kit. Another way to reduce the headroom requirement is to use the Quick-Turn bracket. See Fig. 6-162. The Quick-Turn bracket cannot be used in conjunction

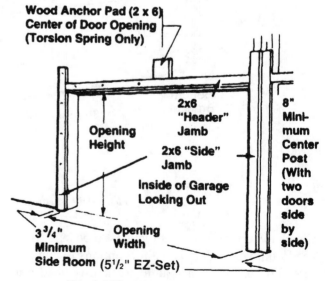

Fig. 6-163 *Preparing the opening.* (Clopay)

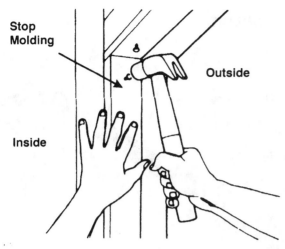

Fig. 6-164 *Door stop molding.* (Clopay)

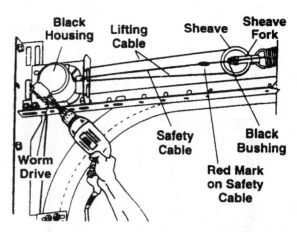

Fig. 6-165 *Adjusting the spring tension with a hand drill.* (Clopay)

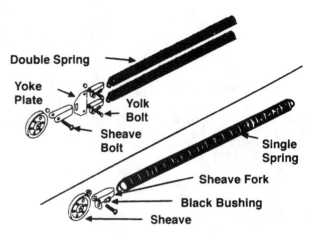

Fig. 6-166 *Single- and double-extension springs.* (Clopay)

rear track hangers into the trusses. A door may fall and cause serious injury if not properly secured.

Attaching springs Lifting cables and springs can be dangerous when installed incorrectly. It is very important to follow the instructions closely when installing the springs. Garage door springs can cause serious injury and property damage if they break under tension and are not secured with safety cables.

Keep your head well below the track when the spring is under tension or being tensioned because springs are dangerous when they are fully or partially wound. The first time the door is opened, make sure the door doesn't fall. This might happen if the tracks are not correctly aligned or the rear track hangers are not strong enough. Proceed slowly and carefully and follow the directions provided by the manufacturer. Both springs should be adjusted equally for proper operation.

The EZ-Set Spring® by Clopay can be adjusted by using a ⅜-inch drill using the ¼-inch hex driver provided in the door kit. Make sure the ¼-inch hex driver is inserted completely into the worm drive. The spring is tensioned by operating the drill in the clockwise direction. See Fig. 6-165.

Extension springs are used on single doors and have two springs, such as in Fig. 6-166. The door with less weight uses a single spring on each side of the door. The heavier door may use the double spring on each side. If your door was supplied with four extension springs, take notice of the color coding on the ends of the springs. If there are two color codes, be sure to use one of each on either side so that the spring tension is equal on both sides.

Torsion spring installation Torsion springs can be very dangerous if they are improperly installed or mis-

handled. Do not attempt to install them alone unless you have the right tools and reasonable mechanical aptitude or experience, and you follow instructions very carefully. It is important to firmly and securely attach the torsion spring assembly to the frame of the garage. See Fig. 6-167.

Attaching an automatic opener When installing an automatic garage door opener, make sure to follow the manufacturer's installation and safety instructions carefully. Remove the pull-down rope and unlock or remove the lock. If attaching an operator bracket to the wood anchor pad, make sure the wooden anchor pad is free of cracks and splits and is firmly attached to the wall. Always drill pilot holes before attaching lag screws.

To avoid damage to the door, you must reinforce the top section of the door in order to provide a mounting point for the opener to be attached. Failure to reinforce the door, as illustrated in Fig. 6-168, will void the manufacturer's warranty. Note: All reinforcing angles are to be attached with #14 × ⅝-inch sheet metal screws at the reinforcement back-up plate locations.

Torsion Spring Installation

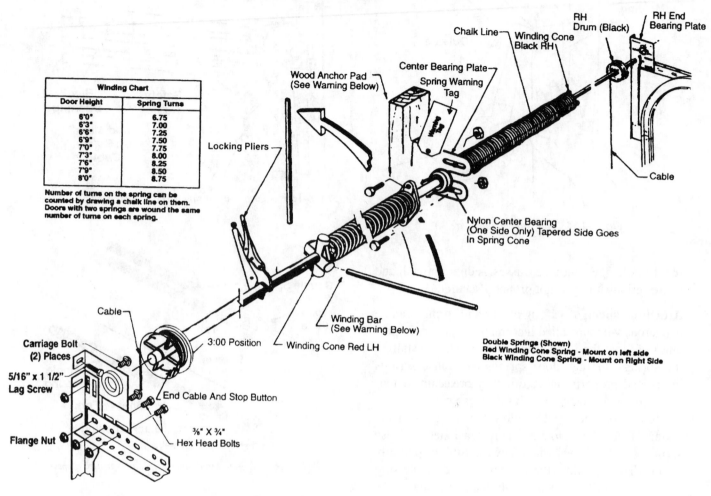

Winding Chart	
Door Height	**Spring Turns**
6'0"	6.75
6'3"	7.00
6'6"	7.25
6'9"	7.50
7'0"	7.75
7'3"	8.00
7'6"	8.25
7'9"	8.50
8'0"	8.75

Number of turns on the spring can be counted by drawing a chalk line on them. Doors with two springs are wound the same number of turns on each spring.

Locking Pliers

Chalk Line

Wood Anchor Pad (See Warning Below)

Center Bearing Plate

Spring Warning Tag

Winding Cone Black RH

RH Drum (Black)

RH End Bearing Plate

Nylon Center Bearing (One Side Only) Tapered Side Goes In Spring Cone

Cable

Winding Bar (See Warning Below)

Winding Cone Red LH

3:00 Position

Cable

Carriage Bolt (2) Places

5/16" x 1 1/2" Lag Screw

Flange Nut

End Cable And Stop Button

⅜" X ¾" Hex Head Bolts

Double Springs (Shown)
Red Winding Cone Spring - Mount on left side
Black Winding Cone Spring - Mount on Right Side

Fig. 6-167 *Torsion spring installation.* (Clopay)

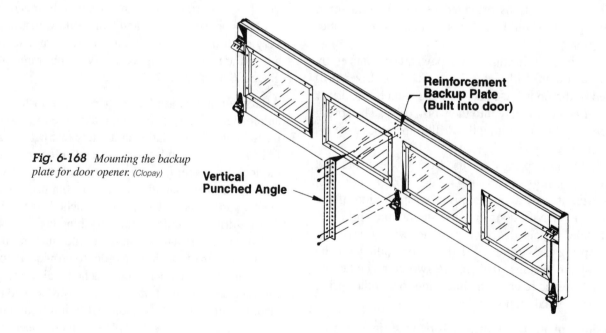

Reinforcement Backup Plate (Built into door)

Vertical Punched Angle

Fig. 6-168 *Mounting the backup plate for door opener.* (Clopay)

Cleaning and painting Before painting the door, it must be free of dirt, oils, chalk, waxes, and mildew. The prepainted surfaces can be cleaned of dirt, oils, chalk, and mildew with a diluted solution of trisodium phosphate. Trisodium phosphate is available over the counter at most stores under the name of Soilax®, in many laundry detergents without fabric softener additives, and in some general purpose cleaners. Check the label for trisodium phosphate content. The recommended concentration is ⅓ cup of powder to 1.5 to 2 gallons of water. After washing the door, always rinse well with clear water and allow it to dry.

The steel door can be painted with a high-quality flat latex exterior grade paint. Because all paints are not created equal, the following test needs to be performed. Paint should be applied on a small area of the door (following the instruction on the paint container), allowed to dry, and evaluated prior to painting the entire door. Paint defects to look for are blistering and peeling. An additional test is to apply a strip of masking tape over the painted area and peel back, checking to see that the paint adheres to the door and not to the tape.

After satisfactorily testing a paint, follow the directions on the container and apply it to the door. Be sure to allow adequate drying time should you decide to apply a second coat.

Window frames and inserts can be painted with a high-quality latex paint. The plastic should first be lightly sanded to remove any surface gloss.

ENERGY FACTORS

There are some coatings for windows and door glass that reflect heat. A thin film is placed over the glass area. This thin film reflects up to 46 percent of the solar energy. Only 23 percent is admitted. That means 77 percent of the solar energy is reflected or turned away. Look at Fig. 6-169 to see the extent of this ability to absorb energy and reflect it.

A thin vapor coating of aluminum prevents solar radiation from passing through glass. It does this by reflecting it back to the exterior. The temperature of the glass is not raised significantly. The coating minimizes undesirable secondary radiation from the glass. Visible light is reduced. However, the level of illumination remains acceptable. During the winter this coating reflects long-wave radiation and keeps the heat in the room.

The film is easily applied to existing windows. See Fig. 6-169. Step 1 calls for spraying the entire window with cleaner. Scrape every square inch of the window with a razor blade. This removes the paint,

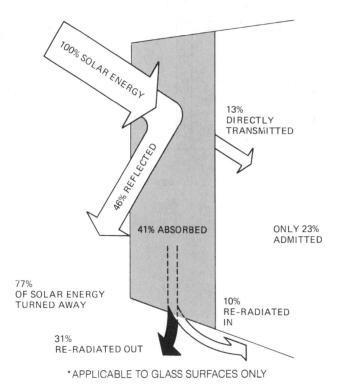

Fig. 6-169 *Energy-saving film for windows.* (Kelly-Stewart)

varnish, and excess putty. Wipe the window clean and dry. Step 2 calls for respraying the window with cleaner. Squeegee the entire window. Use vertical strokes from top to bottom of the window. Use paper towels to dry the edges and corners.

Step 3 calls for clear water to be sprayed onto the glass area. Remove the separator film. See Fig. 6-170. Test for the tacky side of film by folding it over and touching it to itself. The tacky side will stick slightly. Position the tacky side of the film against the top of the glass. Smooth the film by carefully pulling on the edges, or gently press the palms of your hands against the film. Slide the sheet of film easily to remove large wrinkles.

In step 4, spray the film, which is now on the glass, with water. Standing at the center, use a squeegee in gentle, short vertical and horizontal strokes to flatten the film. Slowly work out the wrinkles. Work out the bubbles and remove excess water from under the film. Be careful not to crease or wrinkle the film. Trim the edges with a razor blade. Wipe the excess water from the edges. The film will appear hazy for about two days until the excess water evaporates.

This is shown here to illustrate the possibilities of making your home an energy saver. Films and other devices will be forthcoming as we look forward to conserving energy.

Fig. 6-170 *Step-by-step application of energy-saving film.* (Kelly-Stewart)

7
CHAPTER

Finishing Exterior Walls

EXTERIOR WALLS ARE FINISHED BY TWO basic processes. The first is by covering the wall with a wood or wood product material called siding. The other is to cover the wall with a masonry material such as brick, stone, or stucco. The carpenter will install the exterior siding and will sometimes prepare the exterior wall for the masonry materials. However, people employed in the "trowel trades" install the masonry materials.

Exterior siding is applied over the wall sheathing. It adds protection against weather, strengthens the walls, and also gives the wall its final appearance or beauty. Siding may be made of many materials. Often, more than one type of material is used. Wood and brick could be combined for a different look. Other materials may also be combined. This chapter will help you learn these new skills:

- Prepare the wall for the exterior finish
- Estimate the amount of siding needed
- Select the proper nails for the procedure
- Erect scaffolds
- Install flashing and water tables to help waterproof the wall
- Finish the roof edges
- Install exterior siding
- Trim out windows and doors

INTRODUCTION

Before the exterior is finished, windows and doors must be installed. Also, the roof should be up and the sheathing should be on the walls. Then, to finish the exterior, three things are done. First, the cornices and rakes around the roof are enclosed. Next, the siding is applied. Finally, the finish trim for windows and doors is installed.

The *cornice* is the area beneath the roof overhang. This area is usually enclosed or boxed in. Figure 7-1 shows a typical cornice. In many areas, the cornice is boxed in as part of the roofing job. The cornice is often painted before siding is installed. This is particularly likely when brick or stone is used.

Carpenters install several types of exterior siding. Most types of siding are made of wood, plywood, or wood fibers. Some types of siding are made of plastic and metal. These are usually formed to look like wood siding.

Siding is the outer part of the house that people see. Beauty and appearance are important factors. However, siding is also selected for other reasons. The builder may want a siding that can be installed quickly and easily. This means that it costs less to install. The owner may want a siding that requires little maintenance. Both will want a siding that will not rot or warp. Whether insects will attack the siding is another factor.

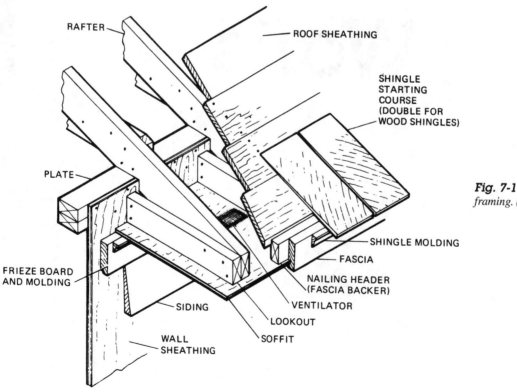

Fig. 7-1 *Typical box cornice framing.* (Forest Products Laboratory)

TYPES OF SIDING

Siding is made of plywood, wood boards, wood fibers, various compositions, metal, or plastics. See Fig. 7-2A to D. However, the shape is the main factor that determines how the siding is applied. Wood fiber is made in two main shapes, "boards" and sheets or panels, which often look like boards. However, the sheets are put up as whole sheets. Fiber panels can also be made to look like other types of siding.

Plywood panels can also be made to look like other types of siding. Figure 7-2B shows a house with plywood siding that looks like individual boards. Plywood can also be made in "board" strips. These are applied like boards.

Siding is also made from shingles and shakes. Shingles are made from either wood or asbestos mineral compounds. Actually, both types of shingle siding are applied in much the same manner.

(A)

(B)

(C)

(D)

Fig. 7-2 *(A) Cedar boards are used both horizontally and diagonally for this siding.* (Potlatch) *(B) Plywood siding has many "looks" and styles. This looks like boards.* (American Plywood Association) *(C) This siding combines rough brick, smooth stucco, and boards. (D) Here stone siding is applied over frame construction.*

SEQUENCE FOR SIDING

Sequence is determined by the type of siding, the type of roof, and the type of sheathing (if any) used on the building. How high the building is also affects the sequence. To work on high places, carpenters erect scaffolds.

In most cases, board siding is put on from bottom to top. However, if scaffolds have to be nailed to the wall, the siding on the bottom could be damaged. In that case the sequence can be changed. The scaffolds can be put up and the siding can be put on the bottom. This way the siding is not damaged by scaffold nails or bolts.

However, many scaffolds can stand alone. They need not be nailed to the wall. The common sequence for finishing an exterior wall is:

1. Prepare for the job. Make sure that the windows are installed, the vapor barrier is in place, the nails are selected, and the amount of siding is estimated.
2. Erect necessary scaffolds.
3. Install the flashing and water tables.
4. Finish the roof edges.
5. Install siding on the upper gable ends and on upper stories.
6. Install siding on the sides.
7. Finish the corners.
8. Trim windows and doors.

PREPARE FOR THE JOB

Several things should be done before siding is put on the wall. Windows and doors should be properly installed. Rough openings should have been moisture-shielded. Then any spaces between window units and the wall frame are blocked in. An air space should be present. Some types of sheathing are also good moisture barriers. However, other types are not. These must have a moisture barrier installed. Figure 7-3 shows a window that has been blocked properly so that siding may now be installed.

Vapor Barrier

One of the reasons for installing a vapor barrier is to protect the outside walls from vapor coming from inside the house. Paint problems on the siding of the house can arise from too much moisture escaping from inside the house. The main purpose of the barrier is to prevent water vapor from entering the enclosed wall space where condensation might occur and cause rot and odor problems. The warm side of the wall should

Fig. 7-3 *A window unit blocked for brick siding.*

be as vapor-tight as possible. The asphalt- or tar-saturated felt papers are not vapor-proof, but can be used on the sheathing for they are water-repellent.

Vapor barriers are available on insulation. This means the barrier side of the insulation should be installed toward the warm or inside of the house. In some instances the entire inside walls and ceiling are covered with a plastic (polyethylene) film to ensure a good vapor barrier. Heated air contains moisture. It travels from inside the building to the outside to blister and bubble paint on the siding. Damp air can also increase the susceptibility to rot and damage.

Sheathing is usually covered with some type of tar-saturated paper to improve on the siding's ability to shed water and at the same time improve the energy efficiency of the house by preventing air leakage through the walls. Reflective energy barriers are also used. See Fig. 7-4. They help keep energy on one side of a wall. They can cause the inside of the house to be warmer in the winter and cooler in the summer.

To put up an air barrier, start at the bottom. Nail the top part in place. Most air-barrier materials will be applied in strips. The strips should shed water to the outside. See Fig. 7-5. To do this, the bottom strip is installed first. Each added strip is overlapped. Moisture barriers should also be lapped into window openings. See Chapter 6 on installing windows for this procedure.

Fig. 7-4 *Foil surfaces reflect energy.*

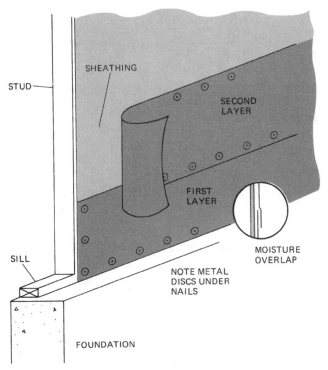

Fig. 7-5 *Barrier is applied from the bottom to the top. The overlap helps shed moisture.*

Vapor barriers are made from plastic films, metal foil, and from builder's felt (also called tar paper). Builder's felt and some plastic films are used most frequently for these purposes.

Nail Selection

Several methods of nailing can be used. Some siding may be put up by any of several methods. Other siding should be put up with special fasteners. The important thing is that the nail must penetrate into something solid in order to hold. For example, nails driven into fiber sheathing will not hold. Siding over this type of sheathing must be nailed at the studs. If the studs are placed on 16-inch centers, the nails are driven at 16-inch intervals. Likewise, splices in the siding should be made over studs. Also, the nail must be long enough to penetrate into the stud.

However, if plywood or hardboard sheathing is used, then nails can be driven at any location. The plywood will hold even a short nail.

Strips of wood are also used to make a nail base. These strips are placed over the sheathing. Then they are nailed to the studs through the sheathing. Nail strips are used for several types of shingled siding. See Fig. 7-6.

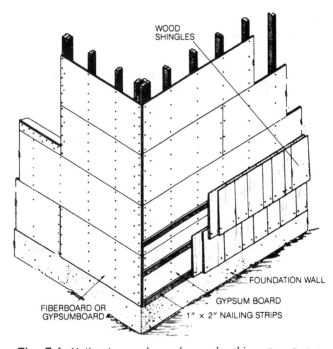

Fig. 7-6 *Nail strips can be used over sheathing.* (Forest Products Laboratory)

The nail should enter the nail base for at least ½ inch. A nonstructural sheathing such as fiber sheathing cannot be the nail base. With fiber or foam sheathing, the nail must be longer. The nail must go through siding and the sheathing and ½ inch into a stud.

Also, the type of nail should be considered. When natural wood is exposed, a finishing or casing-head nail would look better. The head of the nail can be set beneath the surface of the wood. The head will not be seen. This way the nail will not detract from the appearance. However, if the wood is to be painted, a common or a box-head nail may be used. Coated nails are preferred for composition and mineral siding. These nails are coated with zinc to keep them from rusting. This makes them more weather-resistant.

For some siding, the nail is driven at an angle. This means that it must be longer than the straight line distance. This type of nailing is frequently done in grooved and edged siding. The nails are driven in the grooves and edges. This way the siding is put on so that the nails are not exposed to the weather or the eye of the viewer.

Estimate the Amount of Siding Needed

Many people think that "one by six" (1-×-6-inch) boards are 1 inch thick and 6 inches wide. However, a 1-×-6 board is actually ¾ inch thick by 5½ inches wide. Also, most board siding overlaps. Rabbet, bevel, and drop sidings all overlap. When overlapped, each board would not expose 5½ inches. Each board would expose only about 5 inches.

Several things must be known to estimate the amount of siding needed. First, you must determine the height and width of the wall. Then find the type of siding and the sizes of windows or doors. Consider the following example:

Siding: 1-×-8-inch bevel siding
Overlap: 1½ inches
Wall height: 8 feet
Wall length: 40 feet
Windows and Doors: Two windows, each 2 × 4 feet
                     One door, 8 × 3 feet

First, find the total area to be covered. To do this, multiply the length of the wall times the height.

$$\text{Area} = 40 \times 8$$
$$= 320 \text{ square feet}$$

Next, subtract the area of the doors and windows from the area of the wall. To do this:

$$
\begin{aligned}
\text{Area of windows} &= \text{width} \times \text{length} \\
&= 2 \times 4 \\
&= 8 \text{ square feet} \times \text{number of windows} \\
&= 16 \text{ square feet} \\
\text{Door area} &= \text{width} \times \text{length} \\
&= 3 \times 8 \\
&= 24 \text{ square feet} \\
\text{Total opening area} &= 16 \text{ square feet} \\
&\quad + 24 \text{ square feet} \\
\text{Total} &\quad \overline{40 \text{ square feet}}
\end{aligned}
$$

In order to get enough siding to cover this area, the percentage of overlap must be considered. Also, there

is always some waste in cutting boards so that they join at the proper place. When slopes and corners are involved, there is more waste. The amount of overlap for 8-inch siding lapped 1¼ inches is approximately 17 percent. However, it is best to add another 15 percent to this for waste. Thus, about 32 percent would be added to the total requirements. Thus, to side the wall would require 320 − 40 = 280 square feet. However, enough siding should be ordered to cover 280 square feet plus 32 percent (90 board feet). A total of 370 board feet should be ordered. Table 7-1 shows allowances for different types of siding.

Table 7-1 *Allowances that Must Be Added to Estimate Siding Needs*

Type	Size, Inches	Amount of Lap, Inches	Allowance, Percent (Add)
Bevel siding	1 × 4	3/4	45
	1 × 6	1	35
	1 × 8	1¼	32
	1 × 10	1½	30
	1 × 12	1½	25
Drop siding (shiplap)	1 × 4		30
	1 × 6		20
	1 × 8		17
Drop siding (matched)	1 × 4		25
	1 × 6		17
	1 × 8		15

To estimate the area for gable ends, the same procedure is used. Find the length and the height of the gable area. Then multiply the two dimensions for the total. However, since slope is involved, as shown in Fig. 7-7, only one-half of this figure is required. The allowance for waste and overlap is based upon this halved figure.

The amount of board siding required for an entire building may be estimated. First, the total area of all the walls is found. Then the areas of all the gable sections are found. These areas are added together. The allowances are made on the total figure.

Ordering paneled siding Paneled or plywood siding is sold in sheets. The standard sheet size is 4 feet wide and 8 feet long. The standard height of the wall is 8 feet. Thus a panel fully covers a 4-foot length of wall. To estimate the amount needed, find the length of the wall. Then divide by 4, the width of a panel. For example, a wall is 40 feet long. The number of panels is 40 divided by 4. Thus, it would take 10 panels to side this wall with plywood paneling. If, however, the length of the wall is 43 feet, then 11 panels would be

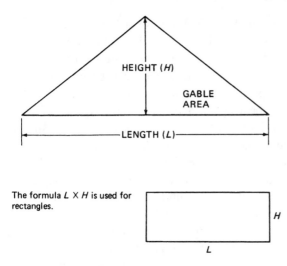

HEIGHT (*H*)

GABLE AREA

←——— LENGTH (*L*) ———→

The formula *L* × *H* is used for rectangles.

H

L

A triangle is half of a rectangle.

$A = \frac{1}{2}(L \times H)$

Its formula is $\frac{1}{2}(L \times H)$
A gable is two triangles.

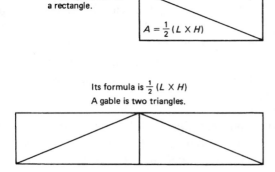

Fig. 7-7 *Estimating gable areas.*

required. Only whole panels may be ordered. No allowances for windows and doors are made. Sections of panels are cut out for windows and doors.

Estimating shingle coverage To find the amount of shingles needed, the actual wall area should be found. The areas of the doors and windows should be deducted from the total area. Shingles vary in length and width. The size of the shingle must be considered when estimating. Also, shingles come in packages called *bundles*. Generally, four bundles are approximately 1 "square" of coverage. This means that four bundles will cover about 100 square feet of surface area. However, the amount of shingle exposed determines the actual coverage. If only 4 inches of a 16-inch-long shingle is exposed, the coverage will be much less. Also, if more of the shingle is exposed, a greater area can be covered. For walls, more of the shingle can be exposed than for roofs. Table 7-2 shows the coverage of 1 square of shingles when different lengths are exposed.

ERECTING SCAFFOLDS

Many areas on a building are high off the ground. When carpenters must work up high, they put up scaffolds. Figure 7-8 shows a typical pump jack scaffold. Scaffolds are also sometimes called *stages*.

A scaffold must be strong enough to hold up the weight of everything on it. This includes the people, the tools, and the building materials. There are several types of scaffolds that can be used. The type used depends upon the number of workers involved and the weight of the materials. Also, how high the scaffold will be is a factor. How long the scaffold will remain up must be considered.

Job-Built Scaffolds

Scaffolds are often built from job lumber. The maximum distance above the ground should be less than 18 feet. Three main types of job scaffolds can be built. Supports for the platform should be no more than 10 feet apart, and less is desirable.

Table 7-2 *Coverage of Wood Shingles for Varying Exposures*

Length and Thickness[b]	Approximate Coverage of Four Bundles or One Carton, [a]Square Feet									
	Weather Exposure, Inches									
	5 1/2	6	7	7 1/2	8	8 1/2	9	10	11	11 1/2
Random-width and dimension 16″ × 5/2″	110	120	140	150[c]	160	170	180	200	220	230
18″ × 5/2 1/4″	100	109	127	136	145 1/2	154 1/2[c]	163 1/2	181 1/2	200	209
24″ × 4/2″	—	80	93	100	106 1/2	113	120	133	146 1/2	153[c]

[a] Nearly all manufacturers pack four bundles to cover 100 square feet when used at maximum exposures for roof construction; rebutted-and-rejoined and machine-grooved shingles typically are packed one carton to cover 100 square feet when used at maximum exposure for double-coursed sidewalls.

[b] Sum of the thickness, e.g. 5/2″ means 5 butts equal 2″.

[c] Maximum exposure recommended for single-coursed sidewalls.

Fig. 7-8 *Pumpjack scaffold for light, low work such as bricklaying, painting, and the application of siding.*

Double-pole scaffolds This style of scaffold is shown in Fig. 7-9. The poles form each support section. They should be made from 2-×-4-inch pieces of lumber. A 1-×-6 board is nailed near the bottom of both poles for a bottom brace. Three 12d nails should be used in each end of the 1-×-6-inch board. The main support for the working platform is called a *ledger*. It is nailed to the poles at the desired height. It should be made of 2-×-4-inch or 2-×-6-inch lumber. It is nailed with three 16d common nails in each end of the ledger. Sway braces should be nailed between each end pole

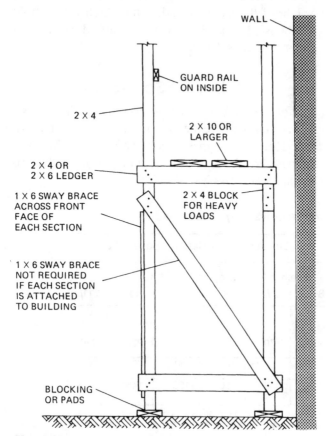

Fig. 7-9 *A double-pole scaffold can stand free of the building. Guard rails are needed above 10'0".*

as shown in Fig. 7-9. The sway braces can be made of 1-inch lumber. They are nailed as needed.

After two or more of the pole sections have been formed on the ground, they can be erected. Place a small piece of wood under the leg of the pole to provide a bearing surface. This keeps the ends of the pole from sinking into the soft earth. The sections should be held in place by the carpenters. Then another carpenter nails braces between each pole section. These are nailed diagonally, as are sway braces. They may be made of 1-×-6-inch lumber. They are nailed with three 14d common nails at each end.

When the ledgers are more than 8 feet above the ground, a guard rail should be used. The guard rail is nailed to the pole about 36 inches above the platform. Note that the guard rail is nailed to the inside of the pole. This way a person can lean against the guard rail without pushing it loose.

The double-pole type of scaffold can be freestanding. That is, it need not be attached to the building. However, carpenters often nail a short board between the scaffold and the wall. This holds the scaffold safely. If this is done, the board should be nailed near the top of the scaffold.

Single-pole scaffolds Single-pole scaffolds are similar to double-pole scaffolds. However, the building forms one of the "poles." Blocks are nailed to the building. They form a nail base for the brace and ledgers. Figure 7-10 shows a single pole scaffold. Sway braces are not needed on the pole sections. However, sway braces should be used between each section of the scaffold.

Wall brackets Wall brackets are often made on the job. See Fig. 7-11. Wall brackets are most often used when the distance above the ground is not very great. They are quicker and easier to build than other scaffold types. Of course, they are also less expensive.

On the ends, the wall brackets are nailed to the outside corners of the wall. On intersections, nail blocks are used, similar to the single-pole scaffold blocks. Wall brackets must be nailed to solid wall members. It is best to use 20d common nails to fasten them in place. Figure 7-12 shows another type of wall bracket. This type is sturdier than the other.

Factory Scaffolds

Today, many builders use factory-made scaffolds. These have several advantages for a builder. They are quick and easy to erect. They are strong and durable and can be easily taken down and reused.

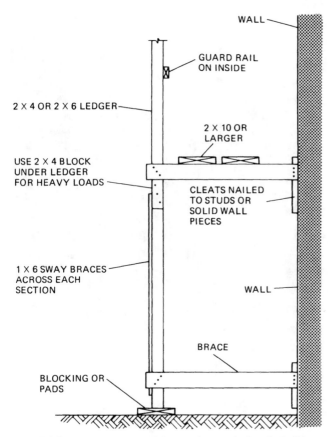

WALL

GUARD RAIL
ON INSIDE

2 X 4 OR 2 X 6 LEDGER

2 X 10 OR
LARGER

USE 2 X 4 BLOCK
UNDER LEDGER
FOR HEAVY LOADS

CLEATS NAILED
TO STUDS OR
SOLID WALL
PIECES

1 X 6 SWAY BRACES
ACROSS EACH
SECTION

WALL

BRACE

BLOCKING OR
PADS

Fig. 7-10 *A single-pole scaffold must be attached to the building. Guard rails are needed above 10'0".*

No lumber is used. Also, no cutting and nailing is needed to erect the scaffold. Thus, no lumber is ruined or made unusable for the building. The metal scaffold parts are easily stored and carried. They are not affected by weather and will not rot. There are several different types of scaffolds.

Double-pole scaffold sections The double-pole type of scaffold features a welded steel frame. It includes sway braces, base plates, and leveling jacks. See Figs. 7-13 and 7-14. Two or more sections can be used to gain greater distance above the ground. Special pieces provide sway bracing and leveling. Guard rails and other scaffold features may also be included.

Wall brackets Wall brackets, as in Figs. 7-15 through 7-17, are common. No nail blocks are needed on the walls for these metal brackets. These brackets are nailed or bolted directly to the wall. Be sure that they are nailed to a stud or another structural member. Use 16 or 20d common nails. After the nails are driven in place, be sure to check the heads. If the nail heads are damaged, remove them and renail the bracket with new nails. Remember that it is the nail head that holds the bracket to the wall. A damaged nail head may break. A break could let the entire scaffold fall.

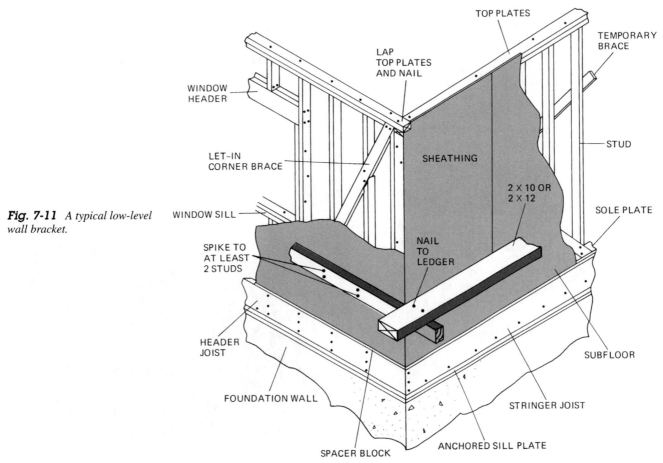

Fig. 7-11 *A typical low-level wall bracket.*

TOP PLATES

TEMPORARY
BRACE

LAP
TOP PLATES
AND NAIL

WINDOW
HEADER

STUD

LET–IN
CORNER BRACE

SHEATHING

WINDOW SILL

2 X 10 OR
2 X 12

SOLE PLATE

SPIKE TO
AT LEAST
2 STUDS

NAIL
TO
LEDGER

HEADER
JOIST

SUBFLOOR

FOUNDATION WALL

STRINGER JOIST

SPACER BLOCK

ANCHORED SILL PLATE

Erecting Scaffolds   239

Fig. 7-12 *A job-made wall-bracket scaffold. It is dangerous to work on this kind of scaffold in high places without safety equipment.*

Fig. 7-14 *Metal sections can be combined in several ways to get the right height and length.*

EASY TO ERECT

SIMPLE INSTALLATION

SIMPLE REMOVAL

EASY STORAGE

SHIPPING/STORAGE CONFIGURATION

Fig. 7-15 *Metal wall brackets are quick to put up and take down.*
(Richmond Screw Anchor)

Fig. 7-13 *Metal double-pole scaffolds are widely used.* (Beaver-Advance)

Fig. 7-16 *Wall brackets can be used on concrete forms.*

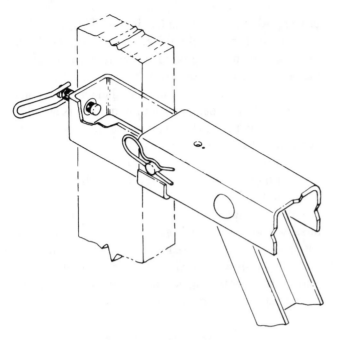

Fig. 7-17 *Guard rails may be added through an end bracket.*
(Richmond Screw Anchor)

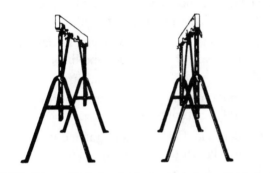

Fig. 7-18 *Two trestle jacks can form a base for scaffold planks.*
(Patent Scaffolding Co.)

Trestle jack Trestle jacks are used for low platforms. They are used both inside and outside. They can be moved very easily. Trestle jacks are shown in Fig. 7-18. Two trestle jacks should be used at each end of the section. A ledger, made of 2-×-4-inch lumber, is used to connect the two trestle jacks. Platform boards are then placed across the two ledgers. Platform boards should always be at least 2 inches thick.

As you can see, it takes four trestle jacks for a single section. Trestle jacks can be used on uneven areas, but they provide for a platform height of only about 24 inches. However, this is ideal for interior use.

Ladder jacks Ladder jacks, as in Fig. 7-19, hang a platform from a ladder. They are most suitable for repair jobs and for light work where only one carpenter is on the job. Two types of jacks are used. The type shown in Fig. 7-19 puts the platform on the outside of the ladder. The type shown in Fig. 7-20 places the platform below or on the inside of the ladder.

Ladder Use

Using ladders safely is an important skill for a carpenter. A ladder to be erected is first laid on the ground. The bottom end of the ladder should be near the building. The top end is raised and held overhead. The carpenter gets directly beneath the ladder. The hands are then moved from rung to rung. As the top end of the ladder is raised, the carpenter walks toward the building. Thus, the ladder is raised higher and higher with every step. Care is taken to watch where the top of the ladder is going. The top of the ladder is guided to its

Fig. 7-19 *Outside ladder jacks.*

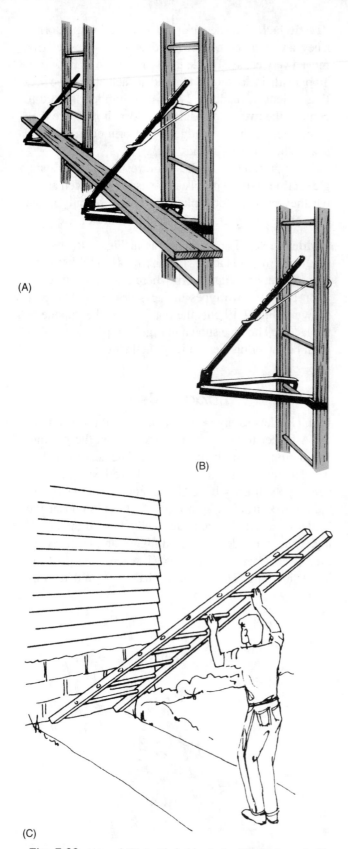

(A)

(B)

(C)

Fig. 7-20 *(A) and (B) Inside ladder jacks. (C) Raising a ladder can be a two-man job. In fact, a heavier ladder should have two men on it. However, if you have a single ladder that's not too heavy, you can place the end of the ladder against the house or some obstruction and walk it up one rung at a time.*

proper position. Then the ladder is leaned firmly against some part of the building. The base of the ladder is made secure on solid ground or concrete. Both ends of the bottom should be on a firm base.

Both wooden and aluminum ladders are commonly used by carpenters. Aluminum ladders are light and easy to manage. However, they have a tendency to sway more than wooden ladders. They are very strong and safe when used properly. Wooden ladders do not sway or move as much, but they are much heavier and harder to handle.

Wooden ladders are sometimes made by carpenters. A ladder is made from clear, straight lumber. It should have been well seasoned or treated. The sides are called rails and the steps are rungs. Joints are cut into the rails for the rungs. Boards should never be just nailed between two rails.

Ladder Safety

1. Ladder condition should be checked before use.
2. The ladder should be clean. Grease, oil, or paint on rails or rungs should be removed.
3. Fittings and pulleys on extension ladders should be tight. Frayed or worn ropes and lines should be replaced.
4. The bottom ends of the ladder must rest firmly and securely on a solid footing.
5. Ladders should be kept straight and vertical. Never climb a ladder that is leaning sideways.
6. The bottom of the ladder should be one-fourth the height from the wall. For example, if the height is 12 feet, the bottom should be 3 feet from the wall.

Scaffold Safety

1. Scaffolds should be checked carefully before each use.
2. Design specifications from the manufacturer should be followed. State codes and local safety rules should also be followed.
3. Pads should be under poles.
4. Flimsy steps on scaffold platforms or ladders should never be used. Height should only be increased with scaffold of sound construction.
5. For platforms, planking that is heavy enough to carry the load and span should be used.
6. Platform boards should hang over the ledger at least 6 inches. This way, when boards overlap, the total overlap should be at least 12 inches.
7. Guard rails and toe boards should be used.

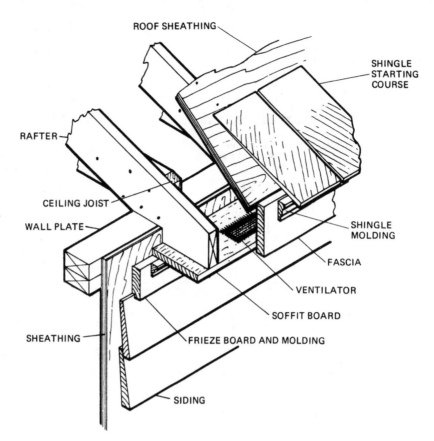

Fig. 7-21 *Narrow box cornice. A closed overhang is called a cornice.* (Forest Products Laboratory)

Labels in figure: ROOF SHEATHING, SHINGLE STARTING COURSE, RAFTER, CEILING JOIST, WALL PLATE, SHINGLE MOLDING, FASCIA, VENTILATOR, SOFFIT BOARD, SHEATHING, FRIEZE BOARD AND MOLDING, SIDING

8. Scaffolds should never be put up near power lines without proper safety precautions. The electric service company can be consulted for advice when a procedure is not known.

9. All materials and equipment should be taken off before a platform or a scaffold is moved.

FINISHING ROOF EDGES

Most roofs have an overhang. This portion is called the *eave* of the roof. If eaves are enclosed, they are also called a *cornice*. See Fig. 7-21. Usually, the edges of roofs are finished when the roof is sheathed. However, building sequences do vary from place to place. Two methods of finishing the eaves are commonly employed. These are the open method and the closed method. There are several versions of the closed method.

Open Eaves

A board is usually nailed across the ends of the rafters when the roof is sheathed. See Fig. 7-22. This board is called the *fascia*. The fascia helps to brace and strengthen the rafters. Fascia should be joined or spliced as shown in Fig. 7-23. However, the fascia is not needed structurally. Some types of open-eave construction do not use fascia.

Open eaves expose the area where the rafters and joists rest on the top plate of a wall. This area should be sealed either by a board or by the siding. See Fig. 7-24. Sealing this area helps prevent air currents from entering the wall. It also helps keep insects and small animals from entering the attic area. However, to com-

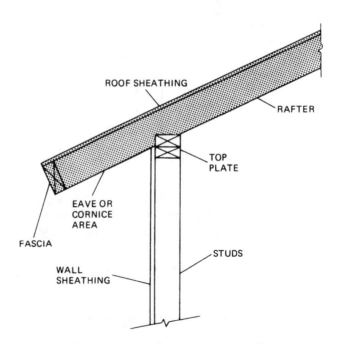

Labels in figure: ROOF SHEATHING, RAFTER, TOP PLATE, EAVE OR CORNICE AREA, FASCIA, STUDS, WALL SHEATHING

Fig. 7-22 *Open overhangs are called eaves.*

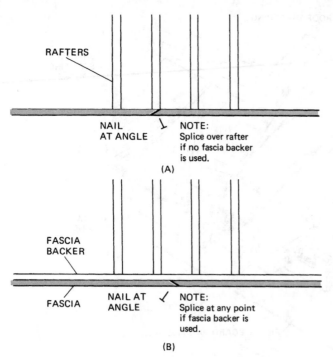

Fig. 7-23 *Splicing the fascia. (A) No fascia backer is used. (B) A fascia backer is used.*

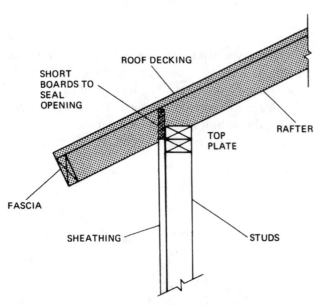

Fig. 7-24 *Open eaves should be sealed and vented properly.*

pletely seal this restricts airflow. This airflow is important to keep the roof members dry. It also prevents rotting from moisture. Airflow will also cool a building. To allow airflow, vents should be installed. These vents are backed with screen or hardware cloth to keep animals and insects from entering.

Enclosed Cornices

Eaves are often enclosed for a neater appearance. Enclosed eaves are called *cornices*. Some houses do not

have eaves or cornices. These are called *close* cornices. See Fig. 7-25.

There are several ways of enclosing cornices. The two most common types are the standard slope cornice and the flat cornice. Both types seal the cornice area with panels called *soffits*. The panels are usually made of plywood or metal. Note that both types should have some type of ventilation. Special cornice vents are used. However, when plywood is used for the soffits, an opening is often left as in Fig. 7-26. This provides a continuous vent strip. This strip is covered with some type of grill or screen.

Standard slope cornice The standard slope cornice is the simplest and quickest to make. It is shown in Fig. 7-27A. The soffit panel is nailed directly to the underside of the rafters. If ¼-inch paneling is used, a 6d box nail is appropriate. Rust-resistant nails are recommended. Casing nails, when used, should be set and covered before painting. Panels should join on a rafter. In this way, each end of each panel has firm support. One edge of the panel is butted against the fascia. The edge next to the wall is closed with a piece of trim called a *frieze*.

Standard flat cornices The standard flat cornice has a soffit that is flat or horizontal with the ground. It is not sloped with the angle of the roof.

A nail base must be built for the soffit. Special short joists are constructed. These are called *lookouts*. See Fig. 7-27B. The flat cornice should be vented, as are other types. Either continuous strips or stock vents can be used.

The lookouts need a nail base on both the wall and the roof. They are nailed to the tail of the rafter for the roof support. On the wall, they can be nailed to the top plate or to a stud. More often than not, neither the top plate nor the stud can be used. Then a ledger is nailed to the studs through the sheathing. See Fig. 7-28.

Either 16d or 20d common nails should be used to nail the ledger to the wall. The bottom of the ledger should be level with the bottom of the rafters.

Next, find the correct length needed for the lookouts. Cut the lookouts to length with both ends cut square. Drive nails into one end of the lookout. The nails are driven until the tips just barely show through the lookout. Then they can be held against the rafter and driven down completely. Butt the other end of the lookout against the ledger. Then toenail the lookout on each side using an 8d or 12d common nail.

Soffit panels should be cut to size. Cornice vents should be cut or spaced next. Cornice vents can be attached to soffits before mounting. After the soffits are

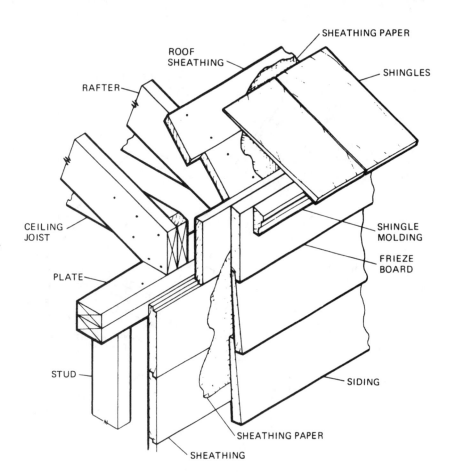

RAFTER

ROOF
SHEATHING

SHEATHING PAPER

SHINGLES

CEILING
JOIST

SHINGLE
MOLDING

FRIEZE
BOARD

PLATE

STUD

SHEATHING PAPER

SHEATHING

SIDING

Fig. 7-25 With close cornices, the roof does not project over the walls. (Forest Products Laboratory)

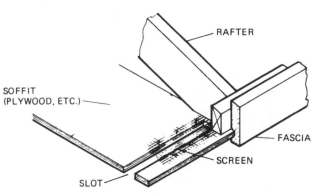

RAFTER

SOFFIT
(PLYWOOD, ETC.)

FASCIA

SCREEN

SLOT

Fig. 7-26 A strip opening may be used for ventilation instead of cornice vents. (Forest Products Laboratory)

ready, they are nailed in place. For ¼-inch panels, a 6d common nail or box nail can be used.

Soffit panels should be joined on a solid nail base. The ends should join at the center of a lookout. One edge is butted against the fascia. The inner edge is sealed with a frieze board or a molding strip.

Many builders use prefabricated soffit panels. Sometimes fascia is grooved for one edge of the soffit panel. When prefabricated soffits are used, a vent is usually built into the panel.

Closed rakes The rake is the part of the roof that hangs over the end of a gable. See Fig. 7-29. When a cornice is closed, the rake should also be closed. However, most rakes are not vented. This is because they are not connected with the attic space as are the cornices.

It is common to add the gable siding before the rake soffit is added. Then a 2-×-4-inch nailing block is nailed to the end rafter. See Fig. 7-30. Use 16d common nails. A fly rafter is added and supported by the fascia and the roof sheathing.

The soffit is then nailed to the bottom of the nail block and the fly rafter. Frieze boards or bed molding are then added to finish the soffit.

Solid rake Today many roofs do not extend over the ends of the gables. In this type of roof, a fascia is nailed directly onto the last rafter. The roof is then finished over the fascia. This becomes what is called a *solid rake*. See Fig. 7-31.

Siding the Gable Ends

On many buildings, cornices are painted before any siding is installed. In other cases the gable siding is installed first, then both the cornice and gables are painted. After these are done, siding is installed. This is common when two different types of siding are used. For example, wood siding is put on the gables

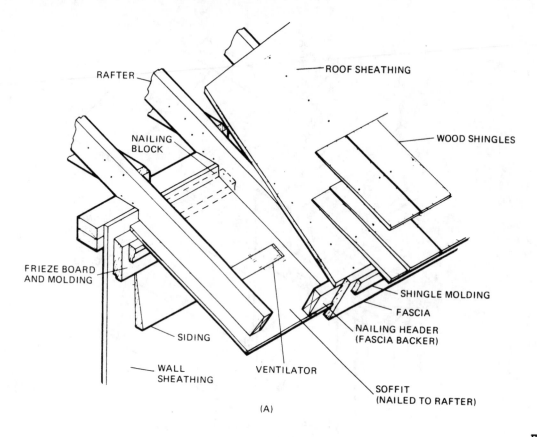

RAFTER

ROOF SHEATHING

NAILING BLOCK

WOOD SHINGLES

FRIEZE BOARD AND MOLDING

SHINGLE MOLDING

FASCIA

NAILING HEADER (FASCIA BACKER)

SIDING

WALL SHEATHING

VENTILATOR

SOFFIT (NAILED TO RAFTER)

(A)

Fig. 7-27 *(A) Standard slope cornice; (B) Standard flat cornice.* (Forest Products Laboratory)

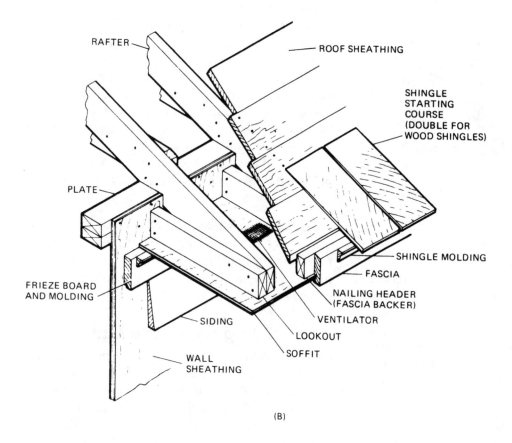

RAFTER

ROOF SHEATHING

SHINGLE STARTING COURSE (DOUBLE FOR WOOD SHINGLES)

PLATE

SHINGLE MOLDING

FASCIA

NAILING HEADER (FASCIA BACKER)

FRIEZE BOARD AND MOLDING

VENTILATOR

SIDING

LOOKOUT

SOFFIT

WALL SHEATHING

(B)

Fig. 7-28 *Lookouts are nailed first to the rafter tail. Then they are toenailed to a solid part of the wall.*

and painted, then a brick siding is laid. See Figs. 7-32 and 7-33.

Gable walls may be sided in several ways. Gable siding may be the same as that on the rest of the walls. In this case, the gable and the walls are treated in the same way. However, gable walls are often covered with different siding. Brick exterior walls topped by wooden gable walls are common. Different types of wood and fiber sidings may also be combined. See Figs. 7-34 and 7-35. Different textures, colors, and directions are used to make pleasing contrasts.

It is important to apply the gable siding in such a way that water is shed properly from gable to wall. For brick or stone, the gable must be framed to overhang the rest of the wall. See Fig. 7-36. This overhang provides an allowance for the thickness of the stone or brick.

For wood, plywood, or fiber siding, special drainage joints are used. Also metal strips and separate wooden moldings are used.

Framed overhangs Gable ends can be framed to overhang a wall. See Fig. 7-36. Remember that this must be done to allow for the thickness of a brick wall.

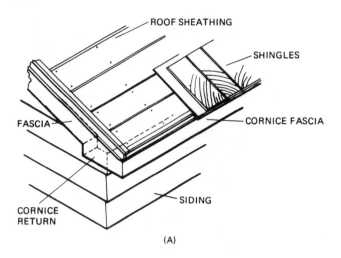

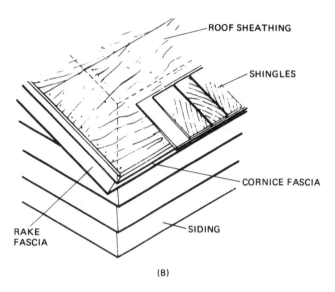

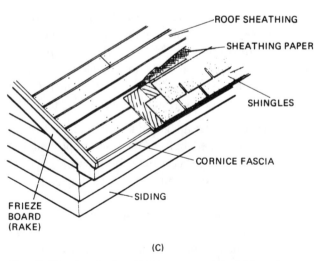

Fig. 7-29 *Closed rakes: (A) Narrow cornice with boxed return; (B) Narrow box cornice and closed rake; (C) Wide overhang at cornice and rake.* (Forest Products Laboratory)

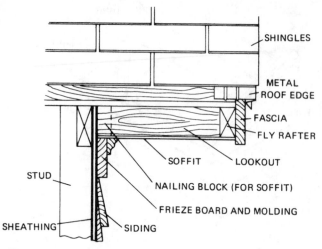

Fig. 7-30 *A nailing block is nailed to the end rafter for the soffit.* (Forest Products Laboratory)

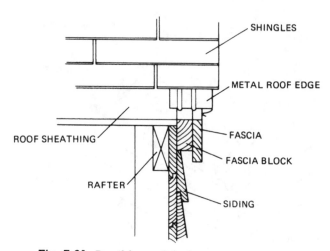

Fig. 7-31 *Detail for a solid rake.* (Forest Products Laboratory)

Fig. 7-32 *Here the gable siding is added and painted before brick siding is added.*

Fig. 7-33 *A wall and gable prepared for a combination of diagonal siding and brick veneer. Note that the wood has been painted before the brick has been applied.*

Several methods are used. One of the most common is to extend the top plate as in Fig. 7-37A. This puts the regular end rafter just over the wall for the overhang. A short (2-foot) piece is nailed to the top plate. The rafter bird's mouth is cut 1½ inch deeper than the others. A solid header is added on the bottom as shown in Fig. 7-37B. The gable is framed in the usual manner. Gable siding is done before wall siding (Figs. 7-38 and 7-39).

Drainage joints Drainage joints are made in several ways. As mentioned before, they are important because water must be shed from gable to wall properly. Otherwise, water would run inside the siding and damage the wall.

Drainage joints are used wherever two or more pieces are used, one above the other. Figure 7-40 shows the most common drainage joints. In one method, molded wooden strips are used. These strips are called *drip caps.*

Some siding, particularly plywood panels, has a rabbet joint cut on the ends. These are overlapped to stop water from running to the inside.

Metal flashing is also used. This flashing may be used alone or with drip caps.

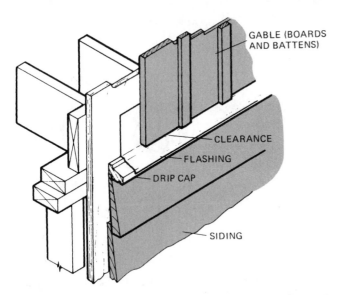

Fig. 7-34 *Different types of siding can be used for the gable and the wall.* (Forest Products Laboratory)

Fig. 7-35 *Smooth gable panels contrast with the brick siding.*

Fig. 7-36 *This gable is framed to overhang the wall.*

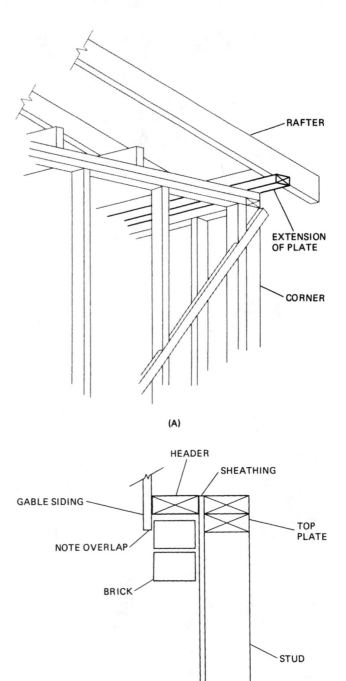

(A)

(B)

Fig. 7-37 *(A) Extending the top plate so that gable siding over-hangs a brick or masonry wall; (B) With a solid header added on the bottom.*

At the bottom of panel siding, a special type of drip cap is used. It is called a water table. It does the same thing, but has a slightly different shape. The water table may also be used with or without metal flashing.

Fig. 7-38 *Gable siding is done before wall siding.* (Fox and Jacobs)

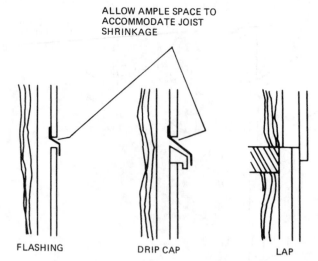

FLASHING   DRIP CAP   LAP

ALLOW AMPLE SPACE TO
ACCOMMODATE JOIST
SHRINKAGE

Fig. 7-40 *Drainage joints between gable and siding.* (Boise-Cascade)

material are installed alike. Panels made of plywood, hardboard, or fiber are installed in a similar manner. Special methods are used for siding made from vinyl or metal.

Putting Up Board Siding

There are three major types of board siding. These are plain boards, drop siding, and beveled siding. Each type is put up in a different manner.

Board siding may be of wood, plywood, or composition material. This material is a type of fiberboard made from wood fibers. Generally, plain boards and drop siding are made from real wood. Composition siding is usually made in the plain beveled shape.

Plain boards Plain boards are applied vertically. This means that the length runs up and down. There are no grooves or special edges to make the boards fit together. There are three major board patterns used. The *board and batten* pattern is shown in Fig. 7-41. In the board and batten style, the board is next to the wall. The board is usually not nailed in place except some-

Fig. 7-39 *Installing gable siding.* (Fox and Jacobs)

INSTALLING THE SIDING

Several shapes of siding are used. Boards, panels, shakes, and shingles are the most common. These may be made of many different materials. However, many of the techniques for installing different types of siding are similar. Shape determines how the siding is installed. For example, wood or asbestos shingles are installed in about the same way. Boards made of any

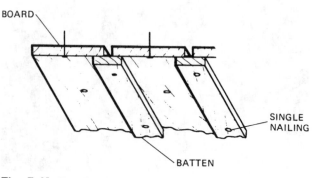

BOARD

SINGLE
NAILING

BATTEN

Fig. 7-41 *Board and batten vertical siding.* (Forest Products Laboratory)

times at the top end. It is sometimes nailed at the top to hold it in place. A narrow opening is left between the boards. This opening is covered with a board called a *batten*. The batten serves as the weather seal. The nails are driven through the batten but not the boards, as shown in Fig. 7-41.

The *batten and board* style is the second pattern. The batten is nailed next to the wall. A typical nailing pattern is shown in Fig. 7-42. The wide board is then fastened to the outside as shown.

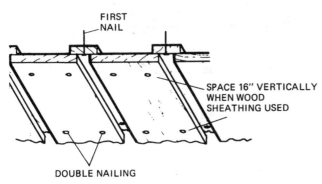

Fig. 7-42 *Batten and board vertical siding.* (Forest Products Laboratory)

Another style is the Santa Rosa style, shown in Fig. 7-43. All the boards are about the same width in this style. A typical nail pattern is shown in the figure. The inner board can be thinner than the outer board.

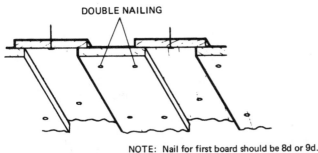

NOTE: Nail for first board should be 8d or 9d. Nail for second board should be 12d.

Fig. 7-43 *The Santa Rosa or board and board style of vertical siding.* (Forest Products Laboratory)

Plain board siding can be used only vertically. This is because it will not shed water well if it is used flat or horizontally. The surface appearance of the boards may vary. The boards may be rough or smooth. For the rough effect, the boards are taken directly from the saw mill. In order to make a smooth board finish, the rough-sawn boards must be surfaced.

Drop siding Drop siding differs from plain boards. Drop siding has a special groove or edge cut into it.

This edge lets each board fit into the next board. This makes the boards fit together and resist moisture and weather. Figure 7-44 shows some types of drop siding that are used. Nailing patterns for each type are also shown.

Beveled siding *Beveled* siding is made with boards that are thicker at the bottom. Figure 7-45 shows the major types of beveled siding. Common beveled siding is also called *lap* siding. The nailing pattern for lap siding is shown in Fig. 7-46.

The minimum amount of lap for lap siding is about 1 inch. For 10-inch widths, which are common, about 1½ inches of lap is suggested. Most lap siding today is made of wood fibers. However, in many areas wood siding is still used. When wood siding is used, the standard nailing pattern shown in Fig. 7-46 should be used. Some authorities suggest nailing the nails through both boards. However, this is not recommended. All wood tends to expand and contract, and few pieces do so evenly. Nailing the two boards together causes them to bend and bow. When this occurs, air and moisture can easily enter.

Another type of beveled siding is called *rabbeted beveled siding*. This is also called *Dolly Varden siding*. The Dolly Varden siding has a groove cut into the lower end of the board. This groove is called a rabbet. When installed from the bottom up, each successive board should be rested firmly on the one beneath it. The nails are then driven into the top of the boards.

Siding Layout

For all types of board siding, two or three factors are important to remember. Boards should be spaced to lay even with windows and doors. One board should fit against the bottom of a window. Another should rest on the top of a window. This way there is no cutting for an opening. Cutting should also be avoided for vertical siding. The spacing should be adjusted so that a board rests against the window on either side.

When siding is installed, guide marks are laid out on the wall. These are used to align the boards properly. To do this, a pattern board, called a story pole, is used just as in wall framing. The procedure for laying out horizontal lap siding will be given. The procedures for other types of siding are similar. This procedure may be adapted accordingly.

Procedure Choose a straight 1-inch board. Cut it to be exactly the height of the area to be sided. Many siding styles are allowed to overlap the foundation 1 to 2 inches. Find the width of the siding and the overlap to

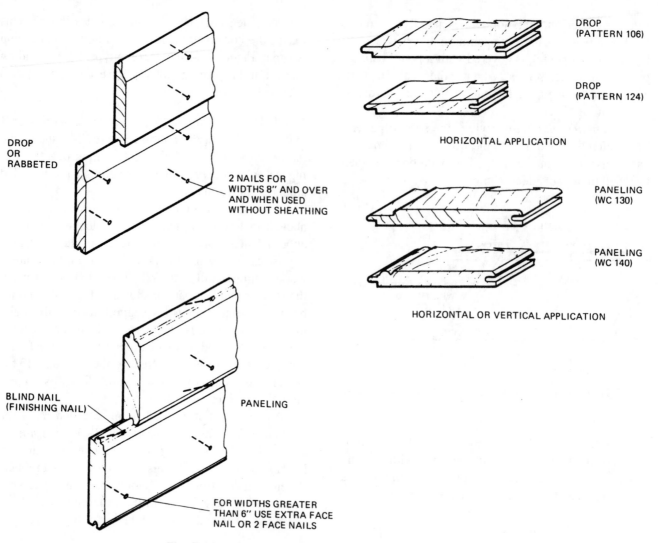

Fig. 7-44 *Shapes and nailing for drop siding.* (Forest Products Laboratory)

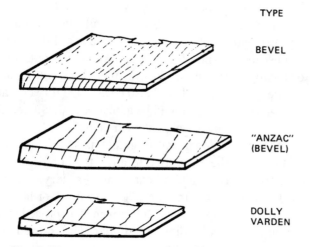

Fig. 7-45 *Major types of bevel siding.* (Forest Products Laboratory)

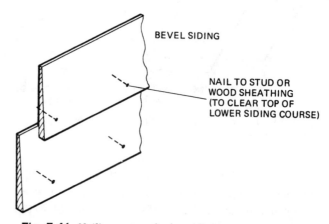

Fig. 7-46 *Nailing pattern for lap siding.* (Forest Products Laboratory)

be used. Subtract the overlap from the siding width. This determines the spacing between the bottom ends of the siding. The bottom end of lap siding is called the *butt* of the board. For example, siding 10 inches wide

is used. The overlap is 1½ inches. Thus, the distance between the board butts is 8½ inches. On the story pole, lay out spaces 8½ inches apart.

Place the story pole beside a window. Check to see if the lines indicating the spacing between the boards line up even with the top and bottom of the windows. The amount of overlap over the foundation wall may be varied slightly. The amount of lap on the siding may also be changed. Small changes will make the butts even with the tops of the windows. They will also make the tops even with the window bottoms. See Fig. 7-47. Story pole markings should be changed to show adjustments. New marks are then made on the pole.

Use the story pole to make marks on the foundation. Also make marks around the walls at appropriate places. Siding marks should be made at edges of windows, corners, and doors. See Fig. 7-48. If the wall is long, marks may also be made at intervals along the wall. In some cases, a chalk line may be used to snap guide lines. A chalk line is often used to snap a line on the foundation. This shows where the bottom board is nailed.

Nailing

Normally, siding is installed from the bottom to the top, with the first board overlapping the foundation at least 1 inch. The first board is tacked in place and checked for level. After leveling, the first board is then nailed firm. Usually bottom boards will be placed for the whole wall first. Then the other boards are put up, bottom to top. The level of the siding is checked after each few boards.

Siding can also be installed from the top down. This is done when scaffolding is used. The layout is the same. The lines should be used to guide the butts of the boards. However, the first board is nailed at the top. Two sets of nails are used on the top board. The first set of nails is nailed near the tops of the board. These may be nailed firm. Then, the bottom nail is driven 1½ inches from the butt of the board. This nail is not nailed down firm. About ¾ inch should be left sticking out. The butt of the board is pried up and away from the wall. A nail bar or the claw of a hammer is used. The board is left this way. Then, the next board is pushed against the nails of the first board. The second board is checked for level. Hold the level on the bottom of the board and check the second level. When the board is level, it is nailed at a place 1½ inches from the butt. Again, the nail in the second board is left out. About ¾ inch should be left sticking out. The board is nailed down at the butt for its length. Then the first board is nailed down firm. The second board is then pried up for the third board. This process is repeated until the siding has been applied.

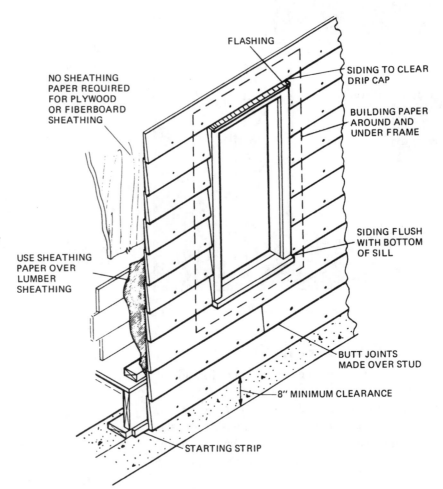

Fig. 7-47 *Bevel siding application. Note that pieces are even with top and bottom of window opening.* (Forest Products Laboratory)

FLASHING

NO SHEATHING PAPER REQUIRED FOR PLYWOOD OR FIBERBOARD SHEATHING

SIDING TO CLEAR DRIP CAP

BUILDING PAPER AROUND AND UNDER FRAME

USE SHEATHING PAPER OVER LUMBER SHEATHING

SIDING FLUSH WITH BOTTOM OF SILL

BUTT JOINTS MADE OVER STUD

8″ MINIMUM CLEARANCE

STARTING STRIP

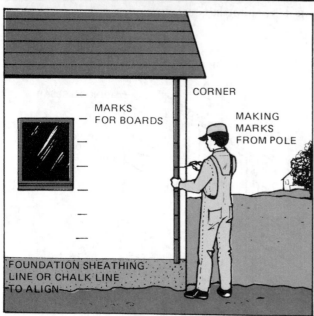

Fig. 7-48 Marks are made to keep siding aligned on long walls.

Corner Finishing

There are three ways of finishing corners for siding. The most common ways are shown in Fig. 7-49. Corner boards can be used for all types of siding. In one style of corner board, the siding is butted next to the boards. In this way, the ends of the siding fit snugly and the corner is the same thickness as the siding. See Fig. 7-49C.

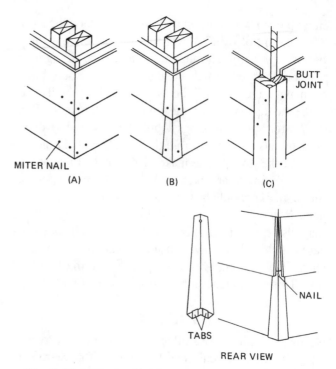

Fig. 7-49 Methods of finishing corners. (Forest Products Laboratory)

In Fig. 7-49B special metal corner strips are used. These are separate pieces for each width of board. The carpenter must be careful to select and use the right-size corner piece for the board siding used.

In other methods, the siding is butted at the corners. The corner boards are then nailed over the siding. See Fig. 7-50. This type of corner board is best used

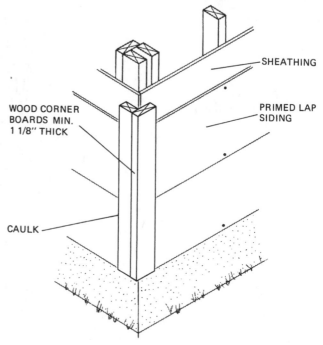

Fig. 7-50 Wood outside corner detail. (Boise-Cascade)

for plywood or panel siding. However, it is also common for board siding.

Another common method uses metal corners for lap siding. Figure 7-49B shows metal corner installation. These corners have small tabs at the bottom. They fit around the butt of each board. At the top is a small tab for a nail. The corners are installed after the siding is up. The bottom tab is put on first. Tabs are then put on other boards from bottom to top. The corners can be put on last. This is because the boards will easily spread apart at the bottoms. A slight spread will expose the tab for nailing.

Another method of finishing corners is to miter the boards. Generally, this method is used on more expensive homes. It provides a very neat and finished appearance. However, it is not as weatherproof. In addition, it requires more time and is thus more expensive. To miter the corners, a miter box should be used. Lap siding fits on the wall at an angle. It must be held at this angle when cut. A small strip of wood as wide as the siding is used for a brace. It is put at the base of the miter box. See Fig. 7-51. This positions the siding at the proper angle for cutting. If this is not handy, an estimate may be made. It is generally accurate enough. This method is shown in Fig. 7-52. A distance equal to the thickness of the siding is laid off at the top. The cut is then made on the line as shown.

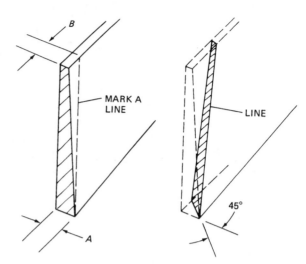

Fig. 7-52 *To cut miters without a miter box, make the top distance B equal to the thickness of A. Then cut back at about 45°.*

Inside corners are treated in two ways. Both metal flashing and wooden strips are used. The most common method uses wooden strips. See Fig. 7-53. A ¾-inch-square strip is nailed in the corner. Then each board of the siding is butted against the corner strip. After the siding is applied, the corner is caulked. This makes the corner weathertight.

Metal flashing for inside corners is similar. The metal has been bent to resemble a wooden square. The strips are nailed in the corner as shown in Fig. 7-54. As before, each board is butted against the metal strip.

PANEL SIDING

Panel siding is now widely used. It has several advantages for the builder. It provides a wider range of appearance. Panels may look like flat panels, boards, or battens and boards. Panels can also be grooved, and they can look like shingles. The surface textures range from rough lumber to smooth, flat board panels. Panels can show diagonal lines, vertical lines, or horizontal lines. Also, panels may be used to give the appearance of stucco or other textures.

Panel siding is easier to lay out. There is little waste and the installation is generally much faster. It is often stapled in place, rather than nailed. See Fig. 7-55. As a rule, corners are made faster and easier with panel siding.

Panel siding can be made from a variety of materials. Plywood is commonly used. Hardboard and various other fiber composition materials are also used. A variety of finishes is also available on panel siding. These are prefinished panels. Panels may be prefinished with paint, chemicals, metal, or vinyl plastics.

Fig. 7-51 *A strip must be used to cant the siding for mitering.*

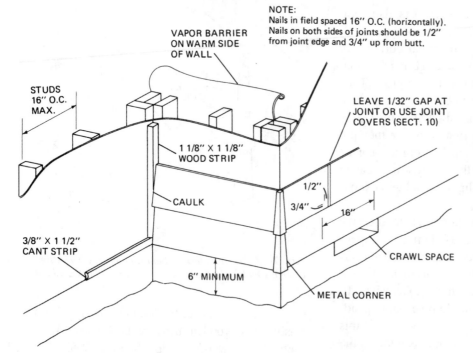

NOTE:
Nails in field spaced 16" O.C. (horizontally).
Nails on both sides of joints should be 1/2"
from joint edge and 3/4" up from butt.

VAPOR BARRIER
ON WARM SIDE
OF WALL

STUDS
16" O.C.
MAX.

1 1/8" X 1 1/8"
WOOD STRIP

CAULK

3/8" X 1 1/2"
CANT STRIP

6" MINIMUM

LEAVE 1/32" GAP AT
JOINT OR USE JOINT
COVERS (SECT. 10)

1/2"
3/4"
16"

CRAWL SPACE

METAL CORNER

Fig. 7-53 *Inside corners may be finished with square wooden strips.* (Boise-Cascade)

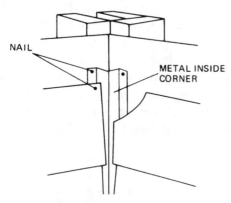

NAIL

METAL INSIDE
CORNER

Fig. 7-54 *Inside corners may be finished with metal strips.*

The standard panel size is 4 feet wide and 8 feet long. The most common panel thickness is ⅜ inch. However, other panel sizes are also used. Lengths up to 14 feet are also available. Thickness ranges from 5/16- to ¾-inch thick in 1/16-inch intervals.

The edges of panel siding may be flat. They may also be grooved in a variety of ways. Grooved edges form tight seams between plywood panels. Seams may be covered by battens. They may also be covered by special edge pieces. Outside corners may be treated in much the same manner as corners of regular board siding. Corners may be lapped and covered with corner boards. They may be covered with metal. They may also be mitered. Inside corners may be butted together. The edge of one panel is butted against the solid face of the first.

As a rule, panels are applied from the bottom up. However, they may be applied from top to bottom.

Fig. 7-55 *Panel siding is often stapled in place.* (Fox and Jacobs)

NAILS AND NAILING

Correct nails and nailing practices are essential in the proper application of wood siding. In general, siding and box nails are used for face nailing, and casing nails are used for blind nailing. Nails must be corrosion-resistant, and preferably rust proof. Avoid using staples. Stainless steel nails are the best choice. High tensile strength aluminum nails are economical and corrosion-resistant, and they will not discolor or cause deterioration of the wood siding. However, aluminum nails will react with galvanized metal, causing corrosion. Do not use aluminum nails on galvanized flashing (nor galvanized nails on aluminum flashing). The hot-dipped galvanized nail is the least expensive, but it might cause discoloration if precautions are not taken.

In some instances, the use of hot-dipped galvanized nails along with clear finishes on Western Red Cedar has resulted in stains around the nails. While this occurrence seems to be limited to the northeastern and north central regions of the country, the combination of hot-dipped galvanized nails with clear finishes on Western Red Cedar is not recommended.

Plastic hammer-head covers can be used when driving hot-dipped galvanized nails. This will reduce the potential for chipping and the subsequent potential for corrosion.

On siding you should not use staples or electroplated nails. These fasteners often result in black iron stains which can be permanent. Copper nails are not suitable for Western Red Cedar as cedar's natural extractives will react with the copper causing the nails to corrode, resulting in stains on the siding.

High-quality nails for solid wood siding are a wise investment. The discoloration, streaking, or staining that can occur with inappropriate nails ruins the appearance of the project, and is very difficult to remove.

Nail Shanks

Many nails have smooth shanks and will loosen as the siding expands and contracts under the extremes of seasonal changes in temperature and humidity. Ring or spiral-threaded nail shanks will increase holding power. Both types of shanks are readily available.

Nail Points

The most commonly used nail points include: blunt, diamond, and needle, as shown in Fig. 7-56. Blunt points reduce splitting while diamond points are the most commonly used. However, needle points should be avoided because they tend to cause splitting.

Recommended penetration into studs or blocking, or into a combination of wood sheathing and these members, is 1.5 inches. Penetration is 1.25 inches with ring shank nails.

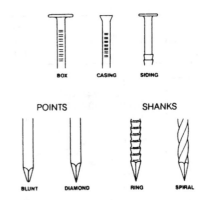

Fig. 7-56 *Nail types. (Western Wood Products Association)*

Vertical siding, when applied over wood bed sheathing, should be nailed to horizontal blocking or other wood framing members not exceeding 35 inches on center when face-nailed, or 32 inches on center when blind-nailed. Vertical siding, when installed without sheathing, should be nailed to wood framing or blocking members at 24 inches on center. Some building codes require 24 inches on center with or without sheathing. Check your local code for the requirements. Horizontal and diagonal siding should be nailed to studs at least 24 inches on center maximum when applied over wood-based, solid sheathing, and 16 inches on center maximum when applied without sheathing.

The siding pattern will determine the exact nail size, placement, and number of nails required. As a general rule, each piece of siding is nailed independently of its neighboring pieces. Do not nail through two overlapping pieces of siding with the same nail, as this practice will restrict the natural movement of the siding and might cause unnecessary problems. Nail joints into the studs or blocking members—nailing into sheathing alone is not adequate.

Drive nails carefully. Hand nailing is preferred over pneumatic nailing because there is less control of placement and driving force with pneumatic nailers. Nails should be snug, but not overdriven. Nails that are overdriven can distort the wood and cause excessive splitting. Predrilling near the ends helps reduce any splitting that can occur with the thinner patterns. Some modern siding patterns with their recommended nailing procedures are shown in Fig. 7-57.

SHINGLE AND SHAKE SIDING

Shingles and shakes are often used for exterior siding. They are very similar in appearance. However, shakes have been split from a log. They have a rougher surface texture. Shingles have been sawn and are smoother in

SIDING PATTERNS		NOMINAL SIZES*	NAILING	
		Thickness & Width	6" & Narrower	8" & Wider
	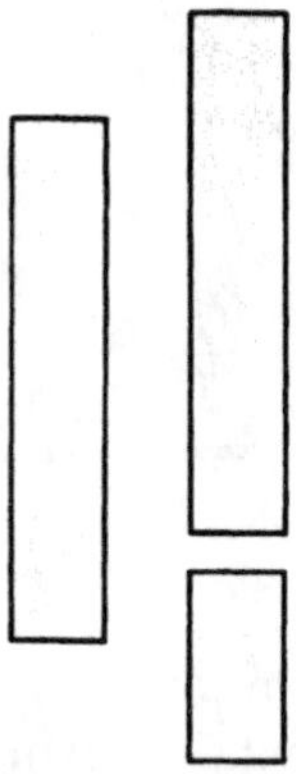 **TRIM** **BOARD-ON-BOARD** **BOARD-AND-BATTEN** Boards are surfaced smooth, rough or saw-textured. Rustic ranch-style appearance. Provide horizontal nailing members. Do not nail through overlapping pieces. Vertical applications only.	1 x 2 1 x 4 1 x 6 1 x 8 1 x 10 1 x 12 1¼ x 6 1¼ x 8 1¼ x 10 1¼ x 12	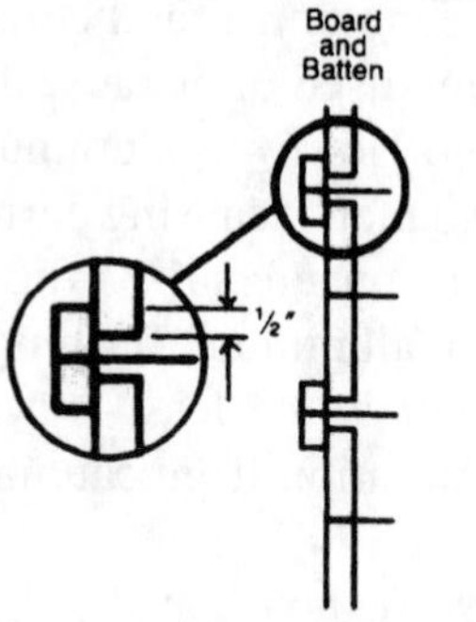 Recommend ½" overlap. One siding or box nail per bearing.	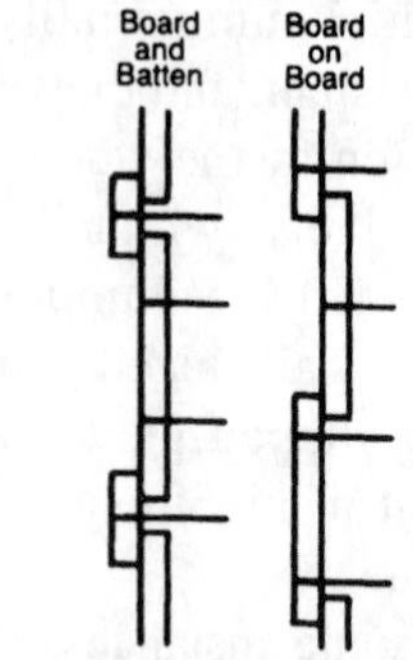 Increase overlap proportionately. Use two siding or box nails, 3-4" apart.
	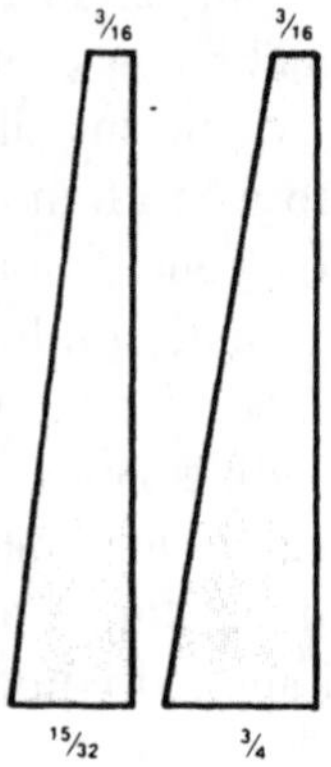 **BEVEL OR BUNGALOW** Bungalow ("Colonial") is slightly thicker than Bevel. Either can be used with the smooth or saw-faced surface exposed. Patterns provide a traditional-style appearance. Recommend a 1" overlap. Do not nail through overlapping pieces. Horizontal applications only. Cedar Bevel is also available in ⅞ x 10,12.	½ x 4 ½ x 5 ½ x 6 ⅝ x 8 ⅝ x 10 ¾ x 6 ¾ x 8 ¾ x 10	 Recommend 1" overlap. One siding or box nail per bearing, just above the 1" overlap.	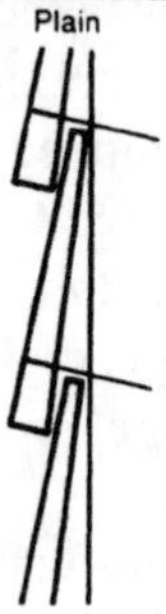 Recommend 1" overlap. One siding or box nail per bearing, just above the 1" overlap.
	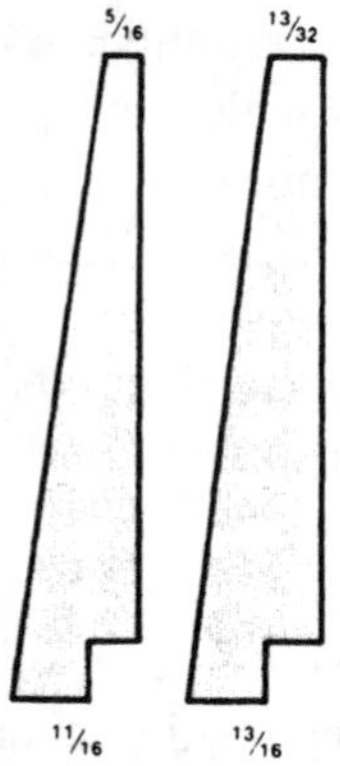 **DOLLY VARDEN** Dolly Varden is thicker than bevel and has a rabbeted edge. Surfaced smooth or saw textured. Provides traditional-style appearance. Allows for ½" overlap, including an approximate ⅛" gap. Do not nail through overlapping pieces. Horizontal applications only. Cedar Dolly Varden is also available ⅞ x 10,12.	Standard Dolly Varden ¾ x 6 ¾ x 8 ¾ x 10 Thick Dolly Varden 1 x 6 1 x 8 1 x 10 1 x 12	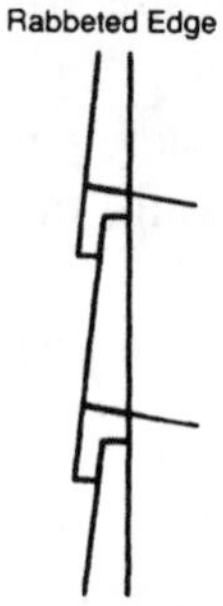 Allows for ½" overlap. One siding or box nail per bearing, 1" up from bottom edge.	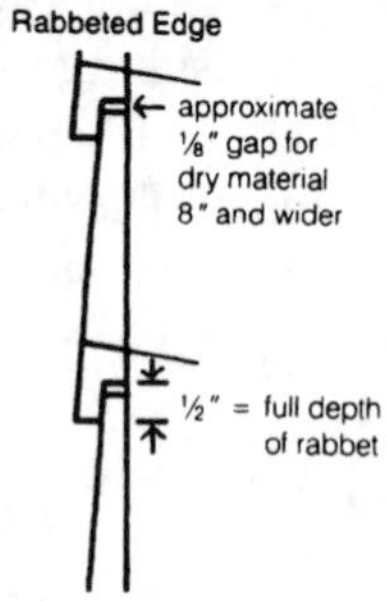 Allows for ½" overlap. One siding or box nail per bearing, 1" up from bottom edge.
	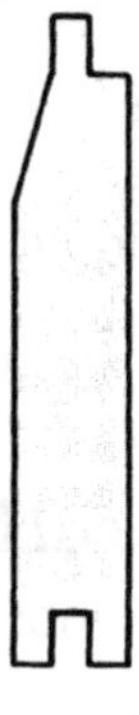 **DROP** Drop siding is available in 13 patterns, in smooth, rough and saw textured surfaces. Some are T&G, others shiplapped. Refer to "Standard Patterns" for dimensional pattern profiles. A variety of looks can be achieved with the different patterns. Do not nail through overlapping pieces. Horizontal or vertical applications. Tongue edge up in horizontal applications.	¾ x 6 ¾ x 8 ¾ x 10	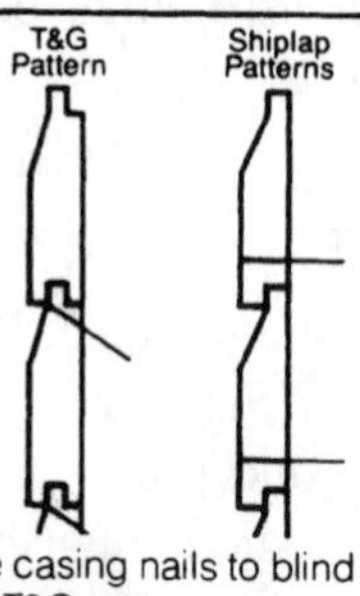 Use casing nails to blind nail T&G patterns, one nail per bearing. Use siding or box nails to face nail shiplap patterns, one inch up from bottom edge.	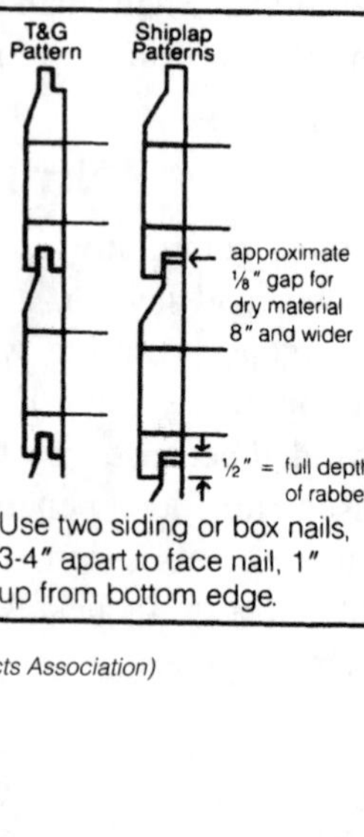 Use two siding or box nails, 3-4" apart to face nail, 1" up from bottom edge.

Fig. 7-57 *Siding patterns, nominal sizes, and recommended nailing.* (Western Wood Products Association)

SIDING PATTERNS	NOMINAL SIZES*	NAILING	
	Thickness & Width	6" & Narrower	8" & Wider
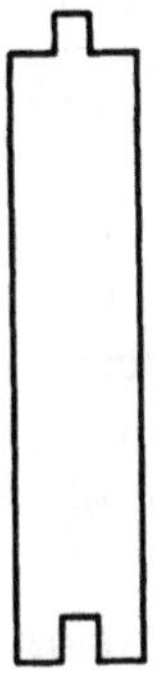 **TONGUE & GROOVE** Tongue & groove siding is available in a variety of patterns. T&G lends itself to different effects aesthetically. Refer to WWPA "Standard Patterns" (G-16) for pattern profiles. Sizes given here are for Plain Tongue & Groove. Do not nail through overlapping pieces. Vertical or horizontal applications. Tongue edge up in horizontal applications.	1 x 4 1 x 6 1 x 8 1 x 10 Note: T&G patterns may be ordered with ¼, ⅜ or ⁷⁄₁₆" tongues. For wider widths, specify the longer tongue and pattern.	Plain 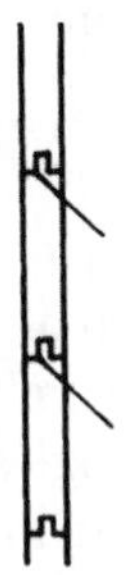 Use one casing nail per bearing to blind nail.	Plain 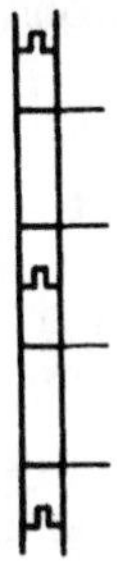 Use two siding or box nails 3-4" apart to face nail.
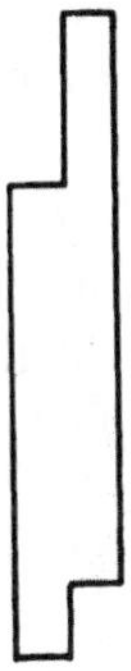 **CHANNEL RUSTIC** Channel Rustic has ½" overlap (including an approximate ⅛" gap) and a 1" to 1¼" channel when installed. The profile allows for maximum dimensional change without adversely affecting appearance in climates of highly variable moisture levels between seasons. Available smooth, rough or saw textured. Do not nail through overlapping pieces. Horizontal or vertical applications.	¾ x 6 ¾ x 8 ¾ x 10	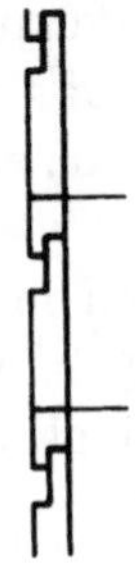 Use one siding or box nail to face nail once per bearing, 1" up from bottom edge.	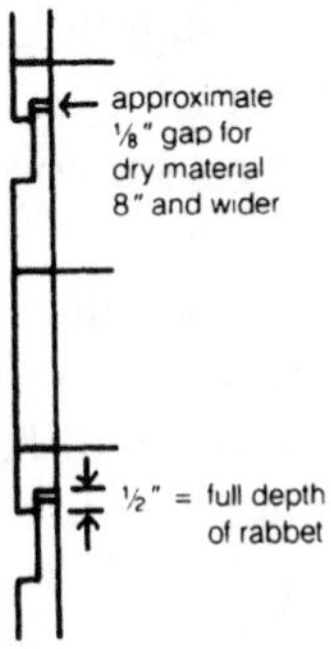 approximate ⅛" gap for dry material 8" and wider ½" = full depth of rabbet Use two siding or box nails 3-4" apart per bearing.
 **LOG CABIN** Log Cabin siding is 1½" thick at the thickest point. Ideally suited to informal buildings in rustic settings. The pattern may be milled from appearance grades (Commons) or dimension grades (2x material). Allows for ½" overlap, including an approximately ⅛" gap. Do not nail through overlapping pieces. Horizontal or vertical applications.	1½ x 6 1½ x 8 1½ x 10 1½ x 12	 Use siding or box nail to face nail once per bearing, 1½" up from bottom edge.	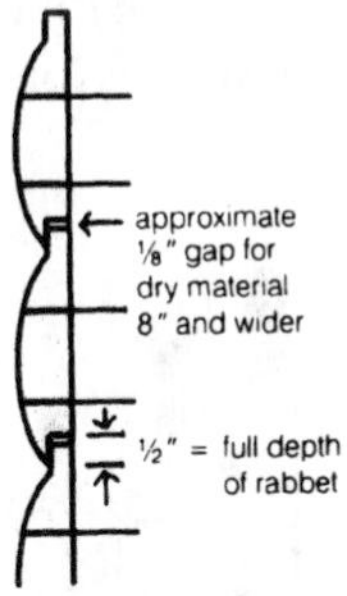 approximate ⅛" gap for dry material 8" and wider ½" = full depth of rabbet Use two siding or box nails, 3-4" apart, per bearing to face nail.

SIDING INSTALLATION TIPS

Do not nail through overlapping pieces. Use stainless steel, high tensile strength aluminum or hot-dipped galvanized nails with ring or spiral-threaded shanks. Use casing nails to blind nail; siding or box nails to face nail.

Horizontal applications only for Bevel, Bungalow and Dolly Varden.

Vertical applications only for Board-and-Board or Board-and-Batten; bevel cut ends of pieces and install so water is directed to outside.

Horizontal or vertical applications for Tongue & Groove, Channel Rustic, Log Cabin or Drop patterns. Tongue edge up in horizontal applications of Drop and T&G patterns

Read the section on Nail Penetration & Spacing to determine nail size.

Read the sections on Moisture Content and Prefinishing before installing siding.

Fig. 7-57 *Continued.*

appearance. The procedure is the same for either shingles or shakes.

Shingles are made from many different materials. Shakes are made only from wood. The standard lengths for wood shakes and shingles are 16, 18, and 24 inches.

Shingles

Shingles may be made of wood, flat composition, or mineral fiber composition. The last is a combination of asbestos fiber and portland cement. Shingles for roofs are lapped about two-thirds. About one-third of the shingle is exposed. When the shingles are laid, or coursed, there are three layers on the roof. However, when shingles are used for siding, the length exposed is greater. Slightly more than one-half is exposed. This makes a two-layer thickness instead of three as on roofs. For asbestos mineral sidings, sometimes less lapping is used.

Wooden shingles are made in random widths. Shingles will vary from 3 to 14 inches in width. The better grades will have more wide shingles than narrow shingles.

Nailing

All types of shingles may be nailed in either of two ways. In the first method, shown in Fig. 7-58, nailing strips are used. Note that a moisture barrier is almost always used directly under shingles. Builder's felt (tar paper) is the most common moisture barrier. Shingles should be spaced like lap siding. The butt line of the shingles should be even with the tops of wall openings. Top lines should be even with the bottoms. For utility buildings such as garages, no sheathing is needed. A moisture barrier can be put over the studs. Nailing strips are then nailed to the studs. However, for most residential buildings, a separate sheathing is suggested. Strips are often nailed over the sheathing. This is done over sheathing that is not a good nail base. Shingles may be nailed directly to board or plywood sheathing. This is shown in Fig. 7-59.

The bottom shingle course is always nailed in place first. Part of the bottom course can be laid for large wall sections. The course may reach for only a part of the wall. Then that part of the wall can be shingled. The shingled part will be a triangle from the bottom corner upward. As a rule, two layers of shingles are used on the bottom course. The first layer is nailed in place, and a second layer is nailed over it. The edges of the second shingle should not line up with the edges of the first. This makes the siding much more weather-resistant. This layout is shown in Fig. 7-59.

Another technique for shingles is called *double coursing*. This means that two thicknesses are applied. The first layer is often done with a cheaper grade of shingle. Again, each course is completed before the one above is begun. The edges should alternate for best weather-resistive qualities. Double coursing is shown in Fig. 7-60. The outside shingle covers the bottom of the inside shingle. This gives a dramatic and contrasting effect.

Shakes

The shake is similar to a shingle, but shakes may be made into panels. Figure 7-61 shows this type of siding. Shake panels are real wooden shakes glued to a plywood base. The strips are easier and quicker to nail than individual shakes. Also, shakes can be spaced more easily. A more consistent spacing is also attained. The individual panels may be finished at the factory. Color, spacing, appearance, and texture can be factory-made. Panels may also be combined with insulation and weatherproofing. The panels are applied in the same manner as shingles.

Corners

Corners are finished the same as for other types of siding. Three corner types are used: corner boards, metal

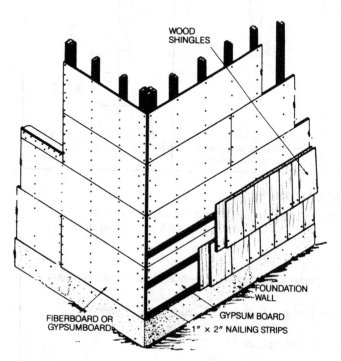

Fig. 7-58 *Nail strips are used over studs and sheathing. Shingles are then nailed to the strips. At least two nails are used for each shingle.*

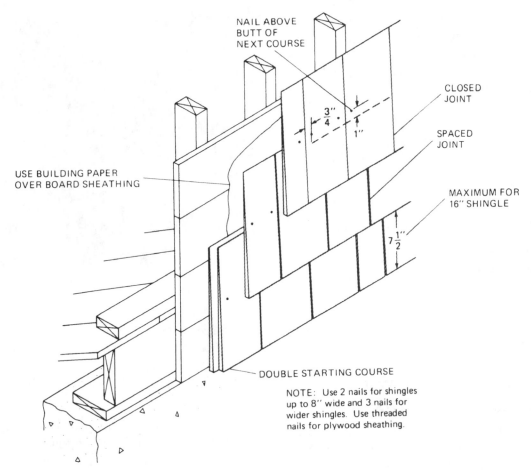

NAIL ABOVE
BUTT OF
NEXT COURSE

CLOSED
JOINT

$\frac{3}{4}''$

1''

SPACED
JOINT

MAXIMUM FOR
16'' SHINGLE

USE BUILDING PAPER
OVER BOARD SHEATHING

$7\frac{1}{2}''$

DOUBLE STARTING COURSE

NOTE: Use 2 nails for shingles
up to 8'' wide and 3 nails for
wider shingles. Use threaded
nails for plywood sheathing.

Fig. 7-59 *Single-course shingle nailed directly to solid sheathing.* (Forest Products Laboratory)

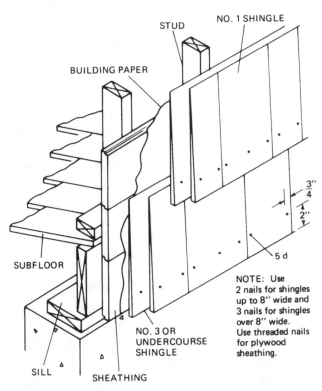

STUD

NO. 1 SHINGLE

BUILDING PAPER

$\frac{3}{4}''$

2''

5 d

SUBFLOOR

NOTE: Use
2 nails for shingles
up to 8'' wide and
3 nails for shingles
over 8'' wide.
Use threaded nails
for plywood
sheathing.

NO. 3 OR
UNDERCOURSE
SHINGLE

SILL

SHEATHING

Fig. 7-60 *Double-coursed shingle siding.* (Forest Products Laboratory)

Fig. 7-61 *Shakes are often made into long panels.* (Shakertown)

Shingle and Shake Siding   261

corners, and mitered corners. Again, mitering is generally used for more expensive buildings.

Corners may also be woven. Shingles may be woven as in Fig. 7-62. This is done on corners and edges of door and window openings.

Fig. 7-63 *A wire mesh is nailed to the wall. Stucco will then be applied over the mesh.*

Fig. 7-62 *Shingle and shake corners may be woven or lapped.*

PREPARATION FOR OTHER WALL FINISHES

There are other common methods for finishing frame walls. These include stucco, brick, and stone. As a rule, the carpenter does not put up these wall finishes. However, the carpenter sometimes prepares the sheathing for these coverings. The preparation depends upon how well the carpenter understands the process.

Stucco Finish

Stucco is widely used in the South and Southwest. It is durable and less expensive than brick or stone. Like brick or stone, it is fireproof. It may be put over almost any type of wall. It can be prepared in several colors. Several textures can also be applied to give different appearances.

Wall preparation A vapor barrier is needed for the wall. The vapor barrier can be made of builder's felt or plastic film. The sheathing may also be used for the vapor barrier. It is a good idea to apply an extra vapor barrier over most types of walls. This includes insulating sheathing, plywood, foam, and gypsum. The vapor barrier is applied from the bottom up. Each top layer overlaps the bottom layer a minimum of 2 inches. Then a wire mesh is nailed over the wall. See Fig. 7-63. Staples are generally used rather than nails. A solid nail base is essential. The mesh should be stapled at 18- to 24-inch intervals. The intervals should be in all directions—across, up, and down. The wire mesh may

be of lightweight "chicken wire." However, the mesh should be of heavier material for large walls.

The mesh actually supports the weight of the stucco. The staples only hold the mesh upright, close to the wall.

Apply the stucco Usually two or three coats are applied. The first coat is not the finish color. The first coat is applied in a very rough manner. It is applied carefully but it may not be even in thickness or appearance. It is called the "scratch" coat. Its purpose is to provide a layer that sticks to the wire. The coat is spread over the mesh with a trowel. See Fig. 7-64.

This scratch coat should be rough. It may be scratched or marked to provide a rough surface. The rough surface is needed so that the next coat will stick. A trowel with ridges may be used. The ridges make grooves in the coat. These grooves may run in any direction. However, most should run horizontally. Another way to get a rough surface is to use a pointed tool. After the stucco sets, grooves are scratched in it with the tool.

One or more coats may be put up before the finish coat is applied. These coats are called "brown" coats. They are scratched so that the next coat sticks better. The last coat is called the finish coat. It is usually white or tan in color. However, dyes may be added to give other colors. The colors are usually light in shade.

Brick and Stone Coverings

Brick and stone are used to cover frame walls. The brick or stone is not part of the load-bearing wall. This means that they do not hold up any roof weight. The covering is called a brick or stone veneer. It adds beauty and weatherproofing. Such walls also increase the resistance of the building to fires.

Fig. 7-64 Stucco is spread over mesh with a trowel.

The preparation for either brick or stone is similar. First, a moisture barrier is used. If a standard wall sheathing has been used, no additional barrier is needed. However, it is not bad practice to add a separate moisture barrier.

The carpenter may then be asked to nail *ties* to the wall. These are small metal pieces, shown in Fig. 7-65. Ties are bent down and embedded in the mortar. After the mortar has set, these ties form a solid connection. They hold the brick wall to the frame wall. They also help keep the space between the

Fig. 7-65 Ties help hold brick or stone to the wall. They are nailed to studs and embedded in the mortar.

walls even. Small holes are usually left at the bottom. These are called *weep* holes. Moisture can soak through brick and stone. Moisture also collects at the bottom from condensation. The small weep holes allow the moisture to drain out. By draining the moisture, damage to the wood members is avoided. The joints are trimmed after the brick is laid (Fig. 7-66).

Fig. 7-66 Joints are trimmed after brick is laid. (Fox and Jacobs)

ALUMINUM SIDING

Aluminum siding is widely used on houses. Aluminum is used for new siding or can be applied over an old wall.

A variety of vertical and horizontal styles are used. Probably the most common type looks like lap siding. However, even this type may come as individual "boards" or as panels two or three "boards" wide. See Fig. 7-67. These types of siding may be hollow-backed or insulated. See Fig. 7-68.

Aluminum siding may also look like shingles or shakes, as in Fig. 7-69. For all types of aluminum siding, a variety of surface textures and colors are available.

Aluminum siding is put up using a special system. Note in Fig. 7-70 that the top edge has holes in it. All nails are driven in these holes. The bottoms or edges of the pieces interlock. In this way, the tops are nailed and the bottoms interlock. Each edge is then attached to a nailed portion for solid support.

To start, a special starter strip is nailed at the bottom. See Fig. 7-71. Then corner strips, as in Fig. 7-72, are added. Special shapes are placed around windows and other features as in Fig. 7-73. For these, the manufacturer's directions should be carefully followed.

Next, the first "board" is nailed in place. Note that it is placed at the bottom of the wall. The bottom of the

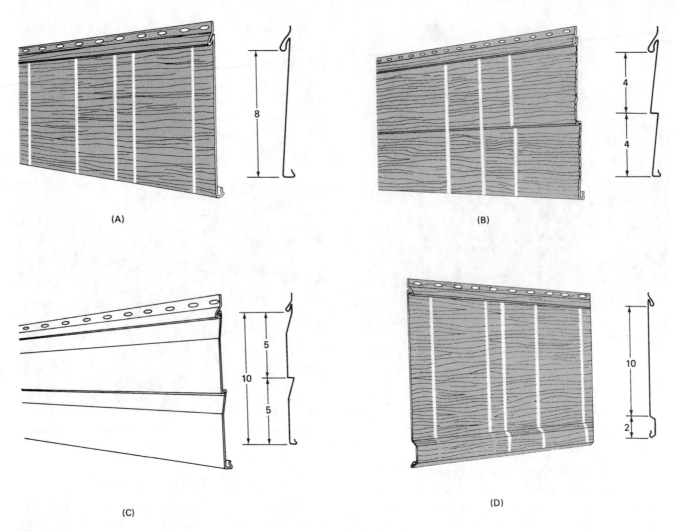

Fig. 7-67 *The most common type of aluminum siding looks like lap siding. (A) Single-"board" lap siding; (B) Two-"board" panel lap siding; (C) Double-"board" drop siding; (D) Vertical "board" siding.*

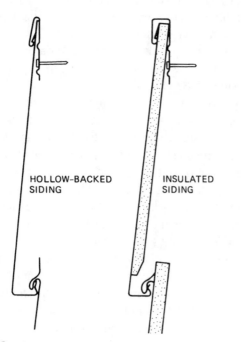

HOLLOW-BACKED SIDING   INSULATED SIDING

Fig. 7-68 *Aluminum siding may be hollow-backed or insulated.* (ALCOA)

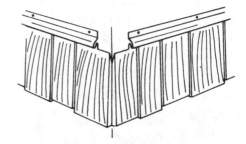

Fig. 7-69 *Aluminum "shake" siding.* (ALCOA)

first piece is interlocked with the starter strip and nailed. It is best to start at the rear of the house and work toward the front. This way, overlaps do not show as much. For the same reason, factory-cut ends should overlap ends cut on the site. See Fig. 7-74. Backer strips, as in Fig. 7-75, should be used at each overlap. These provide strength at the overlaps. Also, overlaps should be spaced evenly, as in Fig. 7-76. Grouping the overlaps together gives a poor appearance.

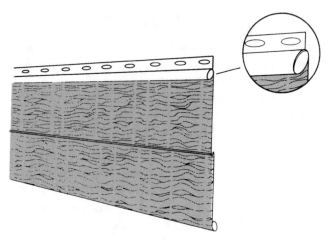

Fig. 7-70 *Panels are nailed at the top and interlock at the bottom.* (ALCOA)

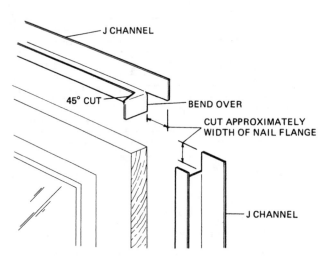

Fig. 7-73 *Special pieces are used around windows and gables.* (ALCOA)

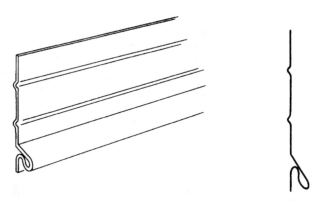

Fig. 7-71 *A starter strip is nailed at the bottom of the wall. The lower edge of the first board can interlock with it for support.* (ALCOA)

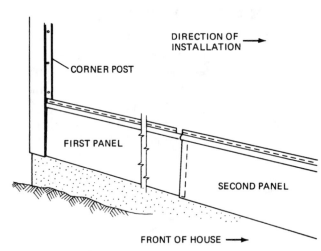

Fig. 7-74 *Siding is started at the back.* (ALCOA)

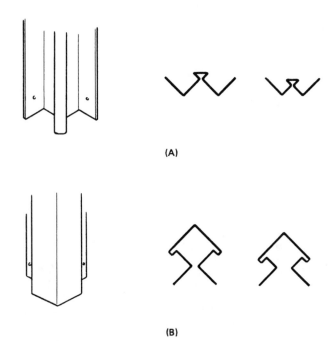

Fig. 7-72 *Corner strips for aluminum siding* (ALCOA). *(A) Inside; (B) Outside.*

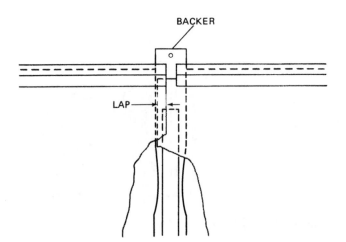

Fig. 7-75 *Backer strips support the ends at overlaps.* (ALCOA)

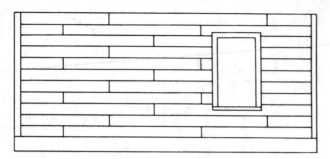

Fig. 7-76 *Overlaps should be equally spaced for best appearance.* (ALCOA)

Allowance must be made for heat expansion. Changes in temperature can cause the pieces to move. To allow movement, the nails are not driven up tightly. It is best to check the instructions that come with the siding.

Vertical Aluminum Siding

Most procedures for vertical siding are the same. Strips are put up at corners, windows, and eaves.

However, the starter strip is put near the center. A plumb line is dropped from the gable peak. A line is then located half the width of a panel to one side. The starter strip is nailed to this line. Panels are then installed from the center to each side. See Fig. 7-77.

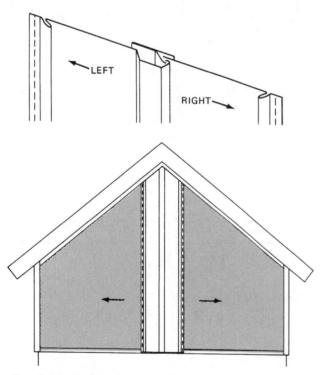

Fig. 7-77 *Vertical aluminum siding is started at the center.* (ALCOA)

SOLID VINYL SIDING

Solid vinyl siding is also used widely. It is available in both vertical and horizontal applications. A wide variety of colors and textures is also available.

As with aluminum siding, vinyl siding is nailed on one edge. The other edge interlocks with a nailed edge for support. See Fig. 7-78. Special pieces are again needed for joining windows, doors, gables, and so forth.

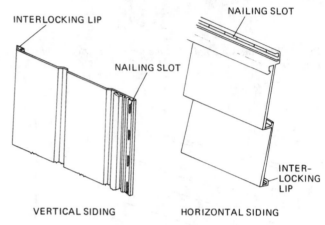

Fig. 7-78 *Solid vinyl siding.*

As with aluminum siding, the finish is a permanent part of the siding. No painting will ever be needed. It may be cleaned with a garden hose.

8
CHAPTER

Alternative Framing Methods

THIS CHAPTER ILLUSTRATES NEW METHODS that are presently being used to frame houses. These newer methods are becoming increasingly popular as the price of lumber continues to increase—while its quality decreases. This chapter covers: the most popular alternative framing methods; the advantages and disadvantages to each of these methods; the tools that are associated with each of these methods; the basic sequence for framing a house with these methods; and installation of utilities inside these framing methods.

WOOD FRAMES PREDOMINATE

Approximately 90% of homes in the United States are built with wood frame walls. Because of this heavy use, almost all of the old-growth forests have been harvested. New-growth lumber is typically the only type available to builders. This presents a present-day problem because new-growth lumber is fertilized and watered to make it grow quickly. This produces a larger cell structure, the end result of which is wood that is not as dense or strong and is more susceptible to warping, cracking, and other defects. In addition, the price of lumber has tripled over the past decade and will continue to rise as quality, old-growth lumber becomes less available.

Because of this trend, many home builders use alternative methods for framing houses. These methods are variations of methods used in commercial buildings and therefore meet most local residential building code requirements. In most cases, these framing methods exceed code specifications and make a home more resistant to hurricanes, earthquakes, termites, and other common nuisances that afflict traditionally framed homes.

STEEL FRAMING

Steel-framed homes are not a new phenomenon. Their debut was in the 1933 Chicago World's Fair. However, it wasn't until the 1960s that the leading steel manufacturers entered the residential housing market. Steel homes have not gained wide popularity in the residential market because of their initial cost. Steel framing had always been more expensive than wood framing. With the rapid increase in the cost of lumber, steel studs are now, in most cases, less expensive than a wooden 2 × 4. In most areas of the country, a steel-framed home costs as much as, or only 15% more than, a wooden-framed home. If the price of lumber continues to rise at the same rate as it has in the past, then steel-framed homes or other methods of framing will become more economical and commonplace.

Advantages and Disadvantages of Steel-Framed Homes

Today, most commercial and industrial buildings are constructed with steel as part of the primary structure. Steel is used because it is the strongest and safest cost-effective building material available. When steel framing is used instead of wood, its advantages include:

- No shrinking, warping, or twisting which allows walls, floors, and ceilings to be straight and square, thereby reducing finish work.
- A higher strength-to-weight ratio so homes can withstand hurricanes, earthquakes, high winds, heavy snow loads, torrential rains, termites, and can span greater distances with less support.
- It is assembled with screws, which eliminates nail-pops and squeaks.
- Steel studs, joists, and trusses are quicker to install.
- Steel-framed homes weigh as much as 30% less, so foundations can be lighter.
- Steel-framed homes provide a better path to ground when struck by lightning, reducing the likelihood of explosions or secondary fires.
- Most steel framing is made to exact size in a factory, so less waste is produced.
- Steel is 100% recyclable.

The main disadvantage to residential steel framing is its initial cost. Other disadvantages include:

- Local building code officials might be unfamiliar with its use.
- Different tools are required.
- Carpenters are not familiar with the building practices associated with steel framing.

Types of Steel Framing

The two main types of steel framing are *red-iron* and *galvanized*. Red-iron framing is based on the standard commercial application in which the structure is built out of a thick red-iron frame at 8-foot spans to distribute the load (Fig. 8-1). Galvanized sheet metal studs, joists, and other framing members are installed at 24 inches O.C. for a screwing base for sheathing.

Galvanized steel framing is based on the traditional wooden "stick-built" method. In this method, the entire home is built using galvanized sheet metal framing members as illustrated in Fig. 8-2. This method is preferred by some builders because its similarity to traditional wooden framing methods makes it easier to

Fig. 8-1 *Red-iron framework at 8-foot intervals with galvanized-steel framing in between to support sheathing.* (Tri-Steel Structures)

retrain carpenters, and cranes are typically not needed to lift heavy red-iron framing members. However, exterior walls require additional horizontal bracing spaced 16–24 inches O.C. and roof framing requires more rafters and purlins.

Tools Used In Steel Framing

Because most metal-framed homes are engineered and sold in kits, most of the framing members are cut to length. However, tin snips or aviation snips are frequently used to cut angles and notches in base track for plumbing manifolds. A chop saw or circular saw with a metal cutting blade is used to make most straight cuts.

Screw guns are used instead of hammers. Powder-actuated nail guns are used to attach base track to concrete foundations. Wrenches and ratchets are used to secure nuts and bolts in red-iron connections. A crane is needed for one day to erect the red-iron framing.

Tape measurers are used to lay out all framing connections and speed squares are used to transfer measurements around corners. Levels and chalk lines are frequently used to keep everything plumb and square. Vice grips are used to clamp framing members together while they are being attached.

Sequence

Most residential steel suppliers are eager for builders to convert to steel framing and will provide extensive training as well as video tapes on framing with steel. In addition, they sell framing kits of precut steel members for standard model floor plans. Detailed plans are provided with each kit that include specifications for stud spacing and other framing members dependent upon the size of the structure. Therefore, only the general sequence for framing with steel is covered in this chapter.

1. Anchor bolts for red-iron framing can only deviate ⅛ inch to ensure proper alignment and accuracy. Therefore, they are installed after the concrete has

Fig. 8-2 *Galvanized-steel framing. Note the horizontal wall bracing and additional rafters that are required for homes made without a red-iron frame.* (Tri-Steel Structures)

hardened. This is done by drilling holes for each anchor bolt with a hammer-drill. Anchor bolts range in length and diameter depending on load specifications. Two anchor bolts are used to fasten each red-iron girder. Anchor bolts are packaged in glass tubes, which break when they are inserted into the hole. When the tube breaks, a two part adhesive mixes together, which bonds the anchor bolt to the concrete foundation. Usually, 24 hours is required for the adhesive to cure before any red-iron framing members are attached.

2. Red-iron framing members are assembled with nuts and bolts on a flat level surface. Each section usually consists of two vertical supports with rafters and bracing. They are erected using a crane.

3. Once the red-iron framing members are attached to the anchor bolts they are straightened by attaching them to come-alongs which are attached to stakes around the perimeter of the home. Come-alongs are tightened and loosened until the framing members are plumb and square. Galvanized sheet metal purlins are then attached every two feet between the eight-foot span of the red-iron rafters to keep them square.

4. Galvanized framing for walls is usually secured to concrete foundations by running two beads of sealant (sill seal) between the concrete and the galvanized base track. A powder-actuated nail gun is then used to drive fasteners every 12 inches O.C. in the base track.

5. Studs are secured by one screw on each side of the base track. Typically, #8 × ½-inch self-drilling low-profile wafer, washer, or pan head screws are used to secure all galvanized framing members. MIG welders also can be used to attach framing members; however, screws are preferred. Studs are attached to red-iron or any other thicker framing members by #12 × ¾-inch self-drilling low-profile head screws. Vice grips should always be used to clamp the framing members together while they are being attached.

6. Furring or hat channel are spaced 16–24 inches O.C. for ceiling framing or for horizontal reinforcement of exterior walls in the galvanized framing method. Steel framing members are identified in Fig. 8-3.

7. Plywood and oriented-strand board (OSB) are commonly used as exterior sheathing for steel-framed homes. Exterior sheathing is staggered every four feet on the roof and on the walls of the galvanized-framing method. Plywood and OSB can be laid horizontally on top of each other without staggering for

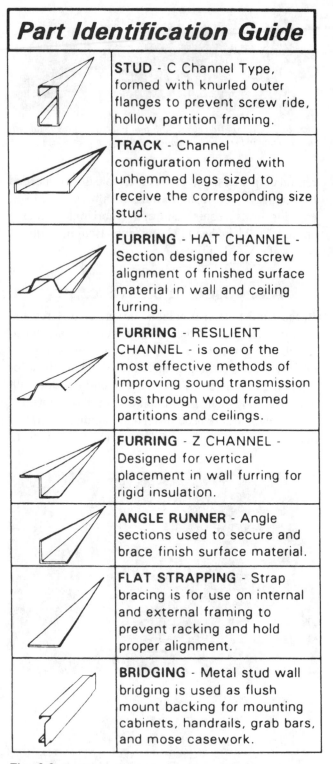

Fig. 8-3 *Steel frame part identification guide.* (Southeastern Metals Manufacturing Co.)

walls in the red-iron method. The basic sequence for framing and sheathing a red-iron home is shown in Fig. 8-4. Sheathing should be attached by screws every 24 inches O.C. with five screws securing a 48-inch width. This equates to a minimum of 25 screws per 4' × 8' sheet.

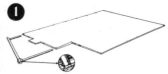

The Foundation (slab, pier & beam, pilings or basement) is prepared and exterior wall sections are bolted together on the ground.

LIFETIME
HOME CONSTRUCTION
Quick, Easy Assembly
Saves Time & Money

Fig. 8-4 *Basic sequence for framing and sheathing using the red-iron method.* (Tri-Steel Structures)

Frame sections are raised, joined together and anchored to the foundation.

Roof trusses are assembled on the ground, lifted into place, attached to the side walls, and braced with furring.

Sheathing is attached to the roof and walls. Doors and windows are installed and dormers are created. Interior framing and insulation are installed.

Exterior finish of your choice is applied and the shell is now complete. The interior can be completed at your own pace.

8. Six screws are used at web stiffening areas such as window sills supported by 6-inch cripple studs. Most other connections require only two screws (one on each side).

9. Utilities, such as electrical wiring and plumbing, can run through the prepunched holes in the studs and joists. However, plastic grommets must first be inserted in the holes so the sharp steel edges of the framing members will not damage the insulation on the wiring. The remaining electrical work usually involves the same techniques used in commercial steel-framing applications.

GALVANIZED FRAMING

When galvanized steel studs are used to frame the entire home, heavier gage studs (12–18 gage)[*] are used for the exterior load-bearing walls and lighter[**] gage studs (20–25 gage) are used for the interior curtain walls. Typically, most houses are engineered by a company in advance; depending on the weight of the roof, tile, or composition shingles, or if there is a second story, the gage of the studs will vary from home to home.

[*]The word *gage* is presently used instead of *gauge*. It is still acceptable to use either form of the word, but it is more "up to date" to use the shorter form.
[**]Keep in mind that the lighter the gage of the steel, the higher the gage number. That means 12 gage is thicker than 15 gage.

There are three main ways to assemble a steel stud wall. One way is to have all the walls fabricated into panels in a factory that is off-site. See Fig. 8-5. This method, typically called *panelization*, is the fastest-growing segment of residential new construction. The advantages to panelization are:

1. Less material is needed because human errors on the job site are prevented.

2. There is no lost time on the job site because of inclement weather.

Fig. 8-5 *Panelized walls and trusses save material and waste on the job site.* (Steel Framing, Inc.)

3. Fewer skilled laborers are needed on the job site. This means a reduction in the builder's liability and in worker's compensation insurance.

4. Panelized walls produce consistent quality. This is because the factory-controlled conditions include preset jigs and layout tables.

5. Panelized walls can be ordered in advance and shipped to the job site for easy installation. This significantly reduces the project construction time. See Fig. 8-6.

Fig. 8-6 *Factory-made panels being unloaded on the construction site.* (Steel Framing, Inc.)

The second way to assemble a steel wall is called *in-place* framing. This method requires the carpenter to first secure the steel-base track to the foundation. It can be done by bolting it down with anchor bolts or by using powder-actuated nailers. When connecting the base track to a slab using powder-actuated nailers, two low-velocity fasteners are spaced a minimum of every 24 inches. Typically, most exterior walls are connected to the slab with anchor bolts or J-bolts. The base track is usually reinforced with an additional piece of track or stud where the anchor bolt extends through the base track, as shown in Fig. 8-7.

Steel studs are attached to the base track every 16 or 24 inches O.C. with a No. 8 self-tapping screw on either side. Spacing depends on the load placed on each wall and the gage of the stud being used. The corner stud is braced to the ground to prevent movement from the wind or from the stresses during assembly. A spacer stud is placed between each stud thereafter to keep the studs in place. Steel channel (which comes in 16-foot lengths) can be placed between the *cut-outs* of each stud to brace the wall instead of the usual spacers, as shown in Fig. 8-8. Additional steps in fabricating the wall will be discussed in the following.

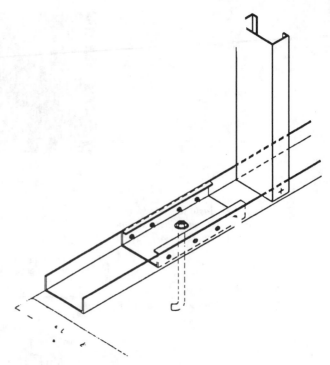

Fig. 8-7 *J-bolt foundation connection.* (Steel Framing, Inc.)

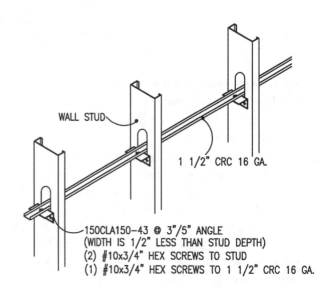

WALL STUD

1 1/2" CRC 16 GA.

150CLA150-43 @ 3"/5" ANGLE
(WIDTH IS 1/2" LESS THAN STUD DEPTH)
(2) #10x3/4" HEX SCREWS TO STUD
(1) #10x3/4" HEX SCREWS TO 1 1/2" CRC 16 GA.

NOTES:
FOR STUD HEIGHTS 10'-0" AND LESS, INSTALL CRC AT MID HEIGHT.
FOR STUD HEIGHTS GREATER THAN 10'-0" INSTALL AT THIRD POINTS.

Fig. 8-8 *Wall bridging detail.* (Nuconsteel Commercial Corp.)

Tilt-up framing is another way to assemble a steel-stud wall. In this method, the wall is assembled flat (horizontally). Then, it is lifted into a vertical position. This method makes it easier for carpenters to assemble a flat straight wall—that is, if the surface they are assembling the wall on is flat. Torpedo levels, which are magnetic, can easily be placed on the steel studs to insure that each is level and square before being installed into the base track with No. 8 self-tapping screws.

Steel-strap blocking, wood blocking, or track blocking can be used between studs for reinforcing. This is particularly helpful where overhead cabinets or other heavy fixtures may be attached. Steel straps are also fastened diagonally in the corners of the framing. This keeps the walls square and protects them from shear forces. However, steel strapping is not necessary if OSB or plywood is fastened to the corners of the steel framing.

Details for corners, window and door openings, wall-to-wall panels, intersecting wall panels, and wall-to-roof trusses are shown in Figs. 8-9 through 8-15.

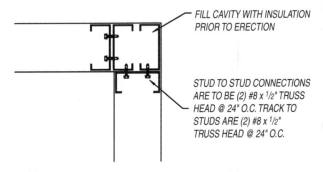

FILL CAVITY WITH INSULATION PRIOR TO ERECTION

STUD TO STUD CONNECTIONS ARE TO BE (2) #8 x 1/2" TRUSS HEAD @ 24" O.C. TRACK TO STUDS ARE (2) #8 x 1/2" TRUSS HEAD @ 24" O.C.

Fig. 8-9 *Corner framing.* (Nuconsteel Commercial Corp.)

INSULATED CONCRETE FORMS

One of the fastest growing framing methods is the insulated concrete form (ICF). In this method, walls are made of concrete reinforced by rebar surrounded by Styrofoam® (expanded polystyrene beads). Most ICFs are hollow Styrofoam® blocks that are stacked together like building blocks with either tongue-and-groove joints or finger joints as shown in Fig. 8-16. Each manufacturer makes the block a different size. Some manufacturers have ties molded in their blocks while others have separate ties that are inserted into the block at the job site. Once the blocks are assembled, reinforced, and braced, concrete is poured into the blocks using a concrete pumper truck. An example of what the concrete and rebar would look like inside the foam is illustrated in Fig. 8-17.

Advantages and Disadvantages to Insulated Concrete Forms

Insulated concrete form homes are framed in concrete, which allows them to have the following advantages over conventional wood-frame homes:

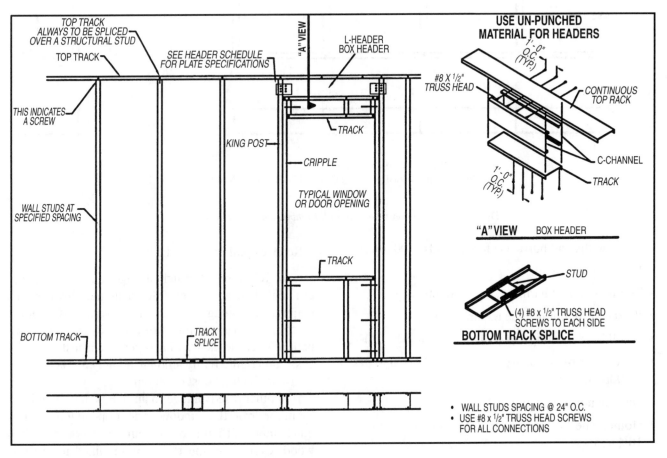

TOP TRACK ALWAYS TO BE SPLICED OVER A STRUCTURAL STUD

TOP TRACK

SEE HEADER SCHEDULE FOR PLATE SPECIFICATIONS

"A" VIEW

L-HEADER BOX HEADER

THIS INDICATES A SCREW

KING POST

CRIPPLE

TRACK

TYPICAL WINDOW OR DOOR OPENING

WALL STUDS AT SPECIFIED SPACING

TRACK

BOTTOM TRACK

TRACK SPLICE

USE UN-PUNCHED MATERIAL FOR HEADERS

1'-0" O.C. (TYP.)

#8 X 1/2" TRUSS HEAD

CONTINUOUS TOP RACK

C-CHANNEL

TRACK

1'-0" O.C. (TYP.)

"A" VIEW   BOX HEADER

STUD

(4) #8 x 1/2" TRUSS HEAD SCREWS TO EACH SIDE

BOTTOM TRACK SPLICE

• WALL STUDS SPACING @ 24" O.C.
• USE #8 x 1/2" TRUSS HEAD SCREWS FOR ALL CONNECTIONS

Fig. 8-10 *Typical structural wall framing.* (Nuconsteel Commercial Corp.)

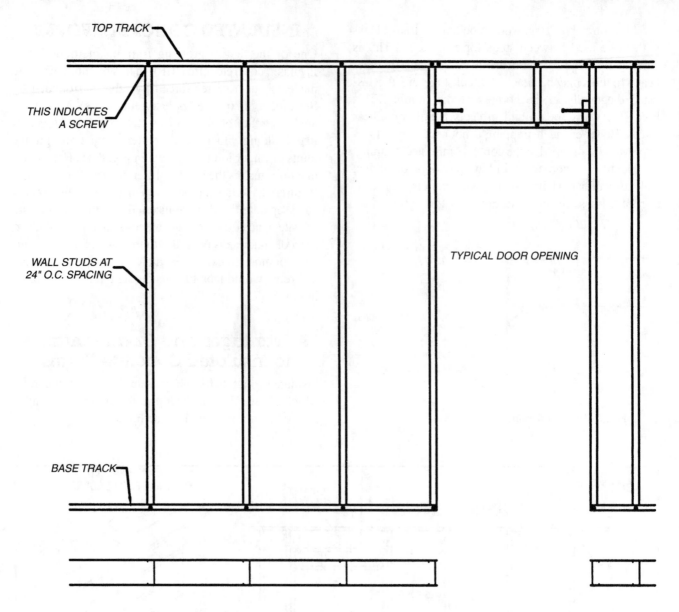

TOP TRACK

THIS INDICATES A SCREW

WALL STUDS AT 24" O.C. SPACING

BASE TRACK

TYPICAL DOOR OPENING

- WALL STUDS SPACING @ 24" O.C.
- USE #6 x 7/16" SHARP POINT SCREWS FOR ALL CONNECTIONS

Fig. 8-11 *Typical non-structural wall framing.* (Nuconsteel Commercial Corp.)

- It can withstand hurricane force winds (200 mph).
- It is bullet resistant.
- Termites cannot harm its structural integrity.
- It exceeds building code requirements.
- It will lower heating and cooling costs by 50%–80%.
- Load bearing capacity is higher (27,000plf vs. 4,000plf).
- Outside noise reduction is higher.
- Homeowner's insurance rates should decrease 10%–25%.
- Fire rating of walls is measured in hours vs. minutes.

- Reduces pollen inside the home.

An ICF home is much stronger and more energy efficient than a wooden-framed home. However, its major disadvantage is cost. The average ICF home costs about twice as much to frame as its chief competitor. Because the framing of a home is just one of its costs, this figure really amounts to only a 5%–15% increase in the total cost of a home.

Another disadvantage to an ICF is that there is no room for error. If a window or door opening is off, it is much more difficult to cut reinforced concrete than wood. Carpenters must also be retrained to use this framing method correctly.

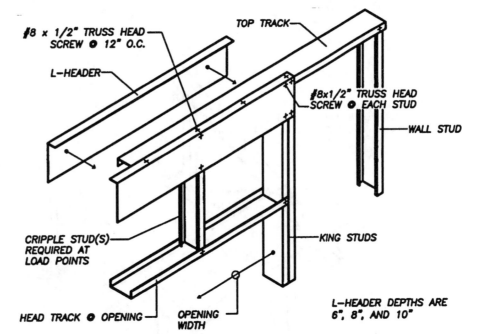

Fig. 8-12 *Double L-header framing.*
(Nuconsteel Commercial Corp.)

#8 × 1/2" TRUSS HEAD SCREW @ 12" O.C.

L—HEADER

TOP TRACK

#8×1/2" TRUSS HEAD SCREW @ EACH STUD

WALL STUD

KING STUDS

CRIPPLE STUD(S) REQUIRED AT LOAD POINTS

HEAD TRACK @ OPENING

OPENING WIDTH

L—HEADER DEPTHS ARE 6", 8", AND 10"

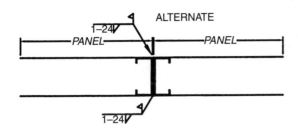

ALTERNATE

1–24∕

PANEL — PANEL

1–24∕

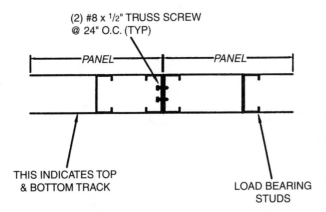

(2) #8 × 1/2" TRUSS SCREW @ 24" O.C. (TYP)

PANEL — PANEL

THIS INDICATES TOP & BOTTOM TRACK

LOAD BEARING STUDS

Fig. 8-13 *Wall panel to wall panel.* (Nuconsteel Commercial Corp.)

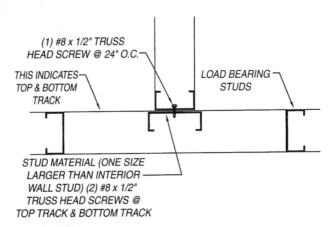

(1) #8 × 1/2" TRUSS HEAD SCREW @ 24" O.C.

THIS INDICATES TOP & BOTTOM TRACK

LOAD BEARING STUDS

STUD MATERIAL (ONE SIZE LARGER THAN INTERIOR WALL STUD) (2) #8 × 1/2" TRUSS HEAD SCREWS @ TOP TRACK & BOTTOM TRACK

Fig. 8-14 *Wall intersection framing.* (Nuconsteel Commercial Corp.)

Tools Used in Insulated Concrete Form Framing

Because concrete can be molded into any shape, it is essential to mold the frame of a home straight and square so walls are plumb and level. In order to achieve this, a carpenter should have a framing square, 2- and 4-foot levels, chalk line, and 30- and 100-foot tape measures. The foam forms can easily be cut with a handsaw or sharp utility knife. In most cases, the foam forms are glued together, so a caulking gun is required.

All the foam forms need to be braced until the concrete hardens. If the construction crew is not using metal braces, then wooden braces need to be constructed. Therefore, a framing hammer, nails, circular saw, and crowbar are needed.

Rebar is run horizontally and vertically throughout the walls of an ICF home. Tools used to work with rebar are rebar cutting and bending tools, metal cutoff saw, rebar twist tie tool (pigtail), 8-inch dikes, 9-inch lineman's pliers, hacksaw, tin snips, and wire-cutting pliers. A hot knife (Fig. 8-18) or router can be used to cut notches in the foam to place electrical wiring.

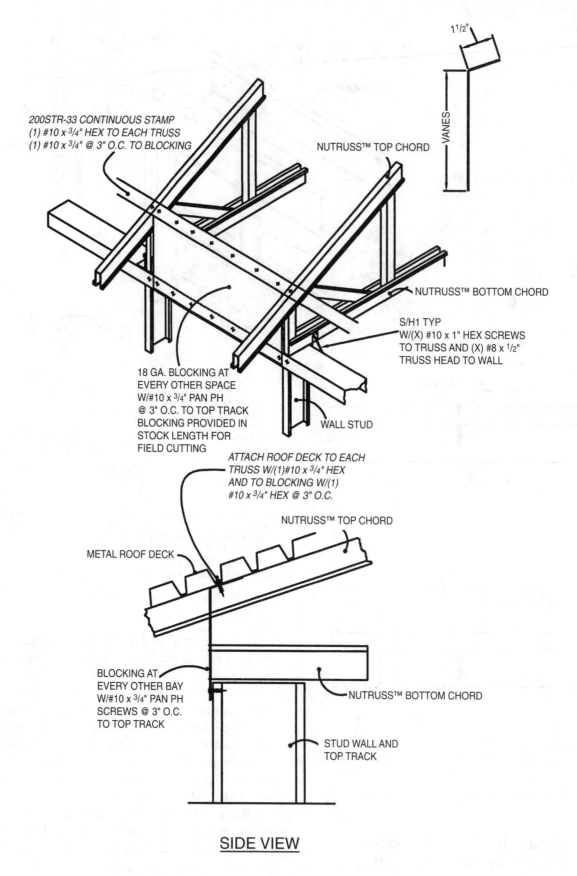

200STR-33 CONTINUOUS STAMP
(1) #10 x 3/4" HEX TO EACH TRUSS
(1) #10 x 3/4" @ 3" O.C. TO BLOCKING

NUTRUSS™ TOP CHORD

1 1/2"

VANES

NUTRUSS™ BOTTOM CHORD

S/H1 TYP
W/(X) #10 x 1" HEX SCREWS
TO TRUSS AND (X) #8 x 1/2"
TRUSS HEAD TO WALL

18 GA. BLOCKING AT
EVERY OTHER SPACE
W/#10 x 3/4" PAN PH
@ 3" O.C. TO TOP TRACK
BLOCKING PROVIDED IN
STOCK LENGTH FOR
FIELD CUTTING

WALL STUD

ATTACH ROOF DECK TO EACH
TRUSS W/(1)#10 x 3/4" HEX
AND TO BLOCKING W/(1)
#10 x 3/4" HEX @ 3" O.C.

NUTRUSS™ TOP CHORD

METAL ROOF DECK

BLOCKING AT
EVERY OTHER BAY
W/#10 x 3/4" PAN PH
SCREWS @ 3" O.C.
TO TOP TRACK

NUTRUSS™ BOTTOM CHORD

STUD WALL AND
TOP TRACK

SIDE VIEW

Fig. 8-15 *Truss to stud wall with blocking.* (Nuconsteel Commercial Corp.)

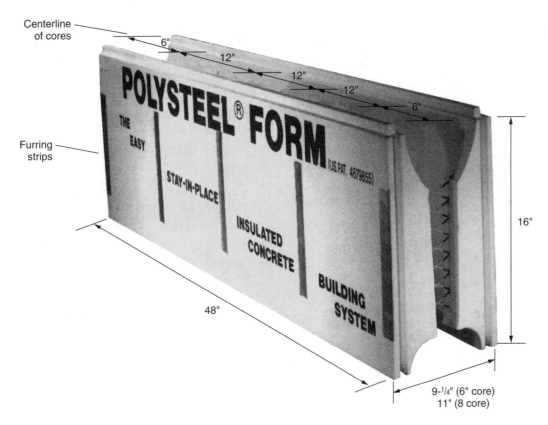

Fig. 8-16 *Insulated concrete form.* (American Polysteel Forms)

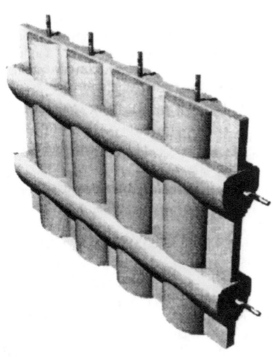

Fig. 8-17 *Concrete and rebar inside an insulated concrete form.*
(American Polysteel Forms)

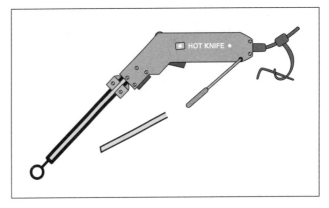

Fig. 8-18 *A hot knife used to cut foam for electrical wiring and plumbing.* (Avalon Concepts)

Sequence

Insulated concrete form framing is a radically new approach for carpenters that are familiar with the traditional wood "stick-built" method. The only wood used in this wall framing method is pressure-treated lumber, which is used for sealing rough door and window openings. Because this method is different and each manufacturer has specific guidelines that must be followed to ensure proper installation, training is offered to all individuals constructing ICFs. Many ICFs also send representatives to the job site to supervise those who are building their first home using this method. Therefore, only general guidelines are covered in the following section.

1. Before starting, rebar should extend vertically 2–6 feet from the foundation every one to two linear feet depending on building code requirements.

2. Most manufacturers suggest placing 2 × 4s or some type of bracing around the perimeter of the foundation as a guide for setting the foam forms.

3. Once the first course of foam form blocks is laid, continue to do so, placing rebar horizontally every one to two feet (or every other course of block) as required.

4. Vertical bracing should be tied to the perimeter bracing at 6-foot intervals. Corners should be braced on each intersecting edge with additional diagonal bracing spaced at 4-foot intervals. See Fig. 8-19 for a typical corner bracing diagram. As forms are stacked, vertical bracing can be screwed into the metal or plastic ties built or inserted into the foam form blocks.

5. Openings for windows and doors should be blocked with pressure-treated lumber.

6. After the foam form blocks are stacked to the correct height, place bracing on the top and secure it to

CORNER BRACE WITH FORMS

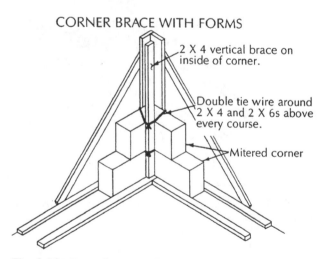

2 X 4 vertical brace on inside of corner.

Double tie wire around 2 X 4 and 2 X 6s above every course.

Mitered corner

Fig. 8-19 *Corner bracing technique for insulated concrete forms.*
(American Polysteel Forms)

the side bracing. Because foam is very light, it will float. Proper bracing is essential to prevent the foam walls from floating and causing blow-outs when the concrete is poured inside them. Additional bracing techniques are illustrated in Fig. 8-20.

7. Walls are usually poured with a boom pump truck in 4-foot increments. Anchor bolts are set in the top of the wall to secure the top plate as a nailing base for roof construction. See Fig. 8-21.

8. Sheathing can be screwed into the metal or plastic ties located every foot in the foam form block, as shown in Fig. 8-22.

9. Electrical wiring and boxes can be installed by gouging out a groove with a router or hot knife. See Fig. 8-23 for electrical wiring tips.

Types of Foam

The most commonly used types of foam utilized in ICFs are the *expanded polystyrene* (EPS), and *extruded polystyrene* (XPS). EPS consists of tightly fused beads of foam. Vending-machine coffee cups, for example, are made of EPS. XPS is produced in a different process and is more continuous, without beads or the sort of "grain" of EPS. The trays in prepackaged meat at the grocery store are made of XPS. The two types can differ in cost, strength, R-value, and water resistance. EPS varies somewhat. It comes in various densities; the most common are 1.5 pounds per cubic foot (pcf) and 2 pounds pcf. The denser foam is a little more expensive, but is a little stronger and has a slightly higher R-value.

Some stock EPS is now available with insect-repellent additives. Although few cases of insect penetration into the foam have been reported to date, some ICF manufacturers offer versions of their product made of treated material. Whether you buy your own foam or you are choosing a preassembled system, you might want to check with the manufacturer about this.

Three Types of ICF Systems

The main difference in ICF systems is that they vary in their unit sizes and connection methods. They can be divided into three types: *plank, panel,* and *block systems* as shown in Fig. 8-24. The panel system is the largest. It is usually 4 by 8 feet in size. That means the wall area can be erected in one step but may require more cutting. These panels have flat edges and are connected one to the other with fasteners such as glue, wire, or plastic channel.

Plank systems are usually 8 feet long with narrow (8- or 12-inch) planks of foam. These pieces of foam are held at a constant distance of separation by steel or

2 X 4 TOP RAIL SUPPORT

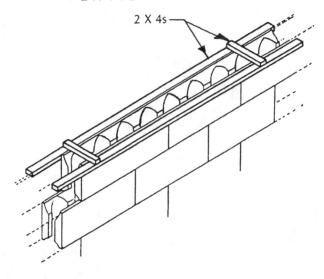

2 X 4s

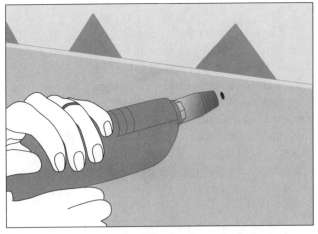

Fig. 8-20 *Additional bracing techniques.* (American Polysteel Forms)

HORIZONTAL TOP BRACE

◆ A horizontal top brace secured to the steel furring strips and supported by a diagonal brace staked to the ground will also provide the alignment and security necessary to keep the Forms in place during the pouring of concrete.

POLYSTEEL FORM® TOP BRACE

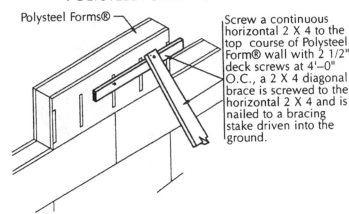

Polysteel Forms®

Screw a continuous horizontal 2 X 4 to the top course of Polysteel Form® wall with 2 1/2" deck screws at 4'–0" O.C., a 2 X 4 diagonal brace is screwed to the horizontal 2 X 4 and is nailed to a bracing stake driven into the ground.

Fig. 8-21 *Anchor bolt placement for roof framing.* (AFM)

plastic ties. The plank system has notched, cut, or drilled edges. The edges are where the ties fit. In addition to spacing the planks, the ties connect each course of planks to the one above and below.

Fig. 8-22 *Attaching sheathing directly to plastic ties in the concrete forms.* (AFM)

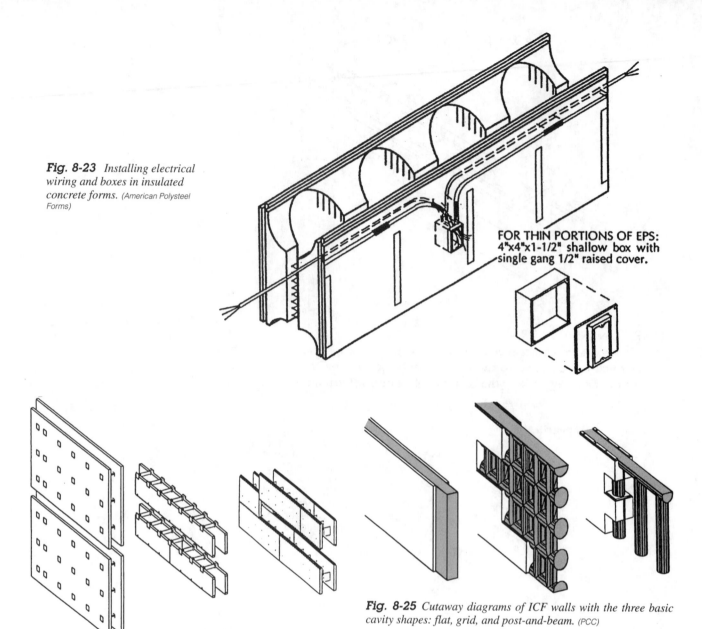

Fig. 8-23 *Installing electrical wiring and boxes in insulated concrete forms.* (American Polysteel Forms)

FOR THIN PORTIONS OF EPS:
4"x4"x1-1/2" shallow box with single gang 1/2" raised cover.

Fig. 8-24 *Diagrams of ICF formwork made with the three basic units: panel on the left, plank in the center, and block on the right.* (PCC)

Fig. 8-25 *Cutaway diagrams of ICF walls with the three basic cavity shapes: flat, grid, and post-and-beam.* (PCC)

Block systems include units ranging from the standard concrete block (8 × 16 inches) size to a much larger 16-inch-high by 4-foot-long unit. Along their edges are teeth or tongues and grooves for interlocking; they stack without separate fasteners on the same principle as children's Lego blocks.

Another difference is the shape of the cavities. Each system has one of three distinct cavity shapes. The shapes are flat, grid, or post-and-beam. These produce different shapes of concrete beneath the foam, as shown in Fig. 8-25.

Note how the flat cavities produce a concrete wall of constant thickness, just like a conventionally poured wall that was made with plywood or metal forms. Cav-

ities are usually "wavy," both horizontally and vertically. If the forms are removed, it can be seen just how the walls resemble a breakfast waffle. The post-and-beam cavities are cavities with concrete only every few feet, horizontally and vertically. In the most extreme post-and-beam systems, there is a 6-inch-diameter concrete "post" formed every 4 feet and a 6-inch concrete "beam" at the top of each story.

Keep in mind that no matter what the shape of the cavity, all systems have "ties." These are the crosspieces that connect the front and back layers of foam that make up the form. If the ties are plastic or metal, the concrete is not affected significantly. However, in some grid systems, they are foam and are much larger. But the forming breaks in the concrete about 2 inches in diameter every foot or so. See Fig. 8-25 for the differences.

Table 8-1 shows eight different systems used in ICF with their dimensions, fastening surface, and various notes with additional details. Note that no matter what the cavity shape is, all systems also have "ties." These ties are the crosspieces that connect the front and back layers of foam. When the ties are metal or plastic, they do not affect the shape of the concrete much. But in some of the grid systems, they are foam and are much larger, forming breaks in the concrete about 2 inches in diameter every foot or so. Figure 8-26 shows the differences.

Table 8-1 *Available ICF Systems*[1] (Portland Cement Assoc.)

	Dimensions[2] (width x height x length)	Fastening surface	Notes
Panel systems			
Flat panel systems			
R-FORMS	8" x 4' x 8'	Ends of plastic ties	Assembled in the field; different lengths of ties available to form different panel widths.
Styroform	10" x 2' x 8'	Ends of plastic ties	Shipped flat and folded out in the field; can be purchased in larger/smaller heights and lengths.
Grid panel systems			
ENER-GRID	10" x 1'3" x 10'	None	Other dimensions also available; units made of foam/cement mixture.
RASTRA	10" x 1'3" x 10'	None	Other dimensions also available; units made of foam/cement mixture.
Post-and-beam panel systems			
Amhome	9 3/8" x 4' x 8'	Wooden strips	Assembled by the contractor from foam sheet. Includes provisions to mount wooden furring strips into the foam as a fastening surface.
Plank systems			
Flat plank systems			
Diamond Snap-Form	1' x 1' x 8'	Ends of plastic ties	
Lite-Form	1' x 8' x 8'	Ends of plastic ties	
Polycrete	11" x 1' x 8'	Plastic strips	
QUAD-LOCK	8" x 1' x 4'	Ends of plastic ties	
Block systems			
Flat block systems			
AAB	11.5" x 16⅝" x 4'	Ends of plastic ties	
Fold-Form	1' x 1' x 4'	Ends of plastic ties	Shipped flat and folded out in the field.
GREENBLOCK	10" x 10" x 3'4"	Ends of plastic ties	
SmartBlock Variable Width Form	10" x 10" x 3'4"	Ends of plastic ties	Ties inserted by the contractor; different length ties available to form different block widths.
Grid block systems with fastening surfaces			
I.C.E. Block	9 1/4" x 1'4" x 4'	Ends of steel ties	
Polysteel	9 1/4" x 1'4" x 4'	Ends of steel ties	
REWARD	9 1/4" x 1'4" x 4'	Ends of plastic ties	
Therm-O-Wall	9 1/4" x 1'4" x 4'	Ends of plastic ties	
Grid block systems without fastening surfaces			
Reddi-Form	9 5/8" x 1' x 4'	Optional	Plastic fastening surface strips available
SmartBlock Standard Form	10" x 10" x 3'4"	None	
Post-and-beam block systems			
ENERGYLOCK	8" x 8" x 2'8"	None	
Featherlite	8" x 8" x 1'4"	None	
KEEVA	8" x 1' x 4'	None	

[1] All systems are listed by brand name.

[2] "Width" is the distance between the inside and outside surfaces of foam of the unit. The thickness of the concrete inside will be less, and the thickness of the completed wall with finishes added will be greater.

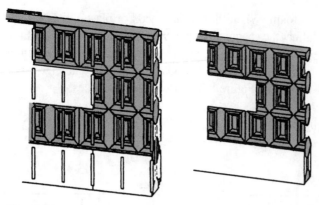

Fig. 8-26 *Cutaway diagrams of ICF grid walls with steel/plastic ties and foam ties.* (PCC)

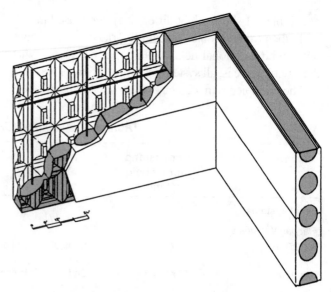

Fig. 8-28 *Cutaway diagram of a grid panel wall.* (PCC)

Another difference is that many of the systems also have a fastening surface which is some material other than foam, embedded into the units that crews can sink screws or nails into the same way as fastening to a stud. Often this surface is simply the ends of the ties; however, other systems have no embedded fastening surface. These units are all foam, including the ties. This generally makes them simpler and less expensive but requires crews to take extra steps to connect interior wallboard, trim, exterior siding, and so on to the walls.

Cutaway views of the wall panels are shown in Figs. 8-27 through 8-34. In Fig. 8-35 the R-Forms panel is being assembled on site. In Fig. 8-36, there is an on-site pile of Ener-Grid panels. Figure 8-37 shows the top view of an Amhome panel that has furring strips embedded. Note how light the panels are. They are easily handled by one person. The worker is carrying a fold-out Fold-Form block before and after it has been spread for adding to the construction.

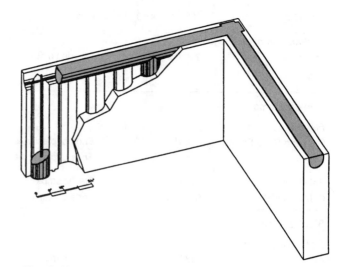

Fig. 8-29 *Cutaway diagram of a post-and-beam panel wall.* (PCC)

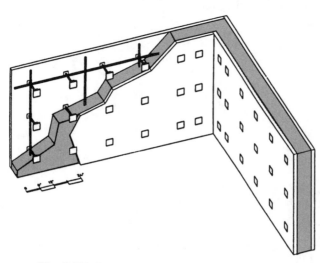

Fig. 8-27 *Cutaway diagram of a flat panel wall.* (PCC)

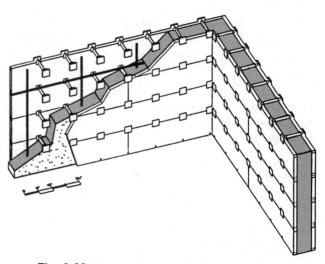

Fig. 8-30 *Cutaway diagram of a flat plank wall.* (PCC)

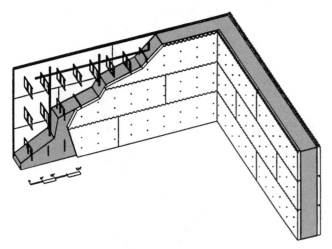

Fig. 8-31 *Cutaway diagram of a flat block wall.* (PCC)

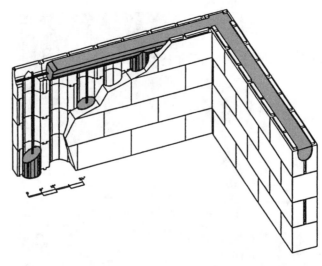

Fig. 8-34 *Cutaway diagram of a post-and-beam block wall.* (PCC)

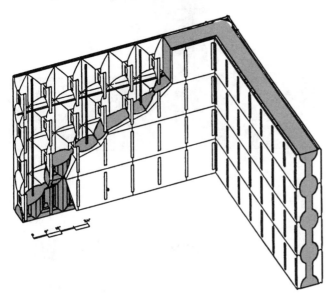

Fig. 8-32 *Cutaway diagram of a grid block wall with fastening surfaces.* (PCC)

Fig. 8-35 *Site-assembly of R-Forms panels.* (PCC)

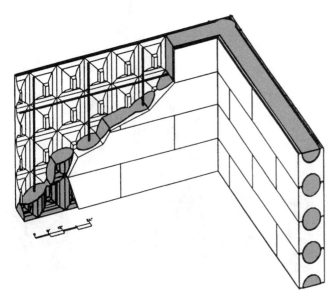

Fig. 8-33 *Cutaway diagram of a grid block wall without fastening surfaces.* (PCC)

Two varieties of SmartBlock are shown in Fig. 8-39. Three of them, A, C, and D, are assembled with plastic ties. A grid block without fastening surfaces is shown in Fig. 8-39B.

Figure 8-40A, B, C, and D shows setting the insulating concrete forms, how the completed forms look, pumping in concrete, and the completed house.

Foam Working Tools

Some of the tools needed for working with foam are not familiar to the usual framer of houses. These tools may include the thermal cutter. This cutter is a new tool that cuts a near-perfect line through foam and plastic units in one pass. Figure 8-41 shows this device. It is made up of a taut resistance-controlled wire mounted on a bench-frame. It is heated with electricity and drawn though the unit while the wire is red hot. It melts a narrow path through foam and plastic

Fig. 8-36 *An ENER-GRID panel. (PCC)*

A

Fig. 8-37 *Top view of an Amhome panel with embedded furring strips. (PCC)*

ties. It is a worthwhile tool to have if you are building a high-volume ICF walled house. It will not cut rough metal ties or the foam-and-cement material of the grid panel systems. However, companies selling thermal cutters are usually located in every community. The grid panel systems can be cut with any of the bladed tools used by a carpenter, but sometimes a chain saw is a handy tool. It goes quickly through the heavier material of these systems and cuts through in one pass, whether cutting on the ground or in place.

Tools needed for cutting and working with foam are shown in Table 8-2.

B

Fig. 8-38 *Folding out a Fold-Form block before use. (PCC)*

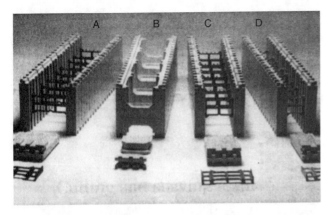

Fig. 8-39 *The two varieties of SmartBlock: a flat block assembled with plastic ties (A,C,D) and a grid block without fastening surfaces in B. (PCC)*

C

Fig. 8-40C *More formwork in place with window backs.*

A

Fig. 8-40A *Bracing corners with window and door inserts.*

D

Fig. 8-40D *Side walls in place with framing to support the concrete pouring operation.*

B

Fig. 8-40B *Garage doors framed, poured, and roof décor applied.*

E

Fig. 8-40E *Garage door framing. Note the steel rebar through the foam blocks and the corner bracing.*

Gluing and Tying Units

ICF units are frequently glued at the joints to hold them down, hold them together, and prevent concrete leakage. Common wood glue and most construction adhesives do the job well. Popular brands are Liquid Nail, PL 200, and PL 400. Some of these can dissolve the foam but, if applied in a thin layer, the amount of foam lost is usually insignificant. Look for an adhesive

Fig. 8-40F *Bracing for a window. Note rebar.*

Fig. 8-40H *Foam blocks filled with concrete; roof is attached and shingled.*

Fig. 8-40I *Foam blocks in place. Roof framing is complete, ready for roofing.*

Fig. 8-40G *Note treatment of the bay window shown here without siding. Also, note the foam being coated where backfill will be placed.*

Fig. 8-40J *Finished house looks like any other type of construction.*

that is "compatible with polystyrene." Figure 8-42 shows an industrial foam gun.

Rebar is often precut to length and pre-bent but, even if it is, the workers generally have to process a few bars in the field. Most ICF systems also have cradles that hold

Fig. 8-40M *Another foam–concrete home with a stucco finish.*

Fig. 8-40L *Finished house with foam foundation and basement to be finished with stucco. This will protect the foam blocks.*

Fig. 8-40N *A smaller foam–concrete home with conventional appearance.*

the bars in place for the pour, but a few bars need to be wired to one another or to ties to keep them in position.

It is possible to bend rebar with whatever tools are handy and cut it with a hacksaw; however, if you process large quantities, you might prefer buying or renting a cutter-bender. A large manual tool is pictured in Fig. 8-43. It makes the job faster and easier. Cutters-benders are available at steel-supply, concrete-supply, and masonry-supply houses. Almost any steel wire can hold rebar in place, but most efficient are rolls of pre-cut wires shown in Fig. 8-44. These wires make the job faster and easier. Wire coils and belt-mounted coil holders are both sold by suppliers of concrete products, masonry, and steel.

Pouring Concrete

Concrete is best poured at a more controlled rate into ICFs than it is into conventional forms. An ordinary

Fig. 8-41 *A thermal cutter. Note the white line that is part of the cutting device. It is attached to the transformer on the left of the framework. (PCC)*

Table 8-2 *Useful Tools and Materials* (Portland Cement Assoc.)

Operation or Class of Material	
Tool or Material	**Comments**

Cutting and shaving foam
Drywall or keyhole saw	For small cuts, holes, and curved cuts.
PVC or miter saw	For small, straight cuts and shaving edges.
Coarse sandpaper or rasp	For shaving edges.
Bow saw or garden pruner	For faster straight cuts.
Circular saw	For fast, precise, straight cuts. For cutting units with steel ties, reverse the blade or use a metal-cutting blade.
Reciprocating saw	For fast cuts, especially in place.
Thermal cutter	For fast, very precise cuts on a bench. Not suitable for steel ties or grid panel units.
Chain saw	For fast cuts of grid panel units.

Lifting units
Forklift, manual lift, or boom or crane truck	For carrying large grid panel units and setting them in place. For upper stories, a truck is necessary.

Gluing and tying units
Wood glue, construction adhesive, or adhesive foam	
Small-gage wire	For connecting units of flat panel systems.

Bending, cutting, and wiring rebar
Cutter-bender	
Small-gage wire or precut tie wire or wire spool	

Filling and sealing formwork
Adhesive foam	

Placing concrete
Chute	For below-grade pours.
Line pump	Use a 2-inch hose.
Boom pump	Use two "S" couplings and reduce the hose down to a 2-inch diameter.

Evening concrete
Mason's trowel	

Damp-proofing walls below grade
Nonsolvent-based damp-proofer or nonheat-sealed membrane product	

Surface cutting foam
Utility knife, router, or hot knife	Heavier utility knives work better. Use a router with a half-inch drive for deep cutting.

Fastening to the wall
Galvanized nails, ringed nails, and drywall screws	For attaching items to fastening surfaces. Use screws only for steel fastening surfaces.
Adhesives	For light and medium connections to foam.
Insulation nails and screws	For holding lumber inside formwork.
J-bolt or steel strap	For heavy structural connections.
Duplex nails	For medium connections to lumber.
Small-gage wire	For connecting to steel mesh for stucco.
Concrete nails or screw anchors	For medium connections to lumber after the pour.

Flattening foam
Coarse sandpaper or rasp	For removing small high spots.
Thermal cutter	For removing large bulges.

Foam
Expanded polystyrene or extruded polystyrene	Consider foam with insect-repellent additives

Concrete
Midrange plasticizer or superplasticizer	For increasing the flow of concrete without decreasing its strength. Can also be accomplished by changing proportions of the other ingredients.

Stucco
Portland cement stucco or polymer-based stucco	

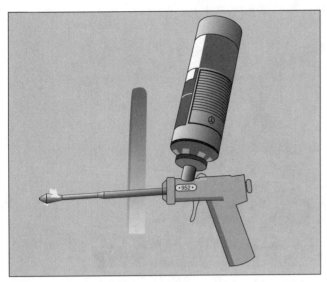

Fig. 8-42 *The industrial foam gun can be used as a glue gun.* (PCC)

Fig. 8-43 *This is a rebar cutter-bender.* (PCC)

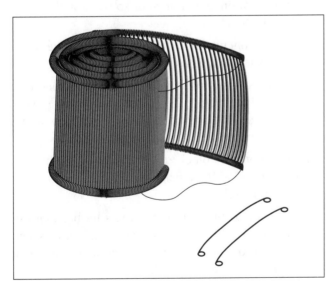

Fig. 8-44 *Roll of precut tie wires.* (PCC)

chute can be used for foundation walls (basement or stem); this is the least expensive option because it comes free with the concrete truck. Precise control is more difficult with a chute. You must pour more slowly than with conventional forms, and you must move the chute and truck frequently to avoid overloading any one section of the formwork.

The smaller line pump pushes concrete through a hose that lies on the ground. See Fig. 8-45. The crew holds the end of the hose over the formwork to drop concrete inside. If possible, use a 2-inch hose. One or two workers can handle it, and it can generally be run at full speed without danger. If only a 3-inch hose is available, you can use it; but pump slowly until you learn how much pressure the forms can take.

Boom pumps are mounted on a truck that also holds a pneumatically operated arm (the boom). The hose from the pump causes the concrete to move along the length of the boom and then hang loose from the end. See Fig. 8-46. By moving the boom, the truck's operator can dangle the hose wherever the crew calls for it. One worker holds the free end to position it over the form-work cavities. The standard hose diameter is 4 inches.

You will need to have the hose diameter reduced to 2 or 3 inches with tapered steel tubes called "reducers." The narrower diameter slows down and smoothes out the flow of the concrete. Figure 8-47 shows how the reducers are arranged on a boom with hoses. Also, ask for two 90-degree elbow fittings on the end of the hose assembly, as seen in Fig. 8-47. These form an "S" in the line that further breaks the fall of the concrete. The concrete can be leveled off on top of the form by using a mason's trowel.

Fig. 8-45 *Line pump for concrete.* (PCC)

Fig. 8-46 *Boom pump for pouring concrete.* (PCC)

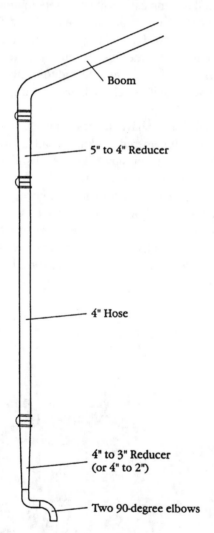

Boom

5" to 4" Reducer

4" Hose

4" to 3" Reducer
(or 4" to 2")

Two 90-degree elbows

Fig. 8-47 *Diagram of boom pump fittings that are suitable for pouring ICF walls.* (PCC)

CONCRETE BLOCK

The second most popular residential framing method is concrete block. Comprised of less than 5% of the residential framing market, concrete block is another traditional method that has been around for a long time. It is mainly used when homeowners are interested in a framing method that is resistant to inclement weather and termites. This type of construction is much more costly and time-consuming. Because the techniques associated with masonry and concrete work are numerous and detailed, only the major steps are reviewed.

1. Snap a chalk line on the footing or foundation where the concrete block should be placed.

2. Spread mortar an inch or two thick on the footing or foundation and an inch or so wider and longer than the block.

3. Place a concrete block on each corner of the wall. Make sure both blocks are level and straight.

4. Stretch a string line tightly between the front, top edge of both blocks. This can be done by wrapping a string around a brick and placing it on top of the concrete block.

5. Place concrete blocks evenly about seven blocks (if using 8" × 16" block) from the corner. Do this for both corners.

6. Continue to build up the corners by backing up each level of block (course) by half a block. When finished, one block should be on the top of each corner. Each corner is called a *lead*.

7. Place concrete blocks between each lead using a string to keep the top and front edge of the block straight and level. Mortar should be placed on one end of each block. A ⅜-inch-thick layer of mortar is common, because the actual size of most 8" × 16" concrete block is 7⅝" × 15⅝".

8. Allowances should be made for rough openings of windows and doors. Lintels or metal ledges are used to support a block over door and window openings.

9. The top course of concrete block is filled with mortar, and anchor bolts are inserted at certain intervals according to building code requirements.

10. Mortar joints are usually smoothed (tooled) with a round bar, square bar, or s-shaped tool.

Estimating Concrete Block

In most instances, you would be laying standard 8- × 8- × 16-inch (height by width by length) concrete block. This is the typical size block for load-bearing walls. Interior walls that are not load bearing would only require 4-inch-thick walls. However, if 4-inch-diameter drain pipes are running through the walls from the second story, then you would use 6- or 8-inch-thick block for that wall. In most cases, only the exterior walls are made of concrete block for protection from hurricanes and other forces in residential construction. Interior walls can be framed from wood or metal studs. If the goal is to protect against termite infestation, metal studs would be the preferred choice. At any rate, the number of blocks required for a wall is determined by the square feet. A 100-square-foot wall made with concrete blocks 8 inches in height and 16 inches in length with 3/8-inch mortar joints would require approximately 113 blocks. If the blocks were only 4 inches in height and 16 inches in length, then you would need 225 blocks.

The amount of mortar required for a 100-square-foot wall made with blocks 8 × 8 × 16 inches would be approximately 8½ cubic feet. A 100-square-foot wall made of 4- × 8- × 16-inch block would require 13½ cubic feet of mortar. Remember not to mix more mortar than you can lay in 1½ hours.

Index

ABOUT THE AUTHORS

Mark R. Miller is Professor of Industrial Technology at The University of Texas in Tyler, Texas. He teaches construction courses for future middle managers in the trade. He is coauthor of several technical books, including the best-selling *Carpentry & Construction,* now in its fourth edition. He lives in Tyler.

Rex Miller is Professor Emeritus of Industrial Technology at State University College at Buffalo and has taught technical curriculum at the college level for more than 40 years. He is the co-author of the best-selling *Carpentry & Construction,* now in its fourth edition, and the author of more than 75 texts for vocational and industrial arts programs. He lives in Round Rock, Texas.